MW00528828

Indoor Radio Planning

Indoor Radio Planning

A Practical Guide for GSM, DCS, UMTS and HSPA

Morten Tolstrup

Technical Director, LGC Wireless — An ADC Company, Denmark

A John Wiley & Sons, Ltd, Publication

This edition first published 2008

Registered office
John Wiley & Sons Ltd, The Atrium, Southern Gate, Chichester, West Sussex,
PO19 8SQ, United Kingdom

For details of our global editorial offices, for customer services and for information about how to apply for permission to reuse the copyright material in this book please see our website at www.wiley.com.

Library of Congress Cataloging-in-Publication Data

Tolstrup, Morten.
Indoor radio planning : a practical guide for GSM, DCS, UMTS and HSPA /
Morten Tolstrup.
p. cm.
Includes bibliographical references and index.
ISBN 978-0-470-05769-8 (cloth)
1. Wireless LANs. 2. Wireless communication systems. 3. Mobile
communication systems. I. Title.
TK5105.78.T65 2008
621.3845'6–dc22 2008006388

A catalogue record for this book is available from the British Library.

ISBN 978-0-470-05769-8 (H/B)

Typeset in 10/12pt Times by Thomson Digital Noida, India.
Printed in Great Britain by CPI Antony Rowe, Chippenham, England.

Contents

Foreword by Professor Simon Saunders

The compelling need for in-building wireless systems derives directly from the needs of the people who use wireless – and that means, increasingly, all of us. We spend most of our time inside buildings, whether in the office or at home, at work or at play. Typically at least two-thirds of voice traffic on cellular networks originates or terminates inside buildings, and for data services the proportion is still higher – probably in excess of 90%.

Yet for too long, most indoor service has been provided from outdoor systems requiring high transmit powers, major civil engineering works and using a relatively large amount of spectrum to serve a given traffic level. This makes great sense for providing economical initial coverage to a large number of buildings and for 'joining the dots' to enable wide area mobility. However, 'outside-in' thinking is 'inside-out', from a technical and practical viewpoint, when attempting to serve users with very high quality and coverage expectations, and for delivering high data rate services within limited spectrum. Buildings offer their own remedy to these challenges, by providing signal isolation from nearby systems and enabling the fundamental principle of cellular systems – that unlimited capacity is available from limited spectrum if the engineering is done right.

Despite these compelling benefits, in-building wireless systems have hitherto been a poor relation of the 'mainstream' macrocellular network operations. With relatively few enthusiasts and a wide range of different favoured techniques for system design and installation, the field has at times resembled a hobby rather than a professional activity. The industry desperately needs best-practice techniques to be shared amongst a wider base of individuals to serve the growing demand – there are not enough engineers for the buildings requiring service – and for these techniques to become standardised in order to drive down costs, improve reliability and drive volumes.

Given this background, I welcome the publication of this book. Morten Tolstrup is a leading practitioner in the field and an engaging and entertaining public speaker. He has written a truly practical and helpful guide to indoor radio planning, which will enable a much wider audience to convert their skills from the old world of two-dimensional networks, comprising macro cells alone, to the new world of three-dimensional hierarchical networks comprising macro, micro, pico and femto cells delivering services to unlimited numbers of users. Following the simple guidelines provided, built on years of real-world experience, will help to avoid some very expensive mistakes.

Most of all, I hope that this book will help to professionalize the industry and encourage sharing of best-practice to the ultimate benefit of end-customers for compelling wireless broadband services.

Professor Simon R. Saunders
Independent Wireless Technologist & Visiting Professor,
University of Surrey
www.simonsaunders.com

Preface

This is Not a Book for Scientists

This book is intended for the RF planners, to serve as a practical tool in their daily work designing indoor radio distribution systems. It is not a complete book about all the deep aspects and corners of GSM, DCS, UMTS and HSPA networks, or all the core network systems. It is dedicated to the last 10–70 m of the network, the indoor air interface between the mobile user and the indoor mobile network.

I have spent the past 20 years working on various parts of the exciting business of cellular communication. During this time I have mostly focused on the planning of the radio interface between the network and the mobile user, with a dedicated focus on indoor radio planning. I have always tried to approach that small part of the systems that involved me, the radio interface, from a practical angle. I have struggled with most of the books available on these subjects mainly due to a theoretical level far beyond my needs. My hope with this book is to present a level of theory that is usable and accessible for a radio planner with basic radio experience.

I also need to emphasize that no matter the radio platform or standard, GSM, UMTS, HSPA or 4G, as long as the interface between the mobile and the network uses radio communication, it will always be a matter of a link calculation with a given signal-to-noise ratio for a given service requirement. After all, it is 'just' radio planning.

The Practical Approach

I am not an expert in cellular, GSM, UMTS or HSPA systems, far from it – but I have gained a lot of experience with RF design, especially with regards to indoor radio planning. An old mountaineering saying is that 'good judgment comes from experience, but experience is often a result of bad judgment'. I have made my share of mistakes along the way, and I will help you avoid making the same mistakes when designing and implementing indoor solutions.

It has been my goal to include what I believe are the most important considerations and design guidelines to enable the RF planner to design and implement a high-performing indoor distributed antenna system.

It was not my intention to provide a deep hardcore mathematical background on RF planning, but to present the most basic calculations of the various parameters that we need to consider when designing a distributed antenna system.

I hope you can use the result – this book. It has been hard but also great fun to write it and to revisit all the background stuff, projects and measurement results that are the basis for this book. I hope you find it to be a useful tool in your daily work. That was my intention.

Morten Tolstrup

Acknowledgments

This book would not have been possible if it was not for my many colleagues and friends who I have spend the better part of the last 20 years with. These friends and colleagues I know mainly from my many indoor projects around the world and from other mobile operators when working on mutual indoor projects. Many hours have I spent with you guys during the design phase, site visits, project implementations and measurements, from the fanciest indoor projects to deep below ground on a tunnel project conducting verification measurements.

I want to thank Simon Saunders for contributing the foreword. Simon is one of the people in this industry I respect the most, for his dedicated work and contribution to so many fields in the industry of telecommunication.

I also thank my friends who have helped me by reviewing the book: Bernd Margotte, Lars Petersen, Kevin Moxsom, Stein Erik Paulsen and Mario Bouchard. In particular, I want to thank Robin Young for his help and inspiration on the section about noise and link budgets. Peter Walther is also acknowledged for his input on the HSPA section, and the link budget example for HSDPA.

Thanks also to the team from Wiley, Mark Hammond, Sarah Hinton and their team, for the support and production of this book.

Last, but not least, thank you Karin my dear wife, for your support on this project and for letting me spend so many late nights, early mornings and weekends on this book. Without you this project would not have been possible. Thanks also for your design concept of the front cover, and the photo for the cover.

Even though I have spent many hours on this project, checking and double-checking everything, there might be an error or two. Let me know, and I will make sure to correct it in any future editions.

You can contact me via: www.ib-planning.com

Morten Tolstrup
Dronninglund, Denmark

1

Introduction

I often think that we have now finally come full circle in the world of radio transmission. We are back to where it all started: after all, the first transmission via radio waves by Marconi in 1895 was digital, using Morse code.

These days we are heading for a fully digitalized form of radio transmission, often using Internet Protocol (IP). Most radio services – broadcast, voice transmission for mobiles and television transmission – are being digitalized and transmitted via radio waves.

Radio waves – what a discovery that truly has changed our world! The effect of electromagnetism was discovered by H. C. Ørsted in 1820. Samuel E. Morse invented his digital system, the 'Morse code', in 1840. Through copper wires the world got connected via the telegraph line, and cross-continental communication was now accessible. Marconi merged both inventions and created the basis of our modern wireless communication systems, performing the first radio transmission over an incredible distance of 1.5 kilometers in 1895. Now we live in a world totally dependent on spin-offs of these basic discoveries.

Marconi struggled to transmit radio signals over a relative short distance: a few kilometers was a major achievement in the early days. Later, radio waves were used to reach several hundred thousand of kilometers into deep space, communicating with and controlling deep space probes and even vehicles on Mars.

Would it not be fair if we could bring back Ørsted, Morse and Marconi, and honor them by showing what we can do today, using the same principles: electromagnetism, digital transmission and radio waves? I am sure that they and the many other scientists who have formed the basis of our modern communication society would be proud. No one today could even consider a world without easy wireless communication; our modern lifestyle is highly dependent on those small devices – mobile telephones.

Things in telecommunications industry are progressing fast. These days we are not happy with anything less than several Mbps over the radio interface, mobile TV, internet, email and mobile media.

Back in the early 1980s I was working on NMT systems. We used analog modems and were able to achieve up to about 300 baud over the mobile phone network. That was truly amazing at the time. People could send a fax from their car and, if they could carry the 18 kg

Indoor Radio Planning: A Practical Guide for GSM, DCS, UMTS and HSPA Morten Tolstrup

mobile cellular phone battery included, they could have a portable phone with up to 30–60 min of talk time. The cost of these types of cellular phones was equivalent to that of a small family car in the early days, so the market was limited to very few professional users. Over a few years the price dropped to about an average month's salary, and mobile phones were getting smaller and smaller. Some were even 'pocket size' – if your pocket was big and able to support a weight of about 1 kg, that was.

At some point I was told about a new futuristic mobile telephone system in the making called GSM. The plan was to convert the voice to data, and the network could support 9600 baud (9.6 kbps), 32 times more that we could do on NMT! This was an amazingly high data speed – higher than we could get over fixed telephone lines at the time. I remember being highly skeptical. Who would ever need such high data rates for mobile use and for what? Mobile TV? Absolutely mad! Man, was I wrong!

These days we are heading for 14 Mbps via HSDPA, more than 4600 times faster than we could perform via NMT in 1980. In reality, we are now able to handle higher mobile data speed to one user than the total data transmission capacity of the whole NMT network in Denmark could handle then for all the users in the network!

The need for data is endless. Data rates via mobile will increase and increase, and actually the radio link is getting shorter and shorter. In order to perform these high data rates, we need a better and better radio link. The radio spectrum is getting more and more loaded, and we are using higher and higher radiofrequencies and more and more complex and quality-sensitive modulation schemes; thus the requirement for the quality of the radio link is getting more and more strict.

It is worthwhile noting that high data rates are not enough on their own. It is also a matter of services; if mobile users are not motivated by an attractive service, even the highest data rate is pointless.

The need for high data rates is motivated by user demand for mobile email, internet and multimedia services. Most UMTS mobile phones are able to support video calling, but it is rarely used. This shows that, even though it is impressive from a technical viewpoint that it is possible at all, the technology has no point if the service is not attractive to mobile users. It is a fact that the most successful mobile data service to date is also the slowest data service in operation over the mobile network, transmitted via a very slow data channel: SMS (Short Message Service). SMS is still the most popular data service and still the 'cash cow' when it comes to data services for most mobile networks. Who would have thought that mobile users of all ages from 8 to 98 would key in long text messages via a 10 digit keyboard on a mobile phone, when they can pick up the phone and talk? Some users in the network have an SMS activity beyond 2000 SMSs per week!

When I was introduced to SMS, I thought it might be a good service to announce voice mails etc. to mobiles, but when the first mobiles arrived that were able to transmit SMS, my thought was 'why?' Wrong again! It clearly shows that it is not only a matter of data speed but also the value of the applications and services offered to the user.

I am happy to note that one thing stays the same: the radio planning of the mobile networks. The air interfaces and especially the modulation schemes are getting more and more complex, but in reality there is no difference when seen from a basic radio planning perspective. The challenges of planning a high-performance HSPA link is the same basic challenge that Marconi faced performing his first radio transmission. It is still a matter of getting a sufficient margin between the signal and the noise, fulfilling the specific requirement for the wanted service, from Morse via long waves to 14 Mbps via UMTS/HSPA. It is still radio planning and a matter of signal-to-noise ratio and quality of the link.

In the old days it was all about getting the radio link transmitted over longer and longer distances. These days, however, the radio link between the network and the mobile user is getting shorter and shorter due to the stricter demands on the quality of the radio link in order to perform the high data rates. Marconi struggled to get his radio transmission to reach a mile. These days we are struggling to get a service range from an indoor antenna in a mobile network to service users at 20–40 m distance with high-speed data and good quality voice service.

We are now moving towards an IP-based world, even on the radio interface, and voice-over-IP. We are now using IP connection to base stations and all other elements in the network. The network elements are also moving closer to the mobile users in order to cater for the requirements for quality of voice and data.

We are now on the brink of a whole new era in the world of telecommunications, an era where the mobile communication network will be an integrated part of any building. The telecommunications industry is just about to start integrating small base stations, 'femto cells', in many residential areas in many countries around the world. People expect mobile coverage and impeccable wireless data service everywhere.

When electricity was invented and became popular, existing buildings had to be post-installed with wires and light fixtures to support the modern technology of electrical apparatus and lighting. Later it was realized that electricity probably was so popular that it was worthwhile pre-installing all the wiring and most of the appliances in buildings from the construction phase. I do believe that, within a few years from now, it will be the same with wireless telecommunications. Wireless services in buildings are one of the basic services that we just expect to work from day one, in our home, in tunnels and surely in corporate and public buildings.

The future is wireless.

2

Overview of Cellular Systems

This book is concentrated around the topic of indoor radio planning from a practical perspective, and it is not the within the scope of this book to cover the full and deep details of the GSM and UMTS systems and structures. This book will only present the most important aspects of the network structure, architecture and system components, in order to provide basic knowledge and information that is needed as a basis for design and implementation of indoor coverage and capacity solutions. For more details on cellular systems in general refer to [2].

2.1 Mobile Telephony

2.1.1 Cellular Systems

The concept of cellular coverage was initially developed by AT&T/Bell Laboratories. Prior to that, the mobile telephony systems were manual systems used only for mobile voice telephony. Typically implemented with high masts that covered large areas, and with limited capacity per mast, they were only able to service few users at the same time – in some cases even only one call per mast! These systems also lacked the ability to hand over calls between masts, so mobility was limited to the specific coverage area from the servicing antenna, although in reality the coverage area was so large that only rarely would you move between coverage areas. Remember that, at that point, there were no portable mobile telephones, only vehicle-installed terminals with roof-top antennas. Over time the use of mobile telephony became increasingly popular and the idea was born that the network needed to be divided into more and smaller cells, accommodating more capacity for more users, implementing full mobility for the traffic and enabling the system to hand over traffic between these small cells.

From this initial concept several cellular systems were developed over time and in different regions of the world. The first of these cellular systems was analog voice transmission, and some 'data transmission' modulated into the voice channel for signaling the occasionally handover or power control command.

Some of the most used standards were/are AMPS, D-AMPS, TACS, PCS, CDMA, NMT, GSM, DCS and UMTS (WCDMA).

Indoor Radio Planning: A Practical Guide for GSM, DCS, UMTS and HSPA Morten Tolstrup

AMPS

AMPS (Advanced Mobile Phone System) is the North American standard and operates in the 800 MHz band. The AMPS system was also implemented outside North America in Asia, Russia and South America. This is an analog system using FM transmission in the 824–849 and 869–894 MHz bands. It has 30 kHz radio channel spacing and a total of 832 radio channels with one user per radio channel.

D-AMPS

D-AMPS (Digital Advanced Mobile Phone System) evolved from AMPS in order to accommodate the increasingly popular AMPS network with fast-growing traffic and capacity constraints. The D-AMPS system used TDMA and thus spectrum efficiency could be improved, and more calls could be serviced in the same spectrum with the same number of base stations.

TACS

TACS (Total Access Cellular System) was also derived from the AMPS technology. The TACS system was implemented in the 800–900 MHz band. First implemented in the UK, the system spread to other countries in Europe, China, Singapore, Hong Kong and the Middle East and Japan.

PCS

PCS (Personal Communications System) is a general term for several types of systems developed from the first cellular systems.

CDMA

CDMA (Code Division Multiple Access) was the first digital standard implemented in the USA. CDMA uses a spread spectrum in the 824–849 and 869–894 MHz bands. There is a channel spacing of 1.23 MHz, and a total of 10 radio channels with 118 users per channel.

NMT

NMT (Nordic Mobile Telephony) was the standard developed by the Scandinavian countries, Denmark, Norway and Sweden, in 1981. Initially NMT was launched on 450 MHz, giving good penetration into the large forests of Sweden and Norway, and later also deployed in the 900 MHz band (the band that today is used for GSM). Being one of the first fully automatic cellular systems in the world (it also had international roaming), the NMT standard spread to other countries in Europe, Asia and Australia.

GSM

GSM (Global System for Mobile communication) was launched in the early 1990s, and was one of the first truly digital systems for mobile telephony. It was specified by ETSI and originally intended to be used only in the European countries. However GSM proved to be a very attractive technology for mobile communications and, since the launce in Europe, GSM has evolved to more or less a global standard.

DCS

Originally GSM was specified as a 900 MHz system, and since then the same radio structure and signaling system have been used for DCS1800 (Digital Cellular Telecommunication System). The GSM basic has also been applied to various spectra around 800–900 and 1800–1900 MHz across the world, the only difference being the frequencies.

UMTS (WCDMA)

After the big global success with the second generation (2G) GSM and the increased need for spectrum efficiency and data transmission, it was evident that there was a need for a third-generation mobile system. UMTS was selected as the first 3G system for many reasons, mainly because it is a very efficient way to utilize the radio resources – the RF spectrum. WCDMA has a very good rejection of narrowband interference, is robust against frequency selective fading and offers good multipath resistance due to the use of rake receivers. The handovers in WCDMA are imperceptible due to the use of soft handover, where the mobile is serviced by more cells at the same time, offering macro-diversity.

However there are challenges when all cells in the network are using the same frequency. UMTS is all about noise and power control. Strict power control is a necessity to make sure that transmitted signals are kept to a level that insures they all reach the base station at the same power level. You need to minimize the inter-cell interference since all cells are operating on the same frequency; this is a challenge.

Even though soft handovers insure that the mobile can communicate with two or more cells operating on the same frequency, one must remember that the same call will take up resources on all the cells the mobile is in soft handover with. The handover zones need to be minimized to well-defined small areas, or the soft HO can cannibalize the capacity in the network.

UMTS has now become the global standard and has been accepted throughout the world. Several upgrades that can accommodate higher data speed HSDPA (High Speed Downlink Packet Access) and HSUPA (High Speed Uplink Data Access) can service the users with data speeds in excess of 10 Mbps.

There are several current considerations about converting the current GSM900 spectrum into UMTS900, giving a much higher spectrum efficiency, and better indoor RF penetration.

2.1.2 Radio Transmission in General

Several challenges need to be addressed when using radio transmission to provide a stable link between the network and the mobile station. These radio challenges are focused around the nature of the propagation of radio waves and especially challenges of penetrating the radio service into buildings where most users are located these days.

The challenges are mainly radio fading, noise control, interference and signal quality. These challenges will be addressed throughout this book, with guidelines on how to design a high-performing indoor radio service.

2.1.3 The Cellular Concept

After the initial success with the first mobile system, it was evident that more capacity needed to be added to future mobile telephony systems. In order to implement more capacity to accommodate more users in the increasingly more popular mobile telephony systems, new principles needed to be applied. The new concept was to divide the radio access network into overlapping 'cells', and to introduce a handover functionality that could insure full mobility throughout the network, turning several masts into one coherent service for the users.

Dividing the network into cells has several advantages and challenges. The advantages are:

- *Frequency reuse* – by planning the radio network with relative low masts with limited coverage area, compared with the first mobile systems, you could design a radio network where the cells will not interfere with each other. Then it is possible to deploy the same radio channel in several cells throughout the network, and at the same time increase the spectrum and radio network efficiency thanks to frequency reuse.
- *Capacity growth* – the cellular network could start with only a few cells, and as the need for better coverage and more capacity grew, these large cells could be split into smaller cells, increasing the radio network capacity even more with tighter reuse of the frequencies (as shown in Figure 2.1).

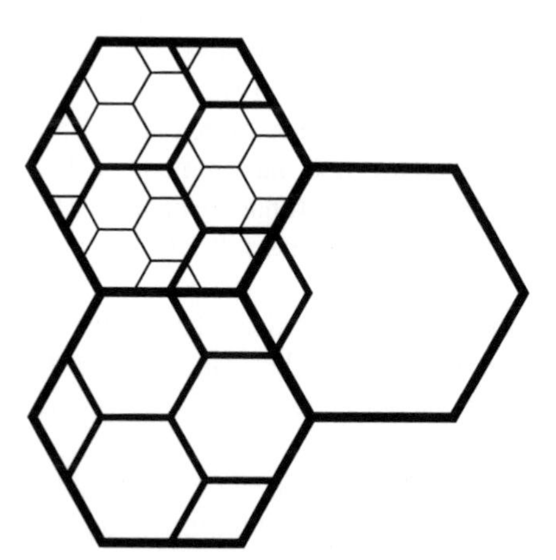

Figure 2.1 The cell structure of a cellular radio network. Cells will be split into smaller cells as the network evolves, and the capacity need grows

- *Mobility* – it is paramount for cellular networks that handovers are possible between the cells, so the users can roam through the network with ongoing connections and no dropped calls. With the advent of the first cellular systems, mobile users could now move around the network, utilizing all the cells as one big service area.

The challenges are:

- *Network structure* – when deploying the cellular structure one needs to design a theoretical hexagonal roll-out, by deploying omni, three-sector or six-sector base stations, and make sure the cells only cover the intended area. It is important that the cells only cover the intended area, and that there are no 'spill' of radio coverage to the coverage area of other cells in the network.
- *Mobility* – to offer full mobility in a cellular network you need to introduce handover among the cells, in order for the call to proceed uninterrupted when a mobile moves from one cell service area to a new one. The handover function creates a need to evaluate the radio quality of a potential new, better cell, compared with the current cell servicing the call. Complex measurements and evaluation procedures of adjacent cells are introduced in order to control the handover decision and mobility management.
- *Power control* – to service the mobile user with coherent radio coverage throughout the network area is a challenge. On the uplink from the mobile to the base station, one of the challenges is for the mobile to reach the base station. However, it is also a challenge that sometimes the mobiles can be close to the serving base station, and at the same time, the same base station using the same radio channel might also serve a mobile at 20 km distance. This can cause a problem known as 'the near–far problem', i.e. the mobile close to the base station will generate a high-level signal in the base station receiver and will simply overpower the low signal from the distant mobile. To overcome this issue, power control of both the downlink (the signal transmitted from the base station to the mobile) and the uplink (the signal transmitted by the mobile to the base station) was introduced. This not only avoids overpowering the receiving end, but also limits the overall interference in the network due to the lower average transmitted power level on both uplink and downlink. It also increases radio quality and spectrum efficiency by tighter frequency reuse among the cells. The challenge with power control is the need to evaluate both the signal strength and the quality of the radio signal received in both ends of the link, in order to adjust the transmitted power to an appropriate level, and first and foremost to make sure that the link is not overpowered.

2.1.4 Digital Cellular Systems

From only performing pure voice connections on the mobile networks around the world, users slowly started to request the possibility of sending and receiving data signals via the mobile connection. The first applications mobile data were the use of analog modems to transmit and receive fax signals over a purely analog mobile connection. Soon it became clear that there was a need for real data transmission via the mobile network.

In order to accommodate the need for mobile data transmission, better voice quality and spectrum efficiency digital technology were applied to the cellular systems. Many competing 2G (second generation) systems were developed, but on a global basis the most successful has been the GSM system

By using digital transmission several advantages were achieved:

- Better spectrum efficiency.
- Lower infrastructure cost on the network side.

- Reducing fraud, by encryption of data and services.
- Lower terminal cost.
- Reduced transmission of power from the mobile, making it possible to have long-lasting batteries for hand-held terminals.

This book mainly focuses on indoor radio planning for 2G (GSM/DCS) and 3G (UMTS) systems. A more thorough introduction to the principles of GSM/UMTS and HSPA is appropriate and is to be found in the following sections.

2.2 Introduction to GSM

To perform indoor GSM radio planning, the radio planner needs some basic information about the network, signaling, etc. This section is not intended as a complete GSM training session, but rather to highlight the general parameters and network functions in GSM networks. For more details on the GSM please refer to [1].

2.2.1 GSM

GSM was launched in the early 1990s, as one of the first truly digital systems for mobile telephony. It was specified by ETSI and originally intended to be used only in Europe. However, GSM proved to be a very attractive technology for mobile communications, and since the launch in Europe, GSM has evolved to be more or less the first truly global standard for mobile communication. Even though GSM is relatively old (most mobile network generations last about 7–10 years), it is still being rolled out all over the world, and in particular, the focus on high indoor usage of mobile telephony has motivated a need for dedicated indoor coverage solutions.

DCS

GSM was originally specified as a 900 MHz system. Since then the same radio structure and signaling have been used for DCS1800. The GSM basic has also been applied to various spectra around 800–900 and 1800–1900 MHz across the world, the only difference being the frequencies.

Security Features

All GSM communication is data; essentially voice is converted from analog to data and back again. Compared with the analog mobile telephony, the GSM system is much more secure and is impossible to intercept. GSM applies digital encryption, authentication of call, checking and validation of calls in order to prevent unauthorized use of the network. For the first time in mobile telephony systems, the identification of the user in the system was not tied to the mobile equipment with an internal telephony number (often an EPROM you had to replace), but to the user by the use of a SIM card (Subscriber Identity Module). The SIM is identified using an IMSI (International Mobile Subscriber Identity) – yes, International – for the first time the mobile user could roam in other countries or networks on a global basis and use their mobiles and supplementary services as if they were in the own home network.

2.2.2 GSM Radio Features

Compared with the analog networks at the time, such as NMT and AMPS, new and sophisticated radio network features were introduced in GSM. One new feature was DTx (discontinuous transmission), which enabled the base station/mobile to only transmit when there was voice activity. This prolongs the battery life of the mobile station, but also minimizes both uplink and downlink interference due to the lower average power level. DTx detects the voice activity on the line, and during the period of no voice activity the radio transmission stops. In order to prevent the users from hanging up when there is complete silence on the link, believing the line to be dead, the system generates 'comfort noise' to emulate the natural background noise of the individual call.

The GSM system can also use frequency hopping, accommodating channel selective fading by shifting between radio frequencies from a predefined list 217 times per second. This hopping is controlled by an individual hopping sequence for each cell.

GSM Data Service

Analog networks at the time were only able to transmit data via analog modems, and only up to a maximum data speed of about 300 baud (0.3 kbps). The possibility of higher data rates on mobile networks with the advent of GSM phase one, enabling data transfer up to 9.6 kbps, was stunning. Today this seems really slow, but at the time most of us were not sure how we should ever be able to utilize this mobile 'broadband' to its full potential – remember, this was before Windows 1.0 was even launched! You were lucky if you could do 300 baud over your fixed telephone line. Over time new more sophisticated packet data transmission modulations have been applied to GSM, GPRS and EDGE, now exceeding 200–300 kbps per data call.

During the specification of the signaling of GSM, it became clear that there was a slight overhead in the data resources, and that this overhead could be used to transmit short portions of limited data – this eventually resulted in a new data service being introduced. The data service that was originally specified and launched as a paging service was SMS. SMS was originally planned to be used by the network to announce voice mails, cell information and other text broadcast services to individuals or all mobiles within an area of the network.

The first GSM mobiles were not even able to transmit SMS; no one was expecting mobile users to struggle with using the small 10-digit keyboard to key in a text message. The concept was that you could dial up a specific support number with a manned message center that could key in a text direct into the network, and charge accordingly for this service. Well, we all know today that, even though SMS has many limitations, the fact remains that this is still the most successful mobile data service and a huge revenue generator for the mobile operators.

Supplementary Services in GSM

With GSM several new services were introduced, such as call forwarding, call waiting, conference calls and voice mails. All this today seems only natural, but it was a major leap for mobile communications. It added new value for mobile users and the network operators, making mobile communication a tool you could finally use as the primary communication platform for professional business use, and a real alternative to fixed line telephone networks, especially if you had good indoor coverage inside your building.

The First Global System, from Europe to the Rest of the World

These new features introduced with GSM – SMS, data service, supplementary services etc. – were a direct result of a good standardizing work by an ETSI/GSM Mou (Memorandum of understanding). It enabled the GSM standard to spread rapidly across the rest of the world, becoming the first truly global standard – even though some countries never implemented the standard.

The GSM Radio Access Structure

The most important part of the GSM network for the indoor radio planner is the GSM radio structure and air interface. In the following the main parameters and techniques are introduced.

Idle Mode

A mobile is considered in idle mode when it is not engaged in traffic (transmitting). The mobile will camp on the strongest cell, monitor system information and detect and decode the paging channel in order to respond to an incoming call.

Dedicated Mode

Mobiles engaged in a call/transmitting data to the network are considered to be in dedicated mode. In dedicated mode a bi-directional connection is established between the mobile station and the network. The mobile will also measure neighboring cells and report this measurement information to the network for handover and mobility management purposes.

FDD + TDD

GSM uses separate frequency bands for the uplink and downlink connection (as shown in Figure 2.2). The two bands are separated by 45 MHz on GSM900, and by 95 MHz on GSM1800. GSM has the mobile transmit band in the lower frequency segment.

By the use of FDD (Frequency Division Duplex), the spectrum is divided into radio channels, each radio channel having 200 kHz bandwidth. Each of these 200 kHz radio channels is then divided in time using TDD (Time Division Duplex) into eight channels (time slots) to be used as logical and traffic channels for the users.

GSM is most commonly assigned to use on the 900 and 1800 MHz frequency bands (as shown in Figure 2.3). However depending on local spectrum assignment in different regions, other band segments are used, such as EGSM or GSM1900.

The GSM Spectrum

The GSM radio spectrum is divided into 200 kHz radio channels. The standard spectrum on GSM900 is 890–915 MHz for the uplink and 935–960 MHz for the downlink spectrum (CH1–CH124), a total of 2 × 35 MHz paired spectra, separated by 45 MHz duplex distance. The DCS spectrum (as shown in Figure 2.4) uses 1710–1785 MHz for the uplink and

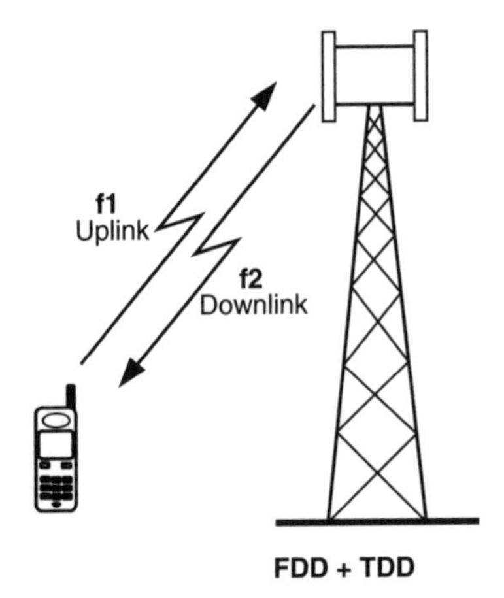

Figure 2.2 GSM uses both frequency division (FDD) in a radio channel of 200 kHz and time division (TDD) in TDMA frames of eight time slots

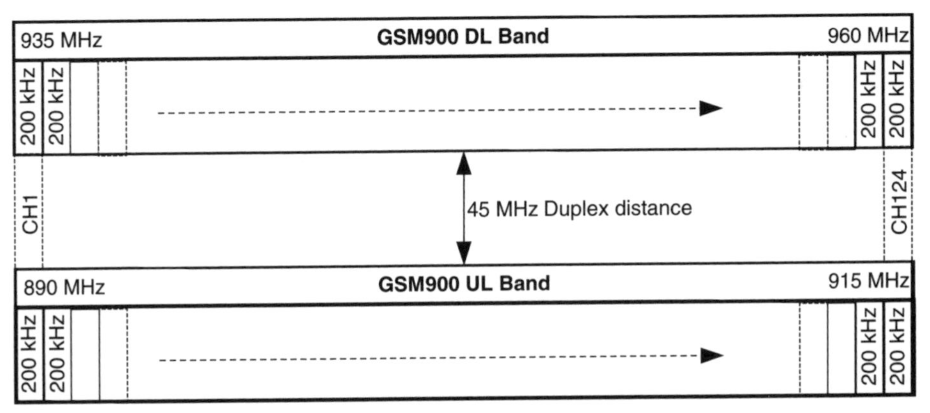

Figure 2.3 The standard GM900 frequency spectrum. GSM uses 124 200 kHz radio channels

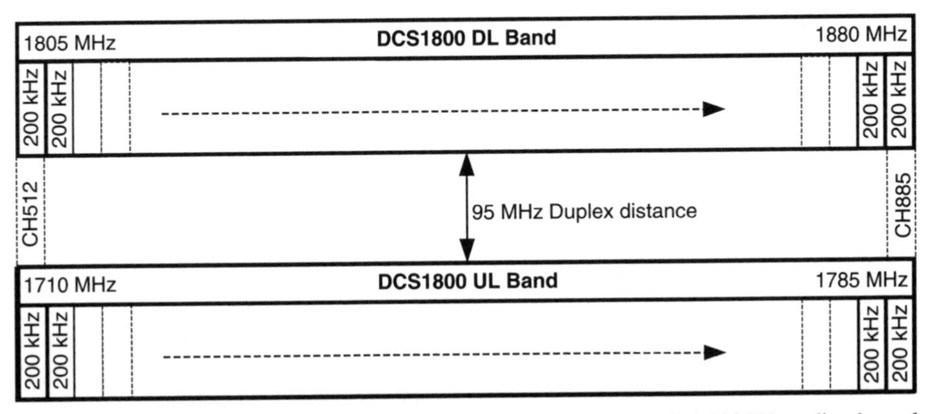

Figure 2.4 The standard DCS1800 frequency spectrum, DCS, uses 374 200 kHz radio channels

1805–1880 MHz for the downlink spectrum (CH512–CH885), a total of 2 × 75 MHz paired spectra separated by 95 MHz duplex distance.

In some regions other spectrum allocations for GSM around 900 and 1800 MHz have been implemented, such as EGSM or DCS1900.

TDMA

In addition to the frequency separation, GSM uses time separation, TDMA (Time Divided Multiple Access); TDMA allows multiple users to occupy the same radio channel. The users will be offset in time, but still use the same 200 kHz radio channel. The radio spectrum (as shown in Figure 2.3) in GSM (900) is separated into 124 radio channels, each of these radio channels then separated into eight time-divided channels called time slots (TSL) (as shown in Figure 2.5).

Each TDMA frame time slot consist of 156.25 bits (as shown in Figure 2.6, 33.9 kbs per TSL or 270.8 kbps per frame), of which 114 (2 × 57) are coded data, including forward error

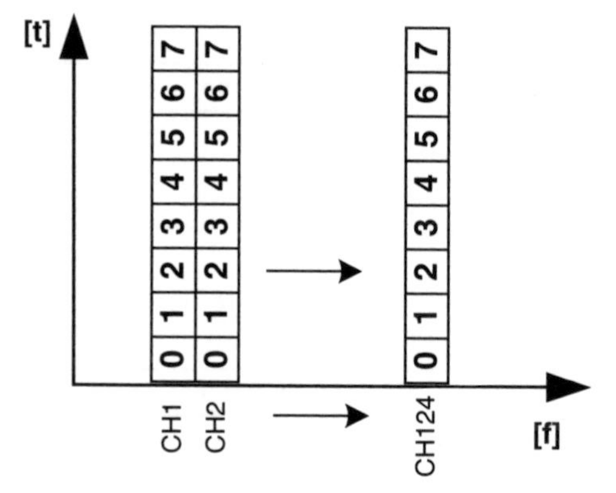

Figure 2.5 Each of the 200 kHz radio channels is time divided into eight channels, using individual time slots, TSL0–7

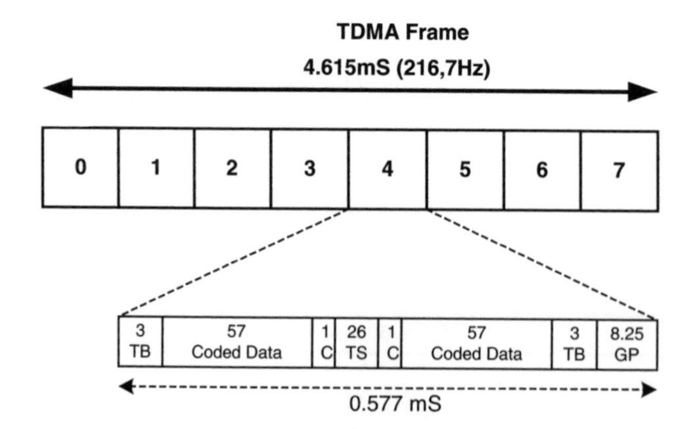

Figure 2.6 The TDMA frame in GSM uses eight time slots, 0–7

correction. All transmitted information is transferred in blocks of 456 bits, divided into four time slot periods ($456 = 4 \times 2 \times 57$). The maximum net bit rate is per time slot is 13 kbps, excluding the error correction.

The GSM Cell Structure

As we know, the GSM network consists of cells (as shown in Figure 2.7), the cells operate on different frequencies and each is identified by the CI (Cell Identity defined in the parameter settings in the network) and the BSIC (Base Station Identity Code). Depending on the geography and the design of the network, the frequencies can be reused by several cells, provided that the cells do not 'spill' coverage into other cells' coverage areas.

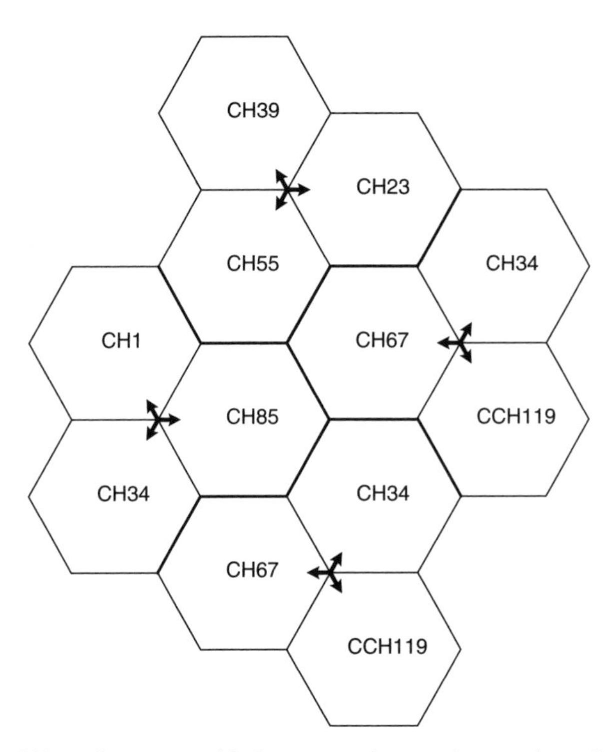

Figure 2.7 GSM cell structure with three macro base stations, each serving three cells

Capacity upgrade for a cell is done by deploying more radio channels in the cell; an upgrade of a cell with one extra radio channel will give eight extra logical channels that can be used for signaling or traffic. In urban areas it is normal to have six to 12 radio channels per cell. In a macro network the typical configuration is to design with three sector sites (as shown in Figure 2.7), where each mast is serving three individual cells.

Most GSM networks were initially deployed as GSM900, but in many countries DCS1800 has been implemented at a later stage, as an extra 'layer' on top of the GSM900 network to add capacity to the network to fulfill the fast growing need for more traffic channels and spectrum.

In most cases (not all), the DCS1800 network has a shorter service range due to the higher free space loss caused by the higher frequency (6 dB). If DCS 1800 is added onto the existing GSM900 network, using the initial mast separation designed for GSM900 might cause the 1800 cells not to overlap adjacent DCS1800 cells; therefore all handovers of traffic between cells have to be done via the GSM900 cells. This puts significant signaling load on GSM900 and increases the risk of dropped calls during a more complex handover. Therefore it is preferred to decrease the inter-site distance between the masts, then the network has coherent and homogeneous coverage on DCS1800. In practice DCS1800 has mostly been implemented to add capacity in urban environments, where the inter-site distance is short even on GSM900 and will provide a good foundation for DCS1800 deployment. In a dual-layer GSM network (900 + 1800) 'Common BCCH' (broadcast control channel) is often used to combine all signaling into one layer and free up more TSLs in the cell for more user capacity.

2.2.3 Mobility Management in GSM

Managing the service quality and mobility of the users in a cellular network like GSM is important. The network must insure the best possible service quality and hand over the mobiles between cells when the users roam through the network. Complex measurements for radio strength and quality have been applied to cater for the mobility management.

Radio Measurements

In dedicated mode (during a call) the base station measures the signal level and radio quality/ BER (bit error rate) of the received signal from the mobile. The mobile will also measure the signal level and the quality of the serving cell, as well as the signal levels of the neighboring cells defined by the network in the 'neighbor list'. The mobile reports these measurements (measurement reports) to the network for quality evaluation, power control and handover evaluation.

Rx-Quality

The radio quality (bit error rate) of both the uplink and downlink connection is divided into eight levels and defined as 'Rx-Qual' 0–7, 0 being the best quality. Rx-Qual 7 will start a network timer that will terminate the call if the quality does not improve within a preset time. If the Rx-Qual rises above 4 it will degrade the voice quality of a speech call to an extent that is noticeable to the user. An Rx-Qual better than 3 must be striven for when planning indoor solutions; preferably, an Rx-Qual of 0 should be insured throughout the building.

The main cause of degradation of the Rx-Qual is interference from cells using the same radio frequency, or by too low a reception level of the servicing signal.

Rx-Level

The radio signal strength is measured as Rx-Level and is reported in a range from 0 to 64. The Rx-Level is the signal level over -110 dBm. Rx-Level 30 is: $-110 + 30\,\text{dB} = -80\,\text{dBm}$.

The mobile will also measure the radio quality and signal strength of the serving cell, as well as the other adjacent cells (specified by the neighbor list received from the base station controller, BSC). The mobile measures and decodes the BSIC (base station identity code) and reports the strongest cells back to the BSC for handover evaluation/execution.

Radio Signal Quality Control

Based on the radio measurements evaluation performed by the BSC and the mobile, the network (the BSC) will control and adjust the various radio parameters, power level, timing and frequency hopping, etc., in order to insure the preset margin for radio signal level and radio quality is maintained

A neighbor list is defined in the network and broadcast to the mobile in service under each cell, with the frequencies and details of the adjacent cells to the serving cell. This list of frequencies (the BCCH of adjacent cells) is used by the mobile to measure the adjacent cells. These measurements are used for the handover evaluation. Normally this is a 'two-way' list, enabling the mobile to handover in both directions, but it is possible to define 'one-way' neighbors, removing the possibility of the mobile handing back to the previous serving cell. This use of 'one-way' neighbors can be useful for optimizing indoor systems, especially in high-rise buildings, in order to make sure that the mobiles inside a building that camp unintentionally on an outdoor cell can hand in to the indoor system, but not hand back to the outdoor network; see Section 5.5 for more details.

Power Control

The network will adjust the transmit power from the mobile and the base station; the downlink power control for the base station is optional. The primary function of the power control is insure that the radio link can fulfill the preset quality parameters, and most importantly to make sure that mobiles close to the base station will not overpower the uplink and desensitize the base station, resulting in degraded service or dropped calls.

Adjustable target levels for minimum and maximum Rx-Level are set in the network, individually for each cell. These Rx-Levels are the reference for the power control function, and will adjust the transmit power to fulfill the requirements and keep the Rx-Levels within the preset window. Separate triggers for Rx-Qual can cause the power to be adjusted up, even if the Rx-Level is already fulfilled in order to overcome interference problems.

The power control in GSM is done in steps of 2 dB, and the power steps can be adjusted in intervals of 60 ms.

GSM Handovers

Based on the measurement reports sent from the mobile to the network, with information on the radio signal level of serving and adjacent cells, if the preset levels for triggering the handover are fulfilled, the network might evaluate that the mobile could be served by a new cell with better radio signal level and command the mobile to hand over to a new serving cell. The handover requires the establishment of a new serving traffic channel on the new cell and, if the handover is successful, the release of the old serving traffic channel. Normally the handover is not noticeable for the user, but there will be 'bit stealing' from the voice channel to provide fast signaling for the handover. This 'bit stealing' can be detected as small gaps in the voice channel, and if indoor users are located in an area where there is constantly handover between two cells ('ping pong handover'), the voice quality on the link will be perceived as degraded by the user, even if the Rx-Qual remains 0.

The handover trigger levels and values can be adjusted in the network individually for each handover relation between the cells. Various averaging windows, offsets, can also be set. For planning and optimizing indoor systems it is very important that the radio planner knows the basics of the parameter set for the specific type of base station network implemented. This will enable the radio planner to optimize the implemented indoor solutions to maximum performance, and compatibility with the adjacent macro network.

The handover in GSM can be triggered by:

- Signal level uplink or downlink, to maintain quality.
- Signal quality uplink or downlink, to maintain quality.
- Distance, to limit the cell range.
- Speed of mobiles.
- Traffic, to distribute load among cells.
- Maintenance – you can empty a cell by network command.

Intracell Handover

If the Rx-Qual is bad but no adjacent cells with higher received signal level are detected and reported by the mobile, the network may initiate an intracell handover, assigning a new TSL or a new radio channel on the same cell. This can improve the quality of the call if interference hits the used time slot or frequency.

Level-triggered Handover

Preset margins for Rx-Level can be set to trigger a handover level. Standard settings will be so that when an adjacent cell measured in the neighbor list becomes 3–6 dB more powerful, the mobile is handed over to the new cell. However, handover margins can actually be set to negative levels, triggering the handover even if the adjacent cell is less powerful.

Quality-triggered Handover

To insure that the Rx-Qual of the radio link is maintained, preset trigger levels for Rx-Qual can be set to 'overwrite' the level handover, initiating a handover if the Rx-Qual degrades below the trigger level. The typical trigger level for the Rx-Quality handover is Rx-Qual > 3. If the quality degrades beyond this level, the network will initiate a handover. It does not matter if the Rx-Level margin is fulfilled or not.

Distance-triggered Handover
The maximum cell size can be set, and handovers triggered when exceeding this distance. This can be useful for indoor cells, making sure that the indoor cell will not service unintended traffic outside the building, but remember to take into account the delay of the DAS, this will offset the cell size.

Speed-triggered Handover
It is also be possible for the network to detect if the mobile is moving fast or slow through the serving cell. This can be utilized by the network by limiting certain cells to serve only slow-moving (hand-held) mobiles. This can be very useful when optimizing indoor GSM systems, in order to avoid vehicular traffic camping on the indoor cell if the indoor cell leaks out signal to a nearby road.

Traffic-triggered Handover
It is possible to predefine a certain traffic level for a cell to start handing over calls to adjacent cells, and thereby offload the cell. This can be very useful for cells that have periodic hot-spot traffic, and might save the expense of deploying extra capacity for servicing only the occasionally peak load. However this is only possible if the adjacent cells are able to service the traffic at the expected quality level.

Maintenance Handover
It is also possible to order the network to empty a specific cell. Similar to the traffic-triggered handover, the cell will push all calls that can be handled by adjacent cells out and wait for the remaining calls to terminate. During this mode the cell will not set up any new calls, and once the cell is empty, the cell can be switched off for service and maintenance. By emptying an operational cell using this method, there will only be limited impact on the users in terms of perceived network quality. This method is to be preferred when doing maintenance in the network rather than just switching off the cell and dropping all the ongoing calls.

The Complexity of the GSM Handovers

In GSM there are basically three handovers seen from the network side (as shown in Figure 2.8), involving several networks depending on the handover type.

Intra-BSC Handovers
Intra-BSC handovers are between cells serviced by the same BSC. This is the least complex handover seen from the network side. Also, there are possibilities for using functions that make the handover smoother and more likely to succeed – pre-sync (chained cells), etc. These functions can be very useful in tricky handover situations, optimizing the handover zone in tunnel systems, for example.

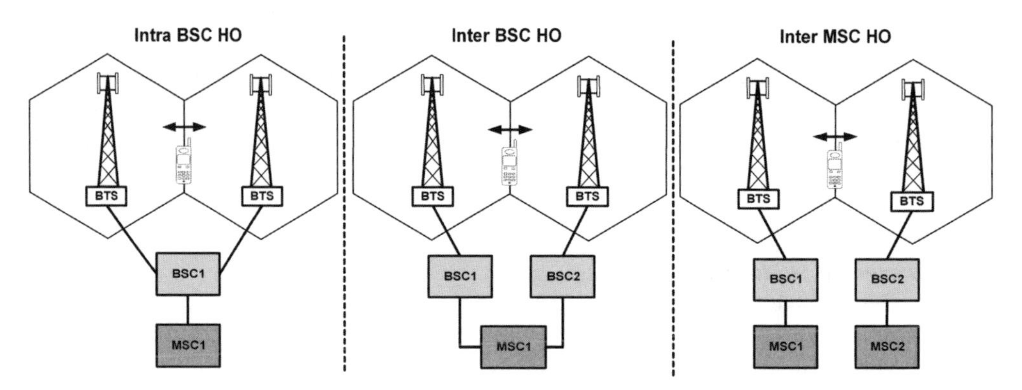

Figure 2.8 Three types of network involvement of the GSM handover: intra-BSC, inter-BSC and inter-MSC handovers (HO)

Inter-BSC Handovers

Inter-BSC handovers are done between cells serviced by different BSCs. The handover function is basically the same as for intra-BSC handovers, but with limited functions like pre-sync, etc.

Inter-MSC Handover

Inter-MSC handovers are controlled by the MSC, and the signaling is more complex due to the implication on several interfaces and network elements. The number of inter-MSC handovers should be limited in high traffic areas (indoors).

Location Management in GSM

The GSM radio network is divided into regions (location areas).The most important motivation for dividing the radio network into several location areas is to limit the paging load. The mobile will monitor and update the network whenever it detects that it is being serviced by a new location area. This is done by the mobile by monitoring the location area code (LAC) broadcast from all the cells in the system information. In the VLR (visitor location register) in the network, the information about the LAC area the mobile was last updated in is stored. Using several location areas, the network can limit the signalling of paging messages by only having to broadcast the paging messages to the mobiles currently serviced by a specific LAC. If the concept of location areas was not implemented, the network would have had to page all mobiles in all cells each time there was an incoming call. This would have severely loaded the paging signalling resources and increased paging response time beyond an acceptable level.

Timing Advance

All the active traffic in a GSM cell is synchronized to the timing and clock of the base station, in order to insure that the time slots are sent and received at the correct time. The

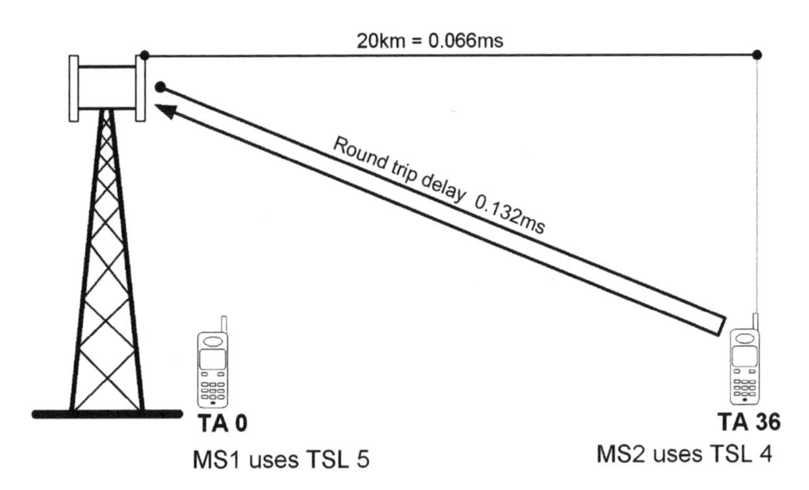

Figure 2.9 Mobiles in traffic on the cell have to compensate for the delay to the base station using timing advance

TDMA frame is 4.615 ms long, and each TSL is 577 μs. Radio waves travel with the speed of light, about 300 000 km/s. Analyzing the example in Figure 2.9 with two mobiles in traffic, one is close to the base station (MS1) and the other (MS2) is 20 km distant. Owing to the propagation delay caused by the difference in distance to the serving base station, the mobiles will be offset in time by about 66 μs (one-way). Both mobiles are locked to the same serving base station's synchronization 'clock'. Actually the distant mobile will in this case be offset by 2×66 μs as the signal has to travel from the base station, to the mobile and back so the synchronization of the distant mobile is offset/delayed in time by about 132 μs. This delay corresponds to about 0.3 TSL, and will cause the transmitted TSL (TSL4) from the distant mobile to overlap with the next TSL (TSL5 in this example), used by the mobile closest to the base station (as shown in Figures 2.9 and 2.10).

To overcome this problem, one could consider including more guard time between the time slots, but this would come at the expense of lower spectrum efficiency. In order to

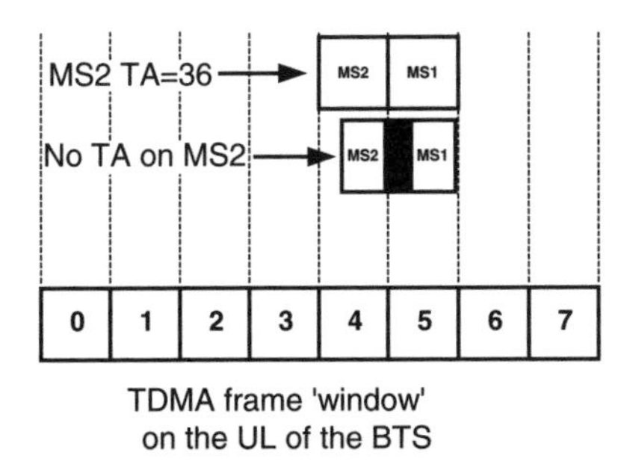

Figure 2.10 If the GSM network did not used timing advance, traffic from mobiles at different distances from the base station would drift into adjacent TSL, due to the propagation delay

preserve spectrum efficiency, the problem was solved using the concept of timing advance (TA). In practice the problem is solved in GSM by having the base station continuously measure the time offset between its own burst schedule and the timing of the received bursts from the mobiles. When the base station detects that the TSL from the mobile has started to drift close to the next or previous TSL, the base station will command a TA step-up or -down. The network simply commands the mobile to send the TSL sooner, and the offset will depend on the distance between the mobile and the serving mast. This timing advance enables the GSM system to compensate for the delay over the air interface.

The use of timing advance insures high spectrum efficiency (low guard times between time slots) and at the same time solves the issue of the timing difference between distant mobiles and mobiles near the base station.

The solution to this problem shown in Figures 2.9 and 2.10 is for the network to command the mobile MS2 to use TA36, causing the mobile to transmit early. In this example, it insures that the transmitted TSL4 from mobile station 2 will not overlap with TSL5 from mobile station 1.

The GSM system is able to handle timing offset up to 233 μs, corresponding to a maximum distance from the base station of 35 km. It has divided this timing offset into 64 timing advance steps, TA0–TA63, with a resolution of 550 m per TA step. However, since the initial GSM phase one specification, several manufacturers are now able to support 'extended cells', thereby supporting traffic up to 70 km and in some cases even further.

TA and Indoor Systems

When designing indoor-distributed antenna systems, you need to keep in mind that all the antenna distribution equipment, passive and active, will delay the signal timing, and offset the TA. Even in a passive cable, the propagation speed and velocity are about 80% of the free space speed of light of 300 000 km/h. Active systems, amplifiers, optical cables etc. might offset the TA by five to eight steps, depending on the equipment, cable distances, etc.

When setting the maximum cell service range, it is very important to remember to compensate for this timing offset, introduced by the indoor antenna system. One must not be tempted to set the maximum TA to one or two per default for indoor systems, as the indoor cell might not pick up any traffic at all. The correct TA setting for the maximum service distance can easily be evaluated on-site by the user of a special test mobile station, where the TA step can be decoded in the display of the mobile, and the maximum cell service range adjusted accordingly.

2.2.4 GSM Signaling

This section will present an overview of the channel types, and basic signaling and mobility management in GSM. Not in a detailed level, but just to introduce the principles.

Logical Channels

Each TDMA frame consists of eight TSL, TSL0–7, which can be used for several types of channels. Traffic and control channels sometimes even combine channels. Several traffic time slots can be assigned to the same user for higher data rates.

Traffic Channels (TCH)

The TCH is the channel that contains all the user data for speech or data. The TCH can be a full-rate channel, or an enhanced full-rate channel with up to 13 kbs user data or good voice quality. It is also possible to divide one 13 kbps TCH into two 6.5 kbps channels, thus doubling the number of channels in the same TDMA frame. One TDMA frame can carry eight full-rate TCHs or voice calls, or 16 calls of 6.5 kbps channels; for voice this is referred to as 'half-rate' and will to some extent degrade the voice quality, although in normal circumstances it is barely noticeable for the voice user.

By the use of half-rate, the GSM network can double the voice capacity, using the same number of radio channels. The half-rate codex can be permanently enabled in the cell, or most commonly the half-rate service will automatic be activated when the traffic on the cell rises close to full load. The mechanics of the half-rate is, for example: if a cell has six ongoing calls, the network starts to convert these calls into half-rate traffic rearranging the users from the 13 kbps full rate channel to two new 6.5 kbp half-rate channels. Now, instead of taking up six TSLs, they will only use three TSLs, thus freeing capacity on the cell for new traffic.

This function for automatic allocation of half-rate channels is referred to as 'dynamic half-rate' and the trigger level for the capacity load needed to start the conversion can be adjusted in the network.

When using half-rate traffic channels; one radio channel can carry up to 16 voice calls. This is a very cost-effective tool for providing high peak load capacity, for big sport arenas, exhibition venues, etc. However it is not recommended to use permanent half-rate for high profile indoor solutions, where users expect perfect voice quality.

Control Channels

In order to handle all the signaling for the mobility management, traffic control, etc., several control channels are needed. Remember these are logical channels and might not be active constantly (most are not), but will be combined into using the same TSL as other control channels.

CCCH (Common Control Channels) and DCCH (Dedicated Control Channels)

One set of CCCH (BCCH + PCH + AGCH for downlink and RACH for uplink) are mapped either alone or together with four SDDCH logical channels into one physical TDMA time slot. The dedicated control channels, SDDCH, SACCH and FACCH, are not mapped together with other control channels. These will have a dedicated TSL or 'steal' some traffic TSL for a period of time when needed.

BCCH

The broadcast control channel (BCCH) is a downlink-only channel, transmitted from the base station to the mobile station. The BCCH broadcasts a lot of different information for mobiles: system information, cell identity, location area, radio channel allocations, etc. The BCCH is assigned using TSL0 of the first radio channel in the cell.

PCH

The paging channel (PCH) is a downlink-only channel, broadcasting all the paging signals to all mobiles currently registered in the location area (LAC). When detecting a page for the actual mobile (IMSI), the mobile will respond via the RACH. Mobiles in the same LAC will be divided into paging groups. The paging group is based on the IMSI (telephone number), so the mobile knows when to listen for specific paging signals belonging to its specific paging group. This enables the mobile to only use resources for paging reception in the short time when its own paging group is active. The mobile can then 'sleep' during the period when the other paging groups are broadcast, thus saving power and extending battery life. This is referred to as a DRX, discontinuous reception.

RACH

The random access channel (RACH) is used when a mobile needs to access the network. It transmits on the RACH as an uplink-only channel. The mobile will perform random access when attaching to the network, when it is switched on, when responding to paging, when setting up a call and when location updating from one LAC to another.

AGCH

The access grant channel (AGCH) is a downlink-only channel that is used by the base station to transmit the access grant signal to a mobile trying to access the network on the RACH.

SDDCH

The stand-alone dedicated control channel (SDDCH) is used for authentication, roaming, encryption activation and general call control. The SDCCHs are grouped into four (SDDCH/4) or eight (SDCCH/8), together with one set of CCCHs and mapped into one time slot.

SACCH

The slow associated control channel (SACCH) is linked with a specific TCH or SDCCH for radio signal control and measurements. One block of SACCH (456 bits) contains one measurement report, sent every 480 ms, reported to the BSC via the base station. The SACCH will use a part of the TCH or SDCCH for the SACCH.

FACCH

The fast associated control channel (FACCH) is for immediate signaling associated with handover commands for the specific call. The FACCH will be transmitted using the TCH, by means of 'bit-stealing' of the TCH (the small gaps in the voice stream that are noticeable during a GSM handover).

GSM Signaling Procedures

Without going into specific details, we will take a quick look at what type of signaling procedures are used in GSM in order to handle mobility and the overall function of the mobile network, such as:

- Location management.
- Paging.
- Accessing the network.
- Authentication and encryption.
- Radio signal quality control.
- Radio measurements.
- Handover.

Location Updating and Roaming

The network is divided into sub-areas called location areas, identified by a specific location area code (LAC). When a mobile detects a new LAC broadcast by the base station, it will access the network and perform a location update, so the network (the VLR) is aware which location area is currently servicing the mobile, and to which location area to send any paging signal for the specific mobile.

Paging

The mobile is not permanently connected to the network. In idle mode the mobile will monitor the PCH in the serving cell. In the case of an incoming call the mobile will respond with a paging response, after connecting to the network via the RACH.

Network Access

Every time the mobile needs to connect to the network for a location update, paging response, call set-up, SMS, data traffic, etc., the mobile has to request a dedicated channel via the RACH.

Authentication and Encryption

When performing a call set-up, the network will verify the identity of the mobile and provide a specific encryption key for the traffic. The network will encrypt the specific call, using a specific key known only to the mobile (SIM) and the network.

2.2.5 GSM Network Architecture

This section presents a short introduction to the elements of the GSM network, in a simplified example (as shown in Figure 2.11). This is not a detailed description of all network elements,

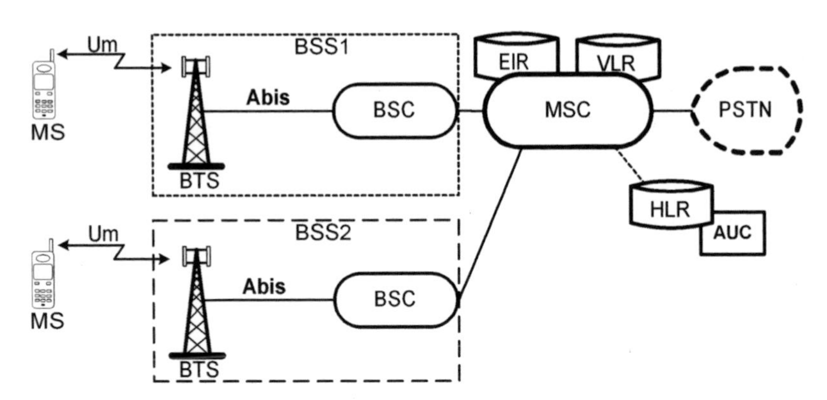

Figure 2.11 The network elements of the GSM network (simplified)

but merely an introduction to the general principle of how the GSM network elements are interconnected and their primary function.

MS

The mobile station (MS) is the user terminal, telephone, data card, etc. The MS is connected to the network (BTS) via the radio interface, the Um interface.

Um, the Radio Interface

The Um is the GSM air interface, the radio channel structure of 200 kHz channels and the TDMA multiplexing presented on the previous pages, combining FDD and TDD. It normally uses the 900 or 1800–1900 MHz spectrum.

Base Transceiver Station

The base transceiver station (BTS) contains all the radio transmitters and receivers that service a specific area (cell). In macronetworks a BTS will typically service three sectors (cells). Each cell can have several radio channels on air in the same cell, serviced by individual transceivers (transmitter/receiver, TRX, TRU, etc.) A standard outdoor base station will use one transmit antenna and two receive antennas (receiver diversity) for each cell. The extra receiver antenna improves the uplink signal from the MS. Mostly, one antenna is used for transmit/receive and one for receiver diversity.

Base Station Controller

The BSC controls large groups of BTSs over a large area. The BSC will perform the power control, etc., of the base stations within its service area. The BSC will also be in charge of handovers and handover parameters, making sure to establish the new connection in the new serving cell, and to close the previously used connection. One BSC and all the BTSs connected are often referred to as a BSS (base sub system). The BSC interfaces to the BTS using the A-bis interface.

Mobile Switching Center

The mobile switching center (MSC) services a number of BSS areas, via the A-interface from the MSC to the connected BSCs. The MSC is in control of all the calls within the BSSs connected, and is responsible for all call switching in the whole area, even two calls under the same cell in the same BTS. All calls are sent back and switched centrally by the MSC. The MSC is also responsible for the supplementary services, call forwarding and charging. To support the MSC with the switching and services, it is connected to supporting network elements; VLR, HLR, EIR and AUC are connected.

Home Location Register

The home location register (HLR) is a large database that contains information about the service profile and current location (VLR address) of the subscribers (SIMs) belonging to the network registered on the particular HLR.

Visitor Location Register

The visitor location register (VLR) is, like the HLR, a database; it contains information about all the mobiles (SIMs) currently camped on the network serviced by the MSC. In the VLR we will find a copy of all the HLR information with regards to the service profile of the user, and information about the current location (location area LAC) of the mobile.

Authentication Register

The authentication register (AUC) contains the individual subscriber identification key (this unique key is also on the SIM) and supports the HLR and VLR with this information, which is used for authentication and call encryption.

Equipment Identity Register

The equipment identity register (EIR) is a database that stores information about mobile station hardware in use. If a mobile is stolen the specific hardware identity (IMEI, International Mobile Equipment Identity) can be blacklisted in the EIR, baring calls to or from the mobile.

Public Service Telephone Network

The public service telephone network (PSTN) is the outside network. It could be the fixed network *a* in this example, or another MSC or mobile network.

2.3 Universal Telecommunication System

This chapter is a short introduction to the universal telecommunication system (UMTS). For more details please refer to References [3] and [4].

UMTS (WCDMA) was specified and selected for 3G system for many reasons, one of the main ones being that it is a very efficient way to utilize the radio resources – the radio spectrum. WCDMA has a very high rejection of narrowband interference, and is very robust against frequency-selective fading. It offers good multipath resistance due to the use of rake receivers. The handovers in WCDMA are smooth and imperceptible due to the use of soft handovers. During handover the mobile is serviced by more cells at the same time, offering macro diversity gain.

However, there are challenges to UMTS radio planning, when all cells in the network are using the same frequency. UMTS radio planning is all about noise and power control. Very strict power control is necessary to make sure that all transmitted signals are kept to a target level that insures that all mobiles reach the base station at the exact same power level. Very good radio planning discipline must be applied and cells must only cover the intended area, and not spill into unintended areas. The reason for this is that you need to minimize the inter-cell interference since all cells are operating on the same frequency; this is one of the biggest radio planning challenges when designing UMTS radio systems.

Even though the concept of soft handovers insures that the mobile can communicate with two or more cells operating on the same frequency, you must remember that the same call will take up resources on all the cells the mobile is in soft handover with, including the backhaul resources of other network elements. Therefore you must try to minimize the handover zones to well-defined small areas, or the soft HO can cannibalize the capacity in the network. This is a particular concern when the two or three cells that are supporting the soft handover are serviced by separate base stations, and different RNCs (radio network controllers) – this will cause added load to the backhaul transmission interfaces.

GSM/DCS and UMTS

UMTS is often added on to existing 2G networks, so in the UMTS specification it has been insured that handovers (hard handover) can be done between 2G and 3G and vice versa.

2.3.1 The Most Important UMTS Radio Design Parameters

UMTS is a very complex mobile system, even if you 'only' need to focus on the radio planning aspects. It can be a challenge to understand the all the deep aspects regarding UMTS radio planning. Many parameters and concerns are important but the most important parameter the RF designer must remember when designing UMTS indoor solutions is that:

- UMTS is power-limited on the downlink (DL).
- UMTS is noise-limited on the uplink (UL).

The RF designer should always strive to design a solution that can secure the most power resources per user, and the least noise load on the UL of the serving cell and other cells in the network.

Remember that all cells in UMTS are on the same frequency; therefore the coverage and noise from one cell will affect the performance on other cells in the network. Do *not* apply 'GSM planning' tactics to UMTS planning, especially not to indoor UMTS radio planning. When planning GSM you can to some extent just 'blast' power from the outside network into a building to provide coverage. In GSM you will also often use few, high-power (exceeding +20 dBm) indoor antennas to dominate the building. What saves you on GSM is the frequency separation; you can assign separate radio channels to the different cells. This is not possible on UMTS. You might be tempted to use a separate UMTS carrier for indoor solutions and by that separate the indoor from the outdoor network, but that is 'GSM thinking' and is a very expensive decision; after all, the networks have typical only two or three UMTS carriers as the total resources for current and future services. The only solution is good indoor radio planning, controlling the noise and the power use to service the users – this is the main topic of the rest of this book.

2.3.2 The UMTS Radio Features

Compared with GSM, the UMTS radio interface is totally different, and can take some time and effort for the hard-core GSM radio planner to understand. Actually it is not complicated at all; if you just focus on the important main parameters it is easy to understand and plan indoor UMTS radio systems. There are in principle two different types of WCDMA – WCDMA-TDD and WCDMA-FDD (as shown in Figure 2.12).

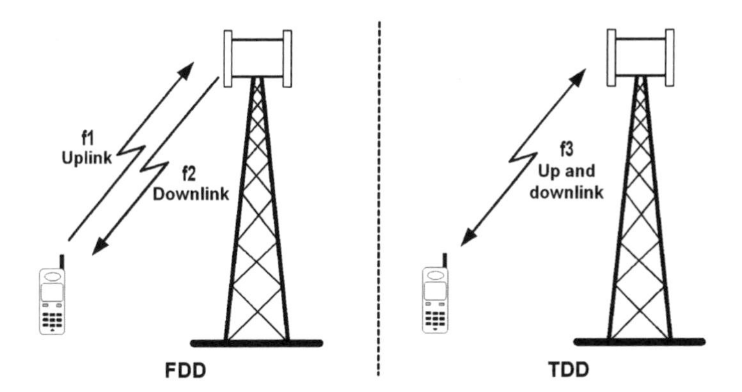

Figure 2.12 The two principle types of WCDMA frequency allocation: FDD and TDD

UMTS TDD

WCDMA-TDD uses the same frequency for the UL and the DL, alternating the direction of the transmission over time. The equipment needs to switch between transmission and reception, so there needs to be a certain guard time between transmission and reception to avoid interference; after all, they are using the same frequency for both links, transmission and reception.

FDD requires a paired set of bands, one for UL and one for DL. TDD can use the same frequency for UL and DL, separated in time. The WCDMA-TDD system can be used in parts of the world where it is not possible to allocate a paired set of bands.

UMTS FDD

The most used (currently only) standard for UMTS is WCDMA-FDD, which transmits and receives at the same time (constantly). Different frequency bands for the UL and DL are allocated as paired bands. The WCDMA-FDD requires a paired set of bands, equal bandwidths separated with the 95 MHz duplex distance throughout the band.

Frequency Allocation

In most regions around the world the band 1920–1980 MHz is assigned for the UL of WCDMA-FDD and the band 2110–2170 MHz is assigned for DL (as shown in Figure 2.13). The frequency allocation for WCDMA-TDD will vary depending on the region, and where it is possible to fit the TDD frequency band assignment into the spectrum.

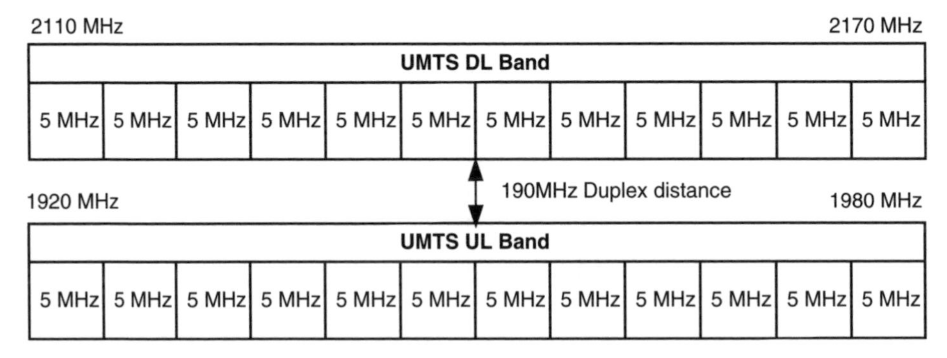

Figure 2.13 UMTS UL and DL frequency bands for the 12 FDD channels

UMTS900

Commonly, most operators are assigned two or three WCDMA-FDD ‘2.1 GHz’ carriers per license. Currently there are considerations on re-using the GSM900 spectrum, by converting the ‘old’ 35 MHz GSM900 band to UMTS, raising the efficiency of the spectrum for better data performance. The 35 MHz GSM spectrum will have the possibility to support seven new UMTS WCDMA channels. Furthermore the 900 MHz band will penetrate buildings much better compared with 2100 MHz, giving better indoor UMTS coverage.

The UMTS/WCDMA RF Channel

Let us have a look at some of the main features of the UMTS radio channel, and how UMTS differs from GSM.

Spread Spectrum, Interference Rejection

UMTS uses WCDMA; this is a spread spectrum signal. The narrow band information from the individual user is modulated, and spread throughout the spectrum. This distributes the energy of the user data over a wider bandwidth (as shown in Figure 2.14). Thus the signal becomes less sensitive to selective interference from narrow-band interference. Typical narrow-band interference will be inter-modulation products from narrowband services at lower frequencies, such as GSM or DCS.

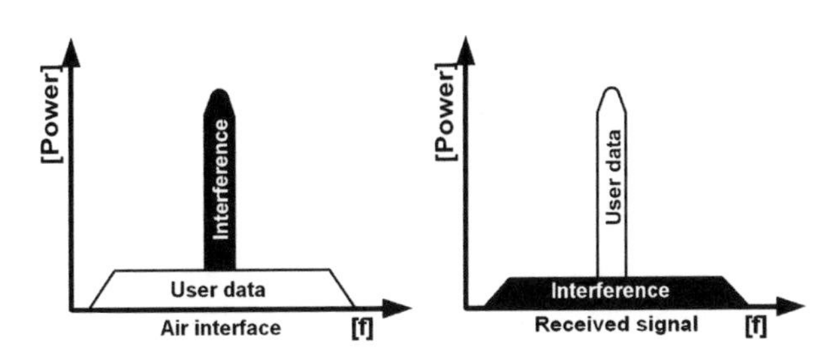

Figure 2.14 Wideband signals are less sensitive to narrowband interference

Fading Resistance

A common problem in radio communication is frequency-selective fading. This occurs when reflections in the environment turn the RF channel into a multipath fading channel. The same signal will have different signal paths, reflections and diffractions, thereby offsetting the phase and timing of the signal arrival at the receiving end. Most of the fading is frequency-selective (as shown in Figure 2.15), but the WCDMA structure solves the problem to some extent. Even if a small portion of the WCDMA fades, most of the energy is maintained and the communication is preserved intact.

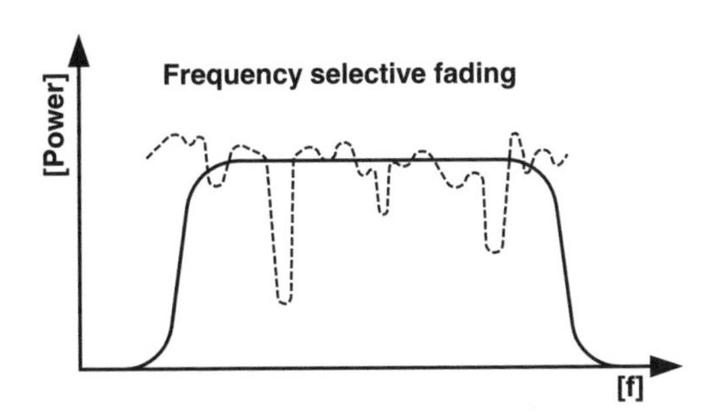

Figure 2.15 The fading channel (dotted line) affects only a small portion of the WCDMA carrier, so the frequency-selective fading is limited

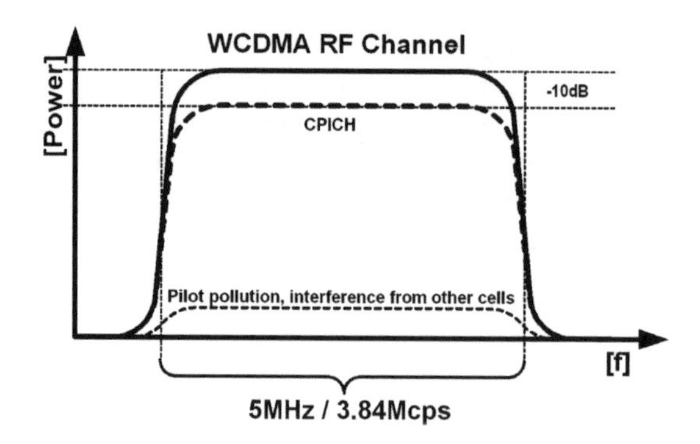

Figure 2.16 The WCDMA air channel, 5 MHz-wide modulated with 3.84 Mchips; a large portion of the power is assigned to the important CPICH channel

The WCDMA RF Carrier

The bandwidth of the UMTS carrier (as shown in Figure 2.16) is about 5 MHz (4.75 MHz), and is divided into 3.84 Mcps. The chips are the raw information rate on the channel, or the 'carrier' if you like. Each user is assigned a specific power according to the service requirement and path loss to that particular user. The more user traffic there is, the more power will be transmitted and the higher the amplitude of the UMTS carrier will be (as shown in Figure 2.17).

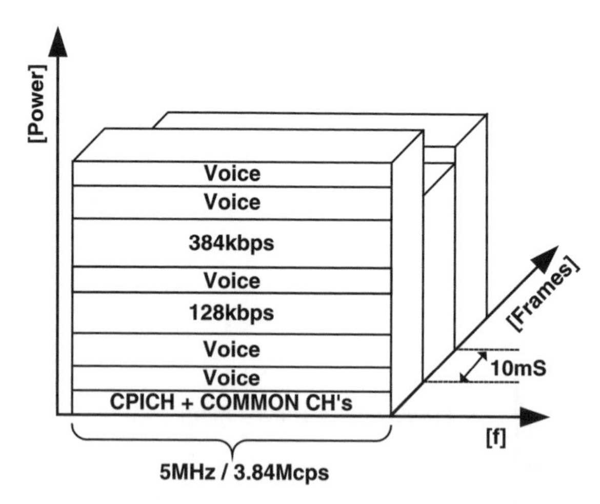

Figure 2.17 The spread WCDMA signal is transmitted in frames of 10 ms, enabling service on demand every 10 ms. One user in a voice session can get a higher data rate assigned in the next frame

Coding and Chips

Each mobile in service is assigned an individual spreading code (as shown in Figure 2.18 and Figure 2.22). All of the users transmit simultaneously using the same WCDMA

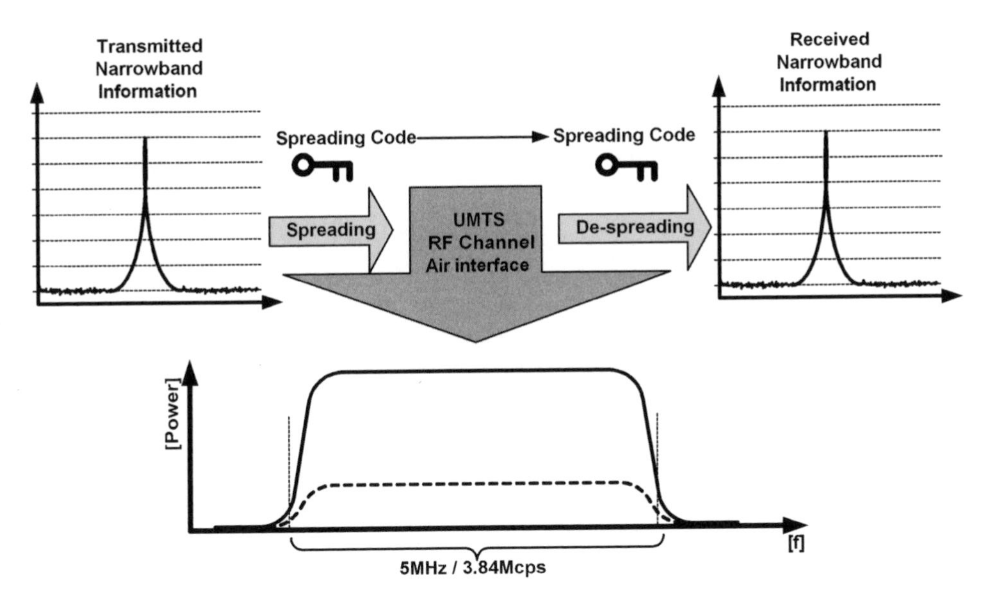

Figure 2.18 The narrowband signal is spread using a dedicated spreading code, modulated into and transmitted over the WCDMA carrier, to be despread and restored in the receiving end using the same dedicated code

frequency. The receiver is able to decode each individual user by applying the same specific spreading code assigned to each user.

The fundamental difference from GSM is that the UMTS system uses the entire 5 MHz spectrum constantly, separating the users only in the code domain, by assigning each mobile a unique code sequence. This unique code is then used to encode the transmitted data; the receiver knows the specific code to use when decoding the signal. This spreading/despreading technique enables the base station to detect the user signal from below the noise.

The coded signal is orthogonal to other users in the cell. The principle is that the code is constructed in such way that one coded signal will not 'spill' any energy to another coded user – if the orthogonality is maintained over the radio channel. Each individual user signal can only be retrieved by applying the same specific code, thereby limiting the interference between users. Essentially all transmitted UMTS information is data; each bit of this data is multiplied by a sequence of code bits, referred to as chips. The number of chips multiplied to each user bit is dependent on the service bit rate of the service assigned to each user. The principle is that the transmitted data is multiplied by the 'raw' channel code rate of a much higher frequency in UMTS 3.84 Mcps (mega-chips per second).

When multiplying the user data rate by a higher-frequency coded bit sequence, the spectrum is spread over the 3.84 Mcps carrier, and becomes a spread spectrum signal. UMTS uses a chip rate of 3.84 Mcps. This is the 3.84 Mcps that takes up the 5 MHz WCDMA radio, but the 'carrier' is 3.84 Mcps.

Frequency Reuse on UMTS

The cells in the UMTS system are operating on the same frequency (as shown in Figure 2.19), using a frequency reuse of 1. UMTS cells are separated only by the use of different primary scrambling codes; this is the 'master key' for the codes in that particular cell.

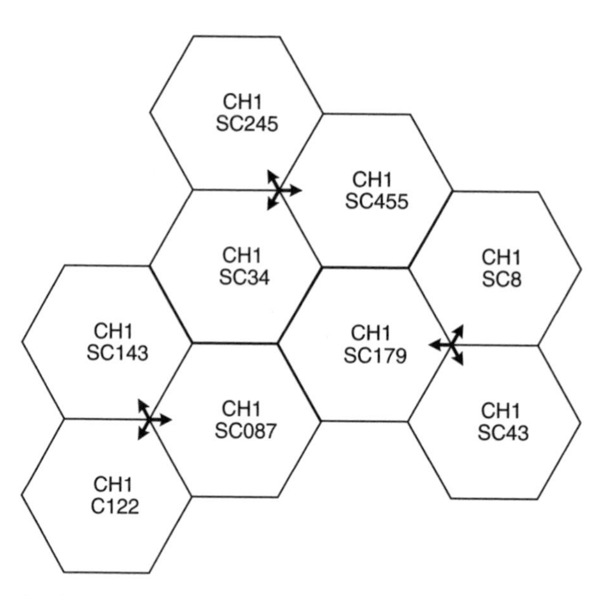

Figure 2.19 Example of three UMTS macro sites with three sectors each; all cells are using the same RF channel, but different scrambling codes

E_b/N_o

The signal quality of the user data on UMTS is defined as the E_b/N_o. The E_b/N_o is defined as the ratio of energy per bit (E_b) to the spectral noise density (N_o). E_b/N_o is the measure of signal-to-noise ratio for a digital communication system. It is measured at the input to the receiver and is used as the basic measure of how strong the signal is over the noise.

$$\frac{\text{Eb}}{\text{No}} = \frac{\text{Bit energy}}{\text{Noise density}} = \frac{\text{Signal bandwidth}}{\text{Bit rate}} * \frac{\text{Signal power}}{\text{Noise power}}$$

The E_b/N_o plays a major role in UMTS radio planning, as the reference point of the link budget calculation. The E_b/N_o defines the maximum data rate possible with a given noise. Different data rates have different E_b/N_o requirements: the higher the data speed, the stricter the E_b/N_o requirements.

The E_b/N_o design level for the various data services are up to the operator, and are used as the basis of the link budget calculation. A typical E_b/N_o planning level for voice will be 7 dB, and about 4 dB for a 384 kps service

Common Pilot Channel

The common pilot channel (CPICH) does not contain any signaling. It is a pure DL channel that serves two purposes only: to broadcast the identity the cell and to aid mobiles with cell evaluation and selection. The CPICH is coded with the primary scrambling code of the cell; it has a fixed channelization code with a spreading factor of 256. The mobiles in the network will measure the CPICH power from the different cells it is able to detect. It will access the cell with the most powerful measured CPICH. The mobile will also measure and evaluate the CPICH levels from other adjacent cells (defined in the neighbor list) for handover evaluation.

A major portion of the total DL power from the base station is assigned to broadcasting the pilot channel. Typically the base station will assign 10% (−10 dB, see Figure 2.16) of the total DL power for the CPICH. For a +43 dBm base station the CPICH is transmitted at 33 dBm.

E_c/I_0

The quality/signal strength of the pilot channel is measured as E_c/I_0, the energy per chip/interference density measured on the CPICH. It is effectively the CPICH signal strength. When the mobile detects two or more CPICH with similar levels, the mobile will enter soft handover. This is essential in order to secure the link. If the mobile did not enter soft handover, the link would break down due to interference between the two cells transmitting the same channel and received at the same power level.

The mobile continuously measures the E_c/I_0 of the serving cell and adjacent cells (defined in the neighbor list/monitor set). The mobile compares the quality (E_c/I_0) of the serving CPICH against the quality of other measured CPICHs. The mobile uses trigger levels and thresholds to add or remove cells from the active set, the cell or cells the mobile is engaged with during traffic. (For more details, refer to the description of the UMTS handover algorithm, Section 2.3.4.)

The CPICH defines the cell size, and the service area and cell border can be adjusted by adjusting the CPICH power of the cell. Thus it is possible to some degree to distribute traffic load to other cells by tuning cell size.

A sufficient quality for the CPICH (not scrambled) will be a E_c/I_0 of −10 to −15 (energy per chip/interference) in the macrolayer. For indoor cells lower CPICH levels of −6 to −8 dB can be considered due to the high isolation of the macrocells (if the indoor system is planned correctly).

Pilot Pollution

When a mobile receives CPICH signals at similar levels from more cells it is supposed to enter soft handover – if it does not the quality will be degraded and the call might be dropped. However there can be cases where a distant cell is not defined in the neighbor list, and thus the mobile is not able to enter soft handover. This will cause interference of the serving cells' CPICH; this is referred to as pilot pollution (as shown in Figure 2.20). This is one of the main concerns when designing UMTS indoor solutions. The problem is often a big concern in high-rise buildings, predominantly in the topmost section of the building, where mobiles are able to detect outdoor cells. The problem is essentially a lack of dominance of one serving cell (see Section 3.5.3).

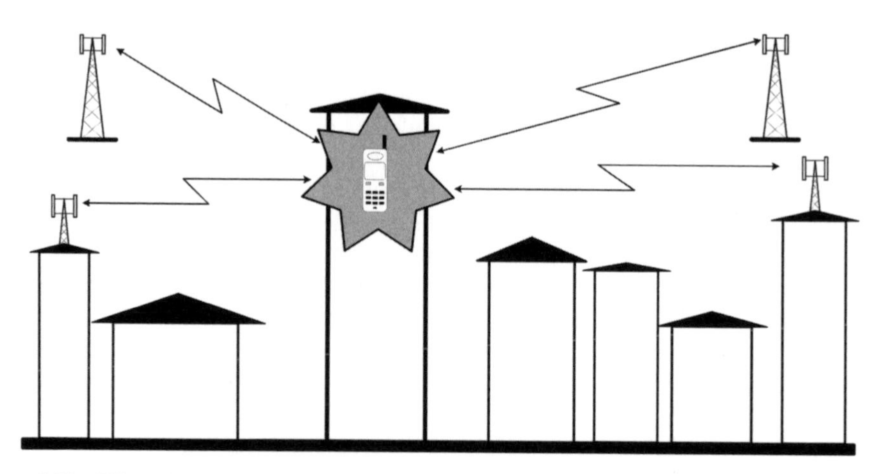

Figure 2.20 Pilot (CPICH) pollution is a big concern, when a mobile receives pilot power from distant base stations without being in handover with these base stations

Geometry Factor

The geometry factor is the power received by the mobile divided by the sum of noise received by the mobile. The sum of noise is thermal noise plus interference (traffic) radiated by other base stations.

$$\text{Geometry factor} = \text{received signal}/(\text{thermal noise} + \text{interference})$$

The closer the mobile is to the servicing base station, the higher the geometry factor will be, and the better the signal quality will be.

Orthogonality

Owing to the unique way the codes are constructed, they are orthogonal to each other. The orthogonal structure of the codes insures that all the users can be active on the same channel at the same time, without degrading each other's service; that is, if the orthogonality is perfect, or 1 (100%)

However, due to multipath propagation on the radio channel, especially in cities and urban environments, the delay spreads of the phases and time of the same transmitted signal in the receiving end will cause the codes to 'spill over', degrading the orthogonality and causing the signals to interfere with each other.

The multipath environment will degrade the orthogonality, depending on how predominant the reflections/diffractions of the signal are. The degraded orthogonality will degrade the effectiveness of the RF channel, degrading the maximum throughput on the WCDMA carrier.

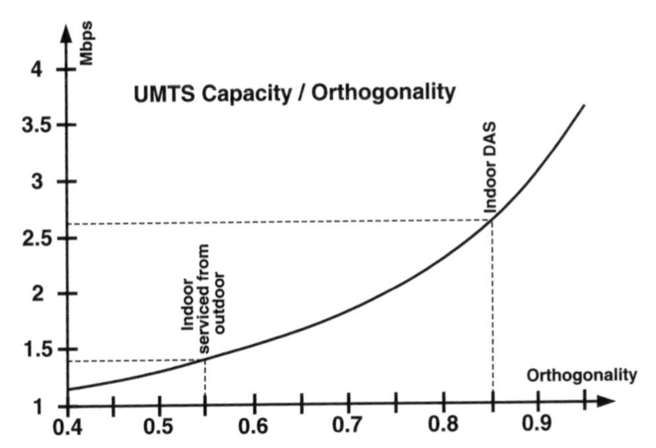

Figure 2.21 The maximum data throughput on the WCDMA carrier is dependent on the orthogonality of the RF channel. Reflections and multipath signals will degrade the orthogonality, and degrade the efficiency of the RF channel

In environments with low orthogonality, the data throughput of the WCDMA carrier is degraded. In the example in Figure 2.21 we can see that users inside a building serviced by a nearby macro base station typically have an orthogonality of less than 0.55. This is due to the fact that most of the signal will arrive at the mobile as reflections, due to the absence of a line-of-sight to the base station.

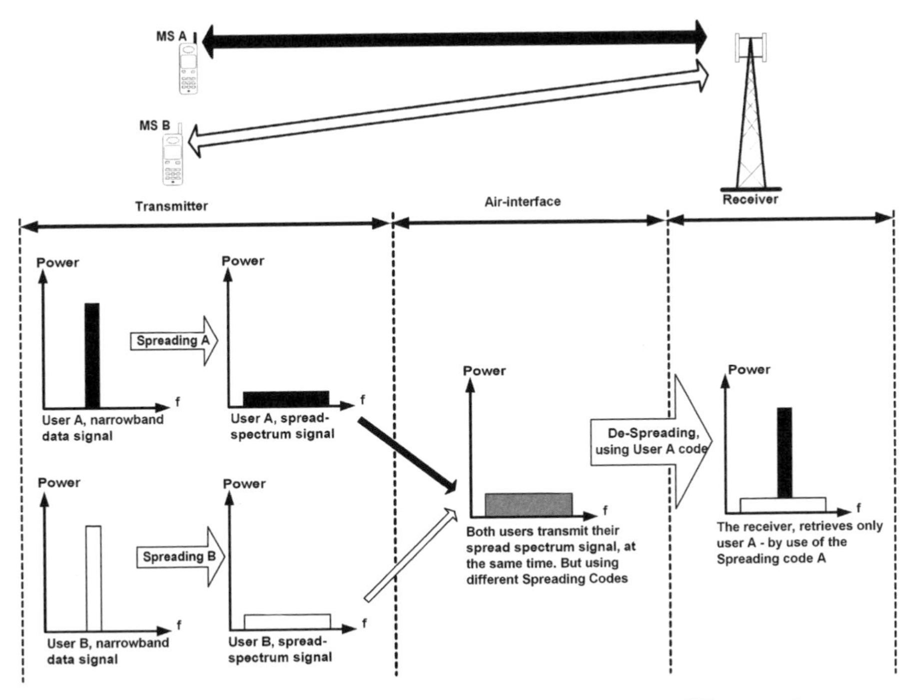

Figure 2.22 Two mobiles serviced by the same cell, using different codes

The degraded orthogonality causes degradation efficiency of the codes, causing the codes to 'spill over', and this 'code interference' will degrade the total throughput of the UMTS carrier. In the example from Figure 2.21, the capacity of the carrier is degraded to less than 1.5 Mbps. However, the same base station connected to an indoor distributed antenna system would have a typical orthogonality of 0.85 and would be able to provide about 2.6 Mpbs, and so is a considerable better use of the same resources. This is due to the fact that the indoor user, serviced by indoor antennas, will mostly be in line-of-sight to the antennas, with only limited reflections. By insuring a high orthogonality and efficient use of the resources, an indoor distributed antenna system can be used. This improves the throughput and efficiency of the UMTS channel, improving the business case.

Processing Gain

When using the technique with spreading and despreading the signal over a wideband carrier, a gain is obtained. This is referred to as the processing gain. The processing gain is dependent on the relation between the carrier chip rate, and the user data bit (chip) rate, as shown in Table 2.1.

Table 2.1 UMTS data rate vs processing gain

User rate	Processing gain
12.2 kbps	25 dB
64 kbps	18 dB
128 kbps	15 dB
384 kbps	10 dB

$$\text{Processing gain} = \text{chip rate}/\text{user data rate}$$

$$\text{Processing gain} = 3.84\,\text{M}/\text{user rate [linear]}$$

$$\text{Processing gain} = 10\log(3.84\,\text{M}/\text{user rate})\ [\text{dB}]$$

Using these formulas you can calculate the processing gain of the specific UMTS data service

Examples

The processing gain for high-speed data, 384 kbps, is 10 log (3.84 Mcps/384 kb) = 10 dB.

The processing gain for speech, 12.2 kbps, is 10 log (3.84 Mcps/12.2 kbps) = 25 dB.

The use of orthogonal spreading codes and processing gain is the main feature in UMTS/WCDMA, giving the system robustness against self-interference. It is the main reason for a frequency reuse factor of 1, due to the rejection of noise from other cells/users.

When you know the required bit power density E_b/N_o the specific service (voice 12.2 kbps + 5 dB, data +20dB) you are able to calculate the required signal-to-interference ratio: voice at 12.2 kbps needs an approximately 5 dB wideband signal-to-interference ratio minus the processing gain, 5 dB − 25 dB = −20 dB. In practice, the processing gain means that the signal for the voice call can be 20 dB lower than the interference or noise, but still be decoded.

2.3.3 UMTS Noise Control

UMTS is very sensitive to excessive noise, and noise control is essential. All traffic is on the same frequency and all signals from active mobiles need to reach the base station at the same level. If one mobile reaches the uplink of the base station at a much higher level, it will interfere with all the traffic from other mobiles in service on the same cell. Every mobile uses the same frequency at the same time.

To control this issue, UMTS uses very strict noise and power control. Traffic in UMTS is essential white noise to other users, and the amount of noise from a cell is directly related to the traffic.

Admission Control

More traffic in a UMTS cell is equal to more radiated noise in the cell. This noise will impact surrounding cells with interference, and therefore it is very important to limit the load in the cells to a preset maximum noise level in order to control the increase in noise.

Loading a UMTS cell by 50% is equal to a 3 dB increase in noise; this is a typical value for the maximum allowable noise increase. The value can be set in the network individually for each cell. Indoor cells are more isolated from the macro layer, and can in principle be loaded relative highly, maybe up to 60–65%, giving a higher capacity.

In order to evaluate the noise increase due to traffic in the cell, node B will constantly measure and evaluate the overall noise power received on the UL in order to evaluate the UL noise increase in the cell. By doing so, the base station can calculate and evaluate how much headroom is left for new traffic, this in order to control the admission of new traffic in the cell (as shown in Figure 2.23).

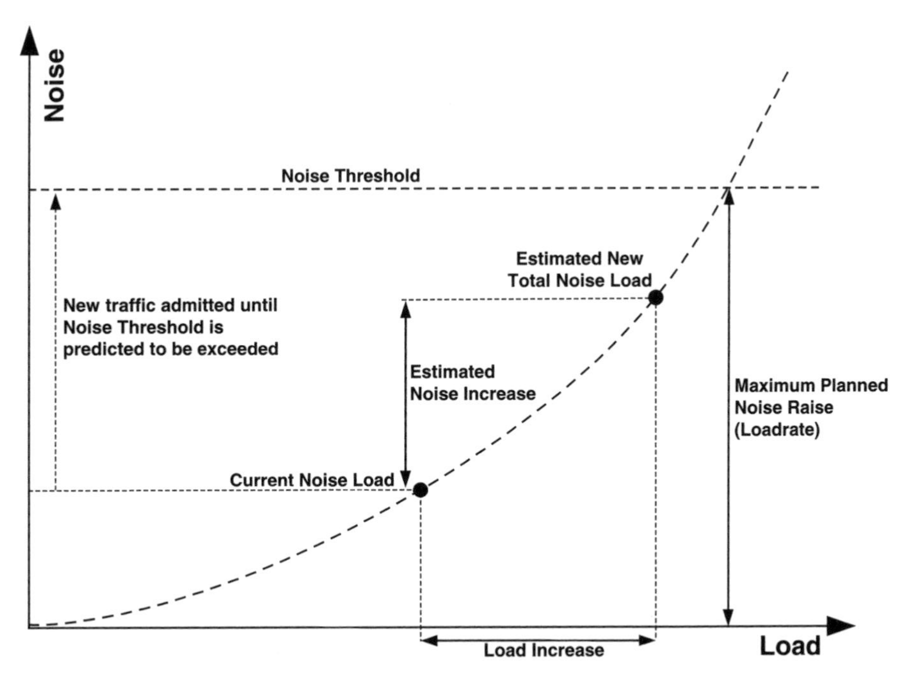

Figure 2.23 Admission control in UMTS will make sure that new admitted traffic in the cell will not cause the noise to increase above a preset level

The total noise power on the UL of node B will be a result of the sum of the noise generated by the traffic in the cell itself, noise (traffic) from other cells and the noise figure from the base station:

$$\text{Noise}_{\text{total}} = \text{own}_{\text{traffic noise}} + \text{other}_{\text{cells' noise}} + \text{node B}_{\text{noise power}}$$

In addition to the noise generated by the traffic in the cell, there will be a contribution from noise from users in other cells and finally the noise figure (NF) of node B and other active hardware in the system.

The base station noise figure (noise floor) is the reference for zero traffic in the cell. If the receiver only measures the noise power generated by the base station itself, the network assumes that there is no traffic in the cell/adjacent cells. The network assumes the load rate to be zero and that the cell has its full potential for servicing new traffic.

Using this noise measurement (including the hardware noise reference of the node B), the network can evaluate if the new admitted traffic will cause a noise increase that exceeds the predefined maximum UL noise increase in the cell, and the base station will keep admitting new traffic in the cell until the max noise rise is predicted or measured.

Impact of Noise Power from External Equipment

It is very important that the base station noise reference level for no traffic is updated when connecting any active equipment to the uplink of the base station. Repeaters, active distributed antenna system, etc., will generate noise power on the uplink, and these systems should be designed so that minimum noise power is injected into the UL port of the base station.

Any noise increase will cause the base station to assume this noise power to be traffic in the cell, or adjacent cells – essentially offsetting the traffic potential for the cell. It may cause the base station to not admit traffic to its full potential, and in severe cases to not admit any traffic at all. Therefore it is crucial to adjust the admission control parameters in the network, with regard to the added hardware (HW) noise power on the UL of the node B.

For most UMTS base station networks these parameters can be found in the parameter set for calibrating the system to connect to mast-mounted amplifiers or low-noise amplifiers (LNA). However the main issue is to design the distributed antenna system so that you minimize the noise load on the base station without compromising the performance of the link; this can be done using uplink attenuators; see Section 7.5 on noise control for more details.

Cell Breathing

Any traffic in UMTS is equal to the noise increase on the radio channel. We know that UMTS is very sensitive to noise increase, and therefore it is important to control the noise increase in the network. When designing UMTS systems you need to decide how much noise increase you will allow in the cell and adjacent cells. Noise from traffic in cells will also increase the noise in adjacent cells and impact the coverage and capacity of those cells.

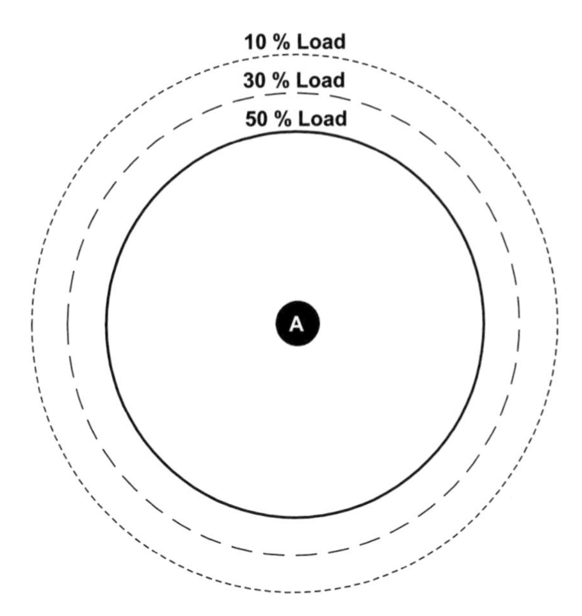

Figure 2.24 Cell breathing is caused by the noise increase owing to the load of the cell; more load will give more noise increase and a smaller coverage footprint

In Figure 2.24 you can see an example of how the noise increase and power load in the cell will affect the footprint of the cell; the greater the load, the more noise, the less the coverage of the cell will be – this is referred to as cell breathing. It will be different for the uplink and the downlink, so a cell can be both downlink and uplink limited, depending on the current load profile (as shown in Figure 2.26).

It is important that you design the system for a maximum allowable noise increase or load. If you do not, the traffic will cause the noise to spin out of control. The cell would collapse, and drop the calls on the cell border due to excessive noise increase.

Noise Increase

You are able to calculate the noise increase in the cell. Using 50% of the remaining capacity in a cell is equal to noise increase of factor 2, corresponding to 3 dB. Loading a cell from 0 to 50% load is a noise increase of 3 dB, and using another 50% of the remaining capacity, from 50 to 75%, is an additional 3 dB, a total of 6 dB, and so on.

You can calculate the noise increase as a function of the load rate:

$$\text{Noise increase} = 10\log[1/(1 - \text{load factor})]$$

Typical cells will be designed to be loaded to a maximum of 50–60%, and the graph in Figure 2.25, based on the noise increase formula, clearly shows why. A 60% load rate, corresponding to a load factor of 0.6, will cause a noise increase of 4 dB. In Figure 2.25 it is very clear what happens with the noise when you exceed 80% – the noise simply rises abruptly, out of control. This excessive noise rise will cause the cell to collapse.

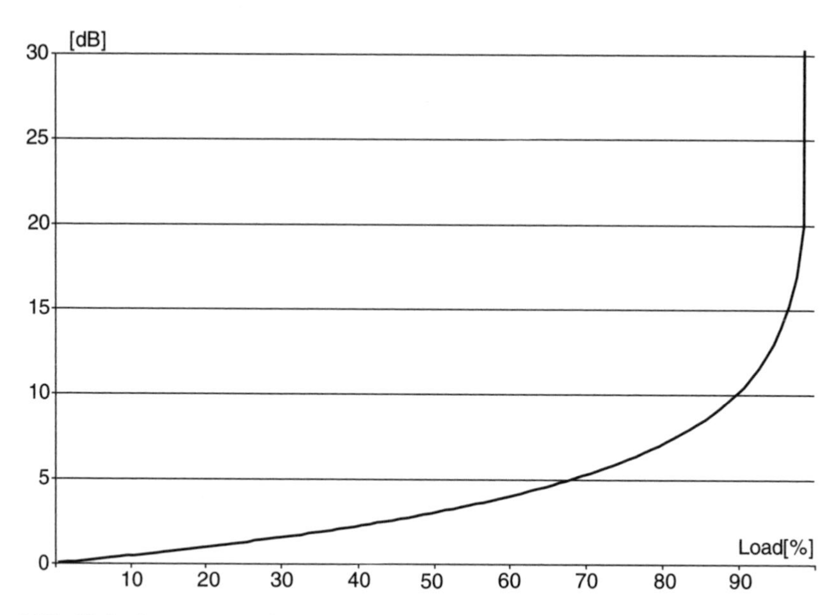

Figure 2.25 Noise increase as a function of the load in the cell; above 80% load the noise increase causes the noise to rise sharply, causing the collapse of the cell

The capacity in UMTS is directly related to the signal-to-noise ratio (SNR; see the geometry factor, Section 2.3.2) in the cell. The capacity is proportional to the interference in the cell. Any added interference in the cell will impact the capacity.

UL and DL Load

UMTS is noise-limited on the uplink and power-limited on the downlink. In a typical cell the link tends to be noise-limited during light traffic load, and power-limited when the traffic load rises (as shown in Figure 2.26).

The service profile of traffic generated by the users in the cell will also affect the balance of the cell between being downlink- or uplink-limited. Typical data users will use higher downlink data speed, for downloads, etc.

The base station's capabilities with regards to processing power, power resources and receiver sensitivity also play a role, but the principle is that the same cell can be both uplink- and downlink-limited depending of the load profile.

2.3.4 UMTS Handovers

There are several handover scenarios in UMTS. The type of handover is dependent on whether the handover is within the same node B, different node B, different UMTS frequencies or even system handover to and from GSM and DCS. As we know, UMTS uses the same frequency for all the cells and the only possible way for a mobile station to hand over from one cell to another is to be connected with both cells in the area where

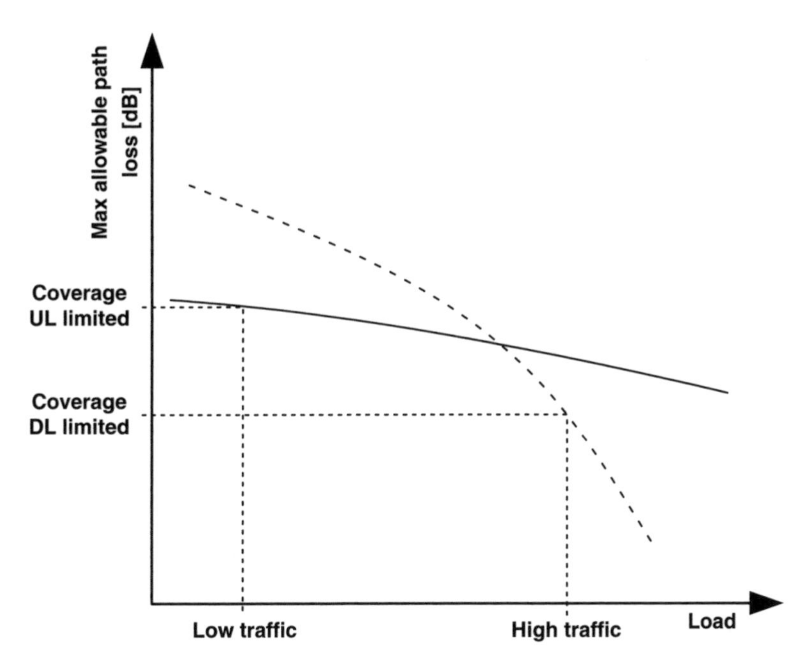

Figure 2.26 The load profile for the cell defines whether the coverage is DL or UL limited

both cells are at equal levels. Typically a handover in UMTS involves two or three cells. This handover function and algorithm is totally different compared with GSM handover zones.

Softer Handover

When a mobile station is within the service area of cells originating from the same node B at the same power level, the mobile station is in softer handover (as shown in Figure 2.27).

The mobile will use both RF links, and the node B will control and combine the signal stream to the RNC (see Figure 2.37 for UTRAN network elements). Two separate codes are used on the downlink, and the mobile will use rake receiving to receive the two downlink paths.

On the UL the base station will use rake receiving for the two signal paths. The mobile will provide some extra load on node B due to the double link for only one call. This extra load is limited to processing power (channel elements) and impacts only the serving base station.

Soft Handover

When a mobile is in the service area of two cells originating from different node Bs (as shown in Figure 2.28) the mobile will use one RF link to both base stations; this is macro diversity and both node Bs will send the call to the RNC (see Figure 2.37 for UTRAN

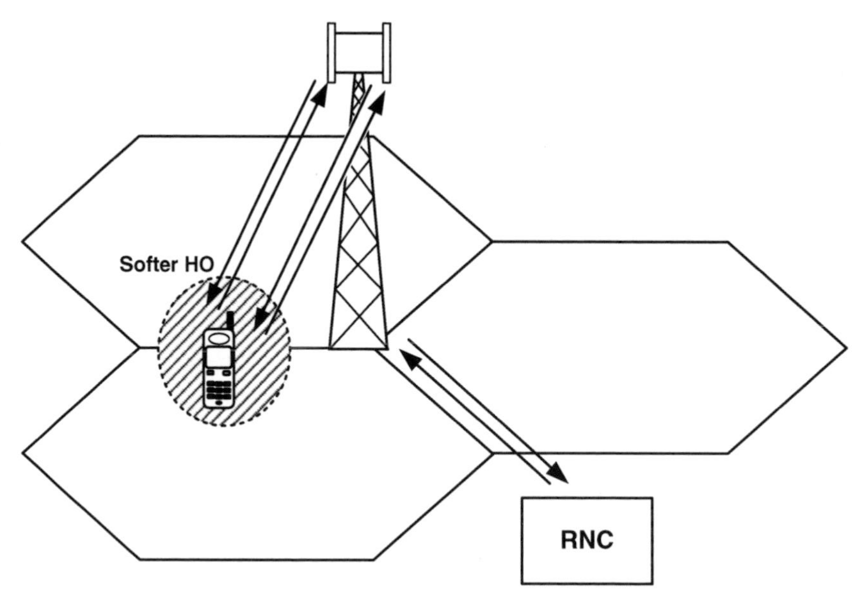

Figure 2.27 The mobile will be in softer handover when it is able to detect more than one cell from the same node B

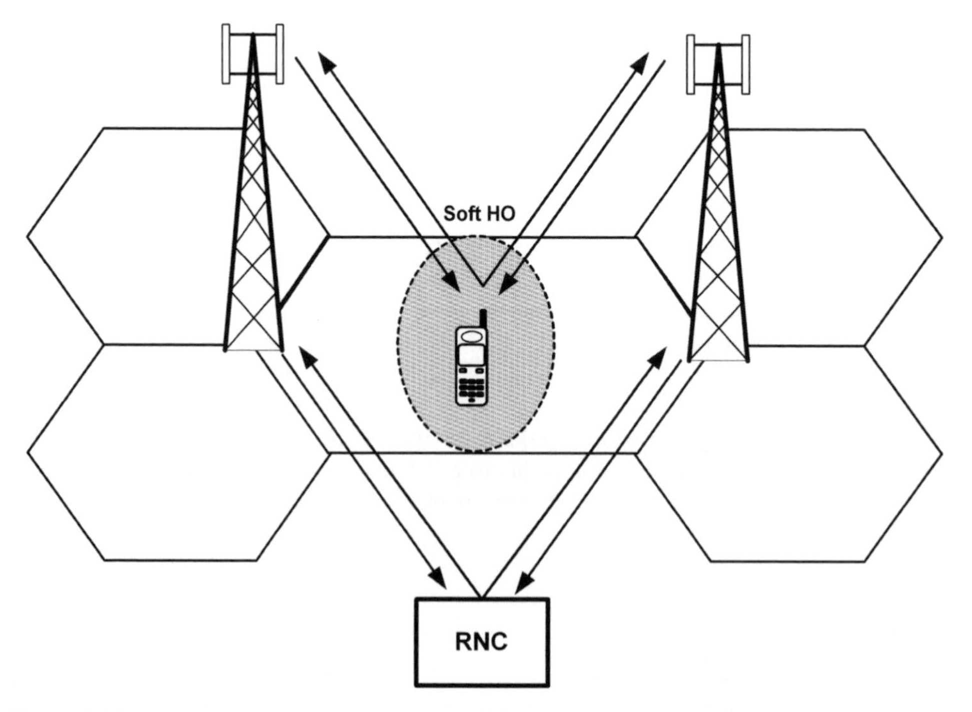

Figure 2.28 Mobiles will be in soft handover when detecting two or more cells from different node Bs

network elements). In this scenario the mobile will not only load both cells with regards to power, noise and processing power (channel elements), but also double the load on the transmission link (IUR interface) back to the RNC/RNCs from both node Bs.

As in softer handover, two separate air interfaces are used; the mobile combines the downlink signals using rake receiving. The main difference is on the uplink, owing to the fact that in soft handover the RNC not the base station will have to perform the combining of the two uplink signal paths.

Typically the RNC will select the uplink signal path with the best frame reliability indicator, which is used for outer loop power control. For indoor radio planning for UMTS, you should aim to keep the soft handover areas to a minimum.

OBS!

Note that if a strong candidate neighbor cell is not monitored (defined in the monitor set/neighbor list), the mobile will not be able to enter soft handover with that nonmonitored cell. If the signal from the nonmonitored cell is high, it will degrade the quality (E_c/I_0) of the serving cell, degrading the data performance, and even causing the call to be dropped.

The Soft Handover Algorithm

The handover algorithm in UMTS (as shown in Figure 2.29) evaluates the radio quality (E_c/I_0) of the serving and adjacent cells. The E_c/I_0 measurement is used by the mobile for evaluating the serving cell and handover candidate cells; the UMTS system uses the following terminology for describing the handovers:

- *Active set* – this is the set of cells that the mobile is in connection with during a soft handover; it is all the cells in the active set currently engaged with traffic to/from the mobile.
- *Monitored set/neighbor set* – these are the neighbors that the mobile must monitor (the neighbor list), but whose pilot E_c/I_0 is not strong enough to be added to the active set and engage in soft handover.

Example

This is an example of how the soft handover algorithm works (shown in Figure 2.29). Specific events are defined in UMTS to describe the status/action.

Event 1A, Radio Link Addition The monitored cell will, if the serving cell pilot $E_c/I_0 >$ best pilot E_c/I_0 – reporting range + hysteresis event 1A, for a period of d*T*, and the active set is not full, be added to the active set, and engage in soft or softer handover.

Event 1B, Radio Link Removal In soft or softer handover, and if one of the serving cells pilot $E_c/I_0 <$ best pilot E_c/I_0 – reporting range – hysteresis event 1B, for a period of d*T*, then this cell will be removed from the active set.

Event 1C, Combined Radio Link Addition/Removal (Replacement) If the active set is full, and no more soft handover links can be added and the best candidate pilot $E_c/I_0 >$ worst old pilot E_c/I_0+ hysteresis event 1C, for a period of d*T*, then the cell with the worst E_c/I_0 in the active set will be replaced with the best candidate cell from the monitored set.

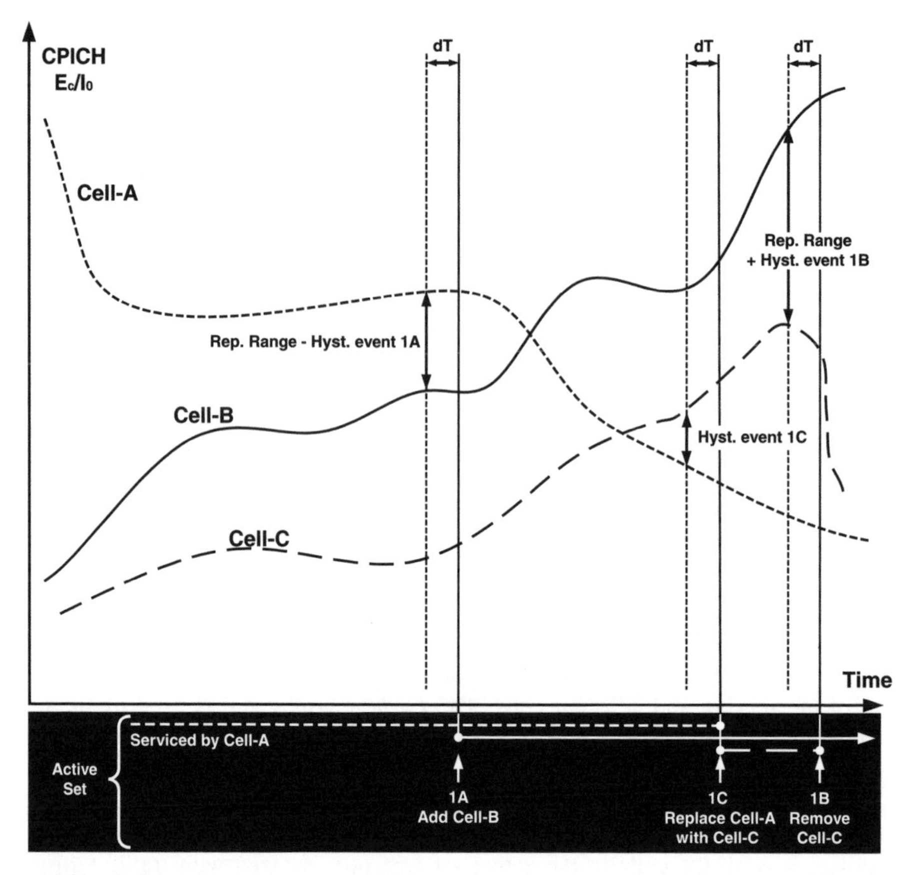

Figure 2.29 Soft handover events in UMTS, example with maximum size of active set assumed to be two cells

UMTS Handover Algorithm Definitions

- Reporting range: threshold for soft handover.
- Hysteresis event 1 A: the cell addition hysteresis.
- Hysteresis event 1 B: the cell removal hysteresis.
- Hysteresis event 1 C: the cell replacement hysteresis.
- d*T*: time delay before trigger.
- Best pilot E_c/I_0: the strongest monitored cell.
- Worst old pilot E_c/I_0: the cell with the worst E_c/I_0 in the active set.
- Best candidate pilot E_c/I_0 : the cell with the best E_c/I_0 in the active set.
- Pilot E_c/I_0: the measured (averaged and filtered) pilot quality.

Hard Handovers

There are a few cases where the mobile station will need to perform a hard handover (like GSM) instead of a soft handover. When performing handover between different UMTS frequencies, the mobile will do a hard handover. This could be different frequencies within the same cell, or between different cells. When the mobile station performs a system handover, i.e. a handover between GSM/DCS and UMTS, it will also be a hard handover. Hard handovers are more difficult to control and more likely to fail, and therefore these events should be limited to a minimum in order to maximize network performance and quality.

2.3.5 UMTS Power Control

Compared with GSM, UMTS power control is complex and extremely strict; the motivation is the need for strict noise control. All traffic on the UL of the base station has to arrive at the exact same level, or else one high signal would overpower the traffic from all other mobiles. They all operate at the same frequency at the same time.

The power control in UMTS has three different stages: no power control, open loop power control (cell access) and closed loop power control, in dedicated (traffic) mode. Figure 2.30 shows the principle of the UMTS power control.

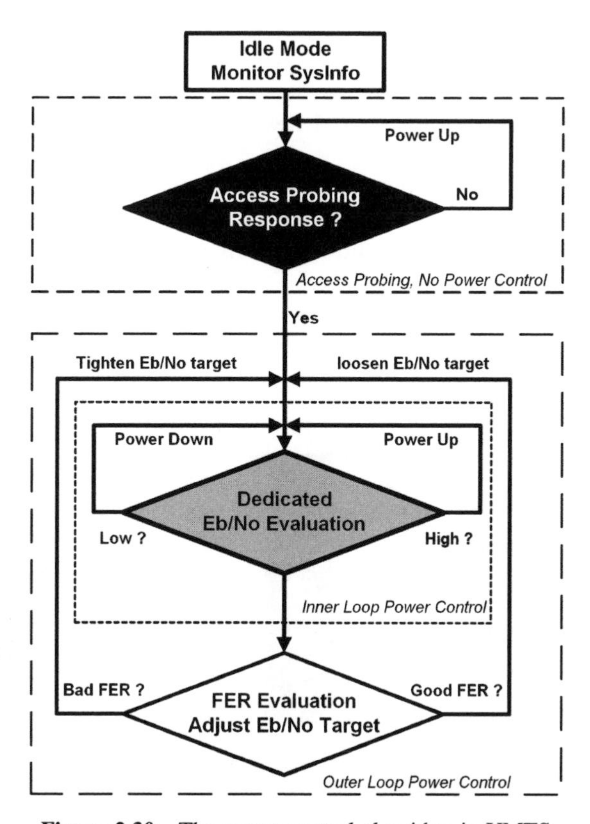

Figure 2.30 The power control algorithm in UMTS

Idle Mode

No power control is needed when the mobile is in idle mode. The mobile will measure the received CPICH power and monitor other power control parameters (PWR_INI and PWR_STEP info). The mobile can thereby calculate what initial power is needed when accessing the base station, in order to make sure it does not overpower the uplink of the base station and degrade the performance for other users.

Open Loop Power Control

Open loop power control (PC) is used when the mobile is in transition from idle mode to dedicated mode (setting up a call or responding to a paging signal). The mobile's initial power for the network attachment is estimated using the downlink pilot (CPICH) signal. The mobile station will send aaccess probes to the base station when accessing the network. The mobile will monitor the system information transmitted by the base station with regards to the reference for the transmitted CPICH power. This enables the mobile station to calculate the path loss back to the base station.

Access Procedure in UMTS

MS starts with low power, to prevent UL 'blocking' the base station (BS), calculating the transmit power for the first access burst (as shown in Figure 2.31):

$$\text{PWR\_INIT} = \text{CPICH\_Tx\_Power} - \text{CPICH\_RSCP} + \text{UL\_Interference} + \text{UL\_Required\_CI}$$

where PWR_INI = initial access power; CPICH_Tx_Power = transmitted CPICH power; CPICH_RSCP = received CPICH signal strength; UL_Interference = uplink interference measured by the node B; and UL_Required_CI = uplink quality margin.

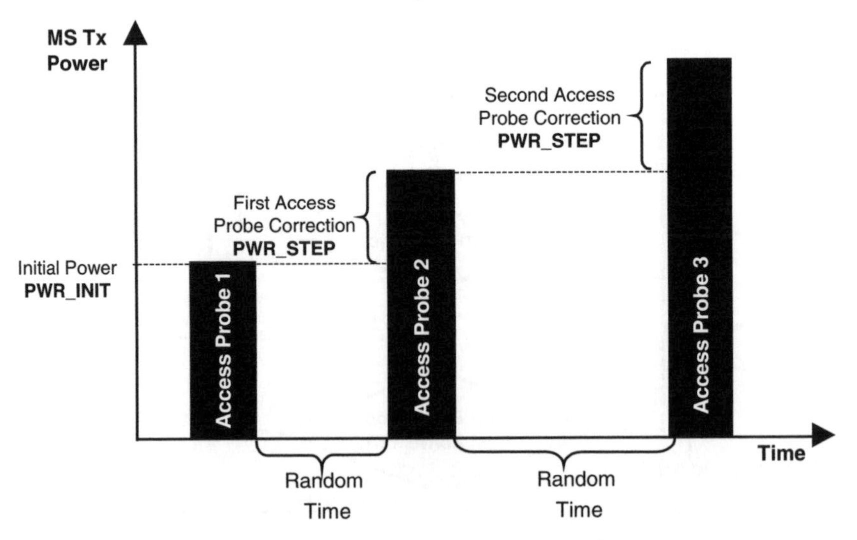

Figure 2.31 The mobile access probe in UMTS

This procedure will insure that the mobile will not use a high initial access power, increasing the noise on the UL, resulting in dropped calls for already ongoing traffic. Then the mobile station waits for response signal (access grant) from node B. If no access grant is received, the mobile assumes that the access probe did not reach the base station. Then the mobile station will increase the transmit power for the next access probe by the PWR_STEP parameter broadcasted by the cell. There will be random intervals for these access bursts, within a specified timing window, in order to prevent the access signals from several mobiles colliding repeatedly if accessing the cell at exactly the same time.

Closed loop Power Control (Inner Loop)

Compared with GSM, where the power adjustment is done every 60 ms, the power control in UMTS is very fast. With a power control rate of about every 666 μs, the mobile transmit power is adjusted; this is a power control of 1500 Hz in closed loop power control. The closed loop power control is active when the mobile station is in traffic mode (dedicated mode). On the downlink, power control on the base station preserves marginal power for mobiles on the edge of the cell.

The Outer loop Power Control

The purpose of the outer loop power control is to adjust the E_b/N_o target in order to fulfil the FER target (no more, no less). Owing to the multipath fading profile, this will be dependant on the speed of the mobile. Mobiles moving at faster speed will have more fading impact, and need more power resources. The FER target will be adjusted accordingly (as shown in Figure 2.32).

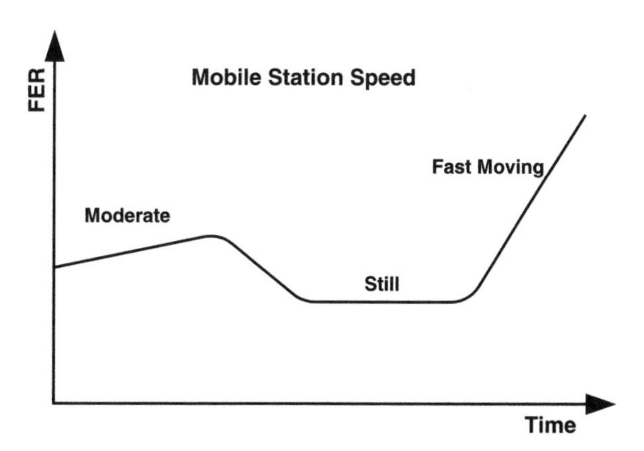

Figure 2.32 The FER target for the UMTS power control is constantly adapted to the fading channel with the speed of the mobile

UMTS Power Control Compensation for Fading

Normally a fading radio channel will affect the signal strength on the uplink. This is dangerous for UMTS because one mobile more powerful than others in the cell will affect the performance – after all, all traffic is on the same frequency at the same time. To compensate for the fading, the UMTS power control is so fast (1500 Hz) that it is able to level out the fading channel and to a large extent compensate for Rayleigh fading, at moderate speeds.

In Figure 2.33 you can see the principle: the fading radio channel (the solid line) will provide a large variance on the path loss, depending on the speed of the mobile. The power control (dotted line) will compensate for this fading, and on the uplink of the base station the topmost graph shows that the resulting radio channel is received compensated by the power control, with a typical maximum variance of ±2–3 dB.

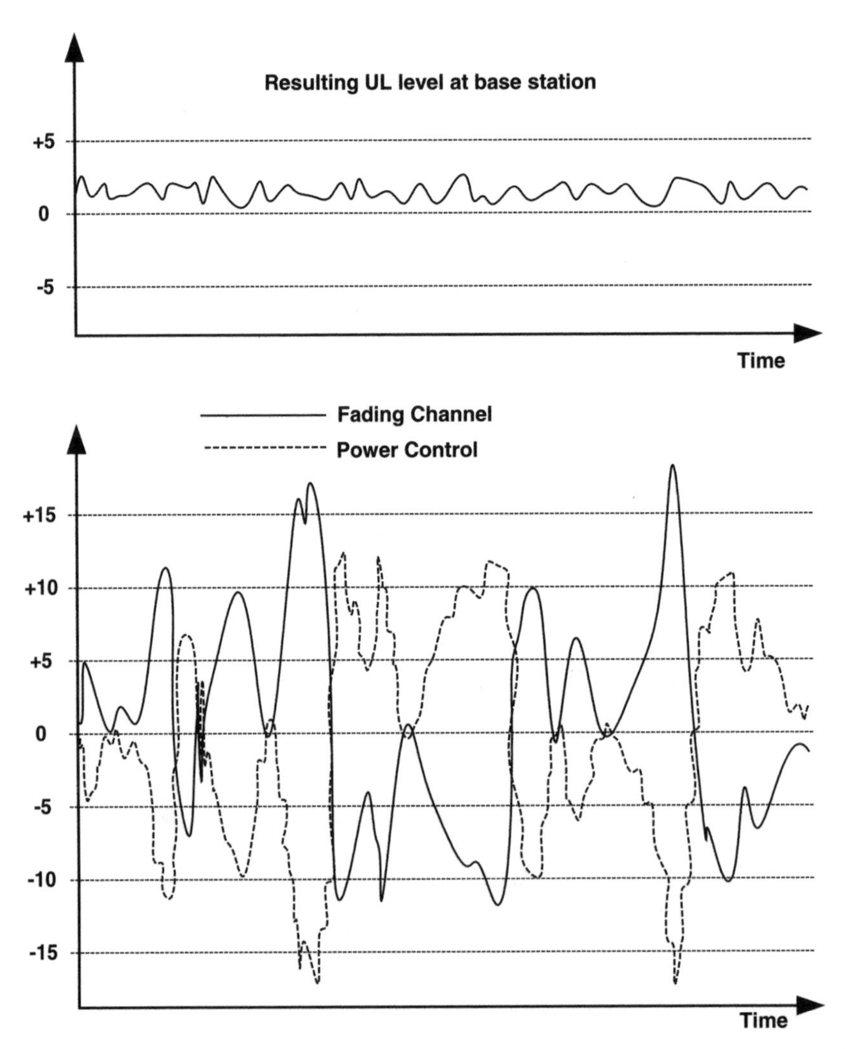

Figure 2.33 The mobile power control in UMTS is so fast and efficient that it will compensate for the fading channel, and all mobiles will reach the uplink of the base station with the same level

2.3.6 UMTS and Multipath Propagation

Ideally the radio channel path should be direct, in line-of-sight (LOS), with no reflections or obstructions between the base station and the mobile. However, in the real world, especially in cities and urban environments, most RF signals reaching the receiver will have been reflected or diffracted by the clutter of the buildings. Typically only a small portion of the RF signal will be the direct signal from the base station; the main signal will derive from reflections. This is called multipath, and the environment creates multipath fading due to the phase shifts and different delays of the signals amplifying and canceling each other. This is the main source for orthogonality degradation, especially when serving indoor mobile users from the macro network.

An example of the fading environment with reflections can be seen in Figure 2.34, where T1 is the direct signal, and T2/T3 are the reflected and delayed signals.

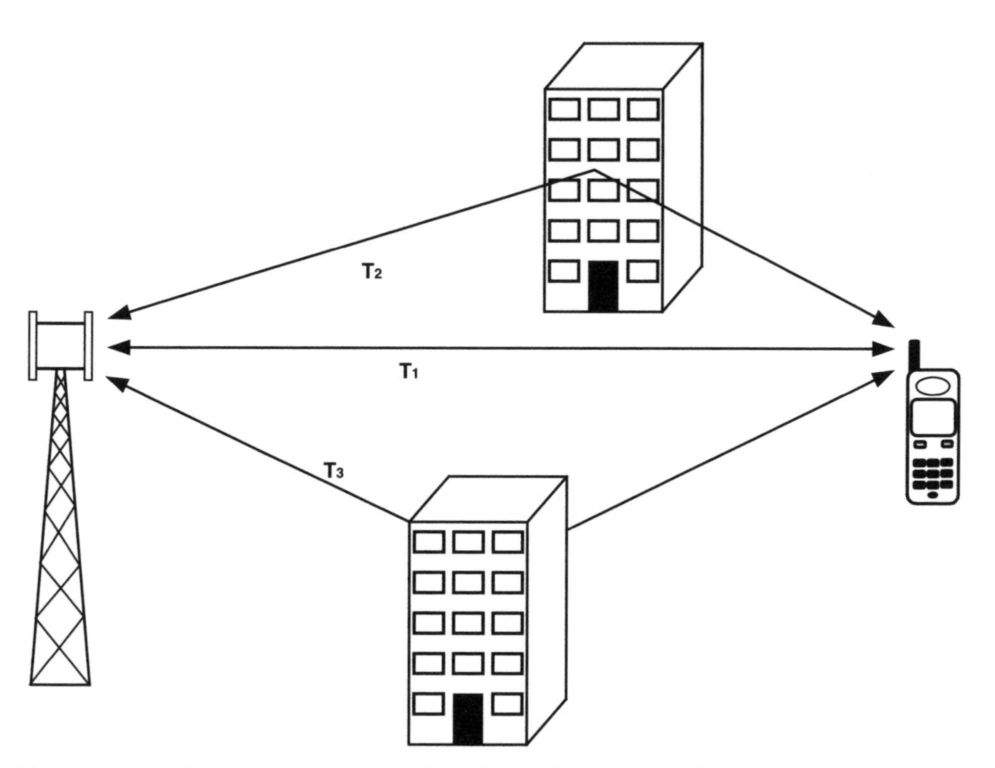

Figure 2.34 In the outdoor environment the RF channel is multipath, having more than one signal path between the mobile and the base station. The signal phase and amplitude of these signals will not be correlated

At the receiver T1, T2 and T3 will be offset in time and amplitude (as shown in Figure 2.35). In a normal receiver this will mean that, depending on the phase and amplitude, the signals will cancel each other out to some extent. However, the UMTS system uses a clever type of receiver, the rake receiver.

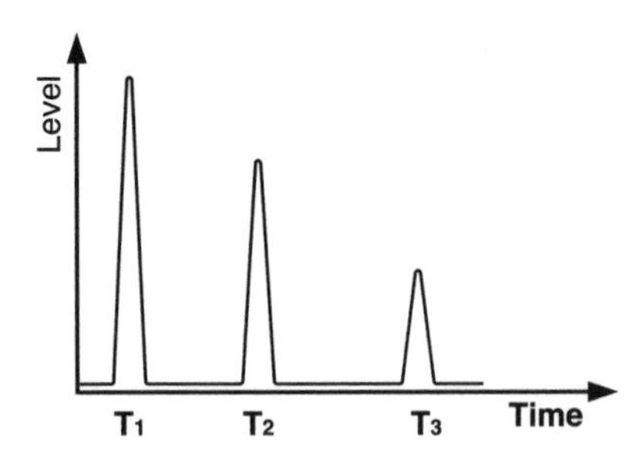

Figure 2.35 Owing to reflections from the environment, the offset signals reaching the mobile will be the different signal components of the multipath signal, with different levels and phases

The Rake Receiver

Like the tines, or fingers, on a rake, the rake receiver has more than one point. Actually the rake receiver could be compared with three or four parallel receivers, each receiving a different offset signal from the same RF channel. The rake receiver in this example (as shown in Figure 2.36) uses three fingers to receive the three differently phased signals, T1, T2 and T3 (as shown Figure 2.35), so these three signals are phase-adjusted, aligned and combined into the output signal by the rake receiver.

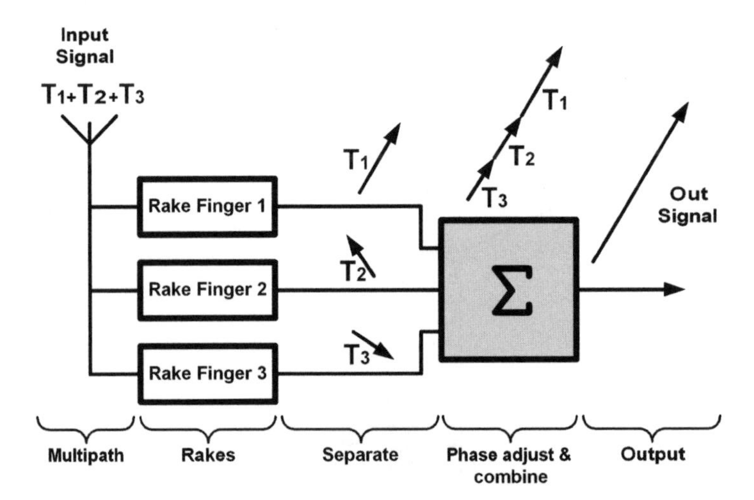

Figure 2.36 The rake receiver will be able to detect the different signals phases and amplitudes in a multipath environment. Separate phases are adjusted and the signals summed back to one signal

In order for the UMTS rake receiver to work, the delay separation between the different signals has to be more than one chip duration (0.26 μs). Delays shorter than one chip cannot be recovered by the rake receiver, and will cause inter-symbol interference.

2.3.7 *UMTS Signaling*

UMTS Transport Channels

The UMTS radio interface has logical channels that are mapped to transport channels. These transport channels are then again mapped to physical channels. Logical-to-transport channel conversion happens in the medium access control (MAC) layer, which is a lower sublayer in the data link layer (layer 2).

Logical Channels

The logical channels are divided into two groups: control channels and traffic channels. The MAC uses the logical channels for data transport. The traffic channels carry the voice and data traffic.

Control Channels

- *BCCH-DL* – this channel broadcasts information to the mobiles for system control, system information about the serving cell and the monitored set (neighbor list). The BCCH is carried by the BCH or FACH.
- *PCCH-DL* – the PCCH is associated with the PICH. The channel is used for paging and notification information. The PCCH is carried by the PCH.
- *DCCH-UL/DL* – this channel carries dedicated radio resource control information, in both directions. Together with the DTCH, the DCCH can be carried by different combinations of transport channels: RACH, FACCH, CPCH, DSCA or DCH.
- *CCCH-UL/DL* – this channel's purpose is to transfer control information between the mobiles and the network. The RACH and FACH are the transport channels used by the CCCH.
- *SHCCH-UL/DL* – this channel is only used in UMTS TDD. It is a bi-directional channel, used to transport shared channel control information.

Traffic Channels

- *DTCH-UL/DL* – this is the bi-directional channel used for user traffic.
- *CTCH-DL* – this is a downlink-only channel, used to transmit dedicated user information to a single mobile or a whole group of mobiles.

Transport Channels

The user data is transmitted via the air interface using transport channels. In the physical layer these channels are mapped into physical channels. There are two types of transport channels, common and dedicated channels. Dedicated channels are for one user only; common channels are for all mobiles in the cell.

Dedicated Transport Channels

- *DCH-UL/DL, mapped to DCCH and DTCH* – this channel is used to transfer user data and control information, handover commands, measurement reports etc. to and from a specific mobile. Each mobile has a dedicated DCH in both UL and DL.

Common Transport Channels

There are six common transport channels:

- *BCH-DL* – this channel transmits identification information about the cell and network, access codes, access slots, etc. The BCH is sent with a low fixed data rate. The BCH must be decoded by all the mobiles in the cell, therefore relative high power is allocated to broadcast the BCH. The BCH is mapped into the PCCPCH (primary common control physical channel).
- *FACH-DL* – this channel transmits control information to the mobiles that are in service on the network. The FACH can also carry packet data. The FACH is mapped into the SCCPCH (secondary control physical channel).
- *PCH-DL* – this is the channel carrying the paging signals within the location area, which alert mobiles about incoming calls, SMS and data connections. The PCH is mapped into the SCCPCH.
- *RACH-UL* – this channel is used by the mobile when accessing the network with access bursts. Control information is sent by the mobile to the network on the RACH. The RACH is mapped into PRACH (physical random access channel).
- *Uplink common packet channel (CPCH)-UL* – this channel is used for fast power control, and also provides additional capacity beyond the capacity of the RACH. The CPCH is mapped into the PCPCH (physical common packet channel).
- *Downlink shared channel (DSCH)-DL* – this channel can be shared by many mobiles and is used for nontime-critical data transmission on the downlink, web browsing, etc. The DSCH is mapped to the PDSCH (physical downlink shared channel).

Physical Channels

The transport channels are mapped into physical channels.

- *CPICH* – this channel identifies the cell, used by the mobile for cell selection, and to measure the quality of serving radio link and the adjacent cells. The CPICH enables the mobile to select the best cell, and the CPICH plays a significant role in cell selection and cell reselection.
- *Synchronization Channel (SCH)* – there is both a primary and a secondary SCH. These channels are used by the mobiles to synchronize to the network.
- *Common control physical channel (CCPCH)* – there are two types of CCPCHs, the primary and the secondary:
  - *PCCPCH* continuously broadcasts the system identification (the BCH) and access control information;

  - *SCCPCH* transmits the FACH (forward access channel), and provides control information, as well as the PACH (paging channel).
- *PRACH* – this channel is used by the mobile to transmit the RACH burst when accessing the network.
- *Dedicated physical data channel DPDCH* – this channel is the user data traffic channel (DCH).
- *Dedicated physical control channel (DPCCH)* – this channel carries the control information to and from the mobile. The channel carries bidirectional pilot bits as well as the transport format combination identifier (TFCI). On the DL channel is included transmit power control and feedback information bits (FBI).
- *PDSCH* – this channel shares control information (DSCH) to all mobiles within the service area of the cell.
- *PCPCH* – this channel is for carrying packet data (CPCH).
- *Acquisition indicator channel (AICH)* – this channel is used to inform the mobile about the DCH is must use to connect to the Node B.
- *Paging indication channel (PICH)* – this channel indicates the paging group the mobile belongs to. This enables the mobiles to be able to monitor the PCH when its group is paged, and in the meantime the mobile can 'sleep' and preserve its battery.
- *CPCH status indication channel (CSICH)* – this DL channel carries information about the status of the CPCH. Can also be used for DL data service.
- *Collision detection channel assignment indication channel (CD/CA-ICH)* – this DL channel is used to indicate to the mobile if the channel assignment is active or inactive.

Power Allocation for Common Control Channels

In the network the power is allocated to the physical common control channels is defined. These power levels can be adjusted, but typical settings could be as shown in Table 2.2. Even with no traffic on the cell, 16% (−8.1dB) of the power resource is allocated for common control channels.

Table 2.2 Typical power levels for common control channels

Channel	Node B power	Activity duty cycle	BS power load
CPICH	10%	100%	−10 dB
CCPCH	5%	90%	−13 dB
SSCH	4%	10%	−14 dB
PSCH	6%	10%	−12.2 dB

Measuring the UMTS Transmit Power

If you need to measure the power level at the base station antenna connector, you will need to know these settings. If the settings are like the settings above, you will measure the maximum power – 8.1 dB, so a 46 dBm base station will give a measurement of 37.9 dBm when there is no traffic on the cell. Most UMTS networks will allow you to preset a specific load in the cell, and then this reference can be used for power measurements.

2.3.8 The UMTS Network Elements

This is a short introduction to the elements of the UTRAN (UMTS terrestrial radio access network) and the key elements of the core network (as shown in Figure 2.37). This book is not intended as a detailed training book about the UTRAN and core network, but it is important that the indoor radio planner understands the general principles of the total network, especially regarding the UTRAN.

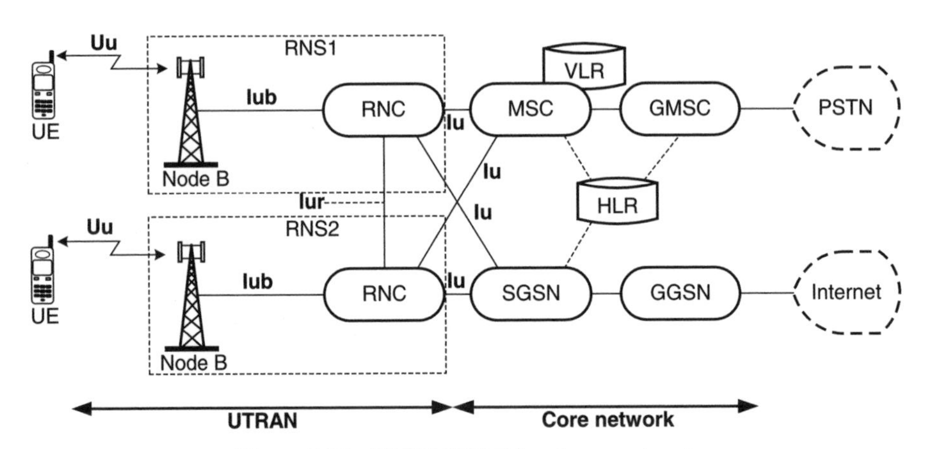

Figure 2.37 UMTS UTRAN and core network

User Equipment

The mobile station is referred to as the user equipment (UE). This indicates that the UE is much more than 'just' a mobile telephone. The UE is a mobile data device (could be a mobile telephone). However, throughout this book, the UE is referred to as the mobile station, in order for the examples to be compatible with both UMTS and GSM.

Node B, Base Station

The base station in the UTRAN network is called node B. Node B consists of transceivers, processing modules that provide the 'channel elements' that service the users. The node B interfaces to the UE over the air interface, the Uu Interface. Interfacing node B to the RNC is the Iub interface.

Node B will also perform important measurements on the Uu interface, and report these measurement results (reports) to the RNC with regards to quality of the link. These are BLER, block error rate, and BER, bit error rate. They are needed in order for the RNC to be able to evaluate the QOS (quality of service), and adjust the power control targets accordingly. Also, the reception level of the mobiles is measured and reported, as well as the signal-to-noise ratio, etc.

Radio Network Controller

The radio network controller (RNC) is controlling the node Bs within its own system (RNS). For speech service, the RNC interfaces to the MSC. For packet-switched data service the RNC interfaces to the SGSN. The RNC is responsible for the load on the individual cells in the RNS, and handles admission (traffic) control, code allocation. Once a connection has been established between UE and node B, signaling to the elements higher up in the network is done by the RNC. This RNC is then referred to as the serving RNC, the SRNC.

The SRNC is responsible for handover evaluation, outer loop power control, as well as the signaling between the UE and the rest of the UTRAN. If the UE signal can be received (in soft handover) by other node Bs controlled by other RNCs, these RNCs are called DRNC, drift RNCs. The DRNC can process the UL signal from the UE and provide macro diversity, transferring the data via the Iub/Iur interfaces to the SRNC.

Radio Network Sub-system

One RNC with all the connected node Bs is defined as a radio network sub-system (RNS). A UTRAN consists of several RNSs; each RNC within the RNS is interconnected with an Iur interface.

- *Uu interface* – this is the radio interface of WCDMA, the air interface between the UE and the node B.
- *Iu interface* – this is the link between the UTRAN and the core network. The interface is standardized so a UTRAN from one manufacture will be compatible with a core network form another manufacture.
- *Iur interface* – the Iur interfaces the different RNCs; it interfaces data from soft handovers between different RNCs.
- *Iub* – between the node B and the RNC the Iub interface is used. Like the Iu interface, this is fully standardized so different RNCs will support different vendors' node Bs.

Core Network

The core network is not of interest of the radio planner on a daily basis, but a short description and overview is appropriate.

- *Mobile switching center* – the mobile switching center (MSC) controls the circuit switched connections, speech and real-time data applications for an active UE in the network. Often the VLR will be co-located with the MSC.
- *Gateway mobile switching center* – the gateway mobile switching center (GMSC) interfaces the MSC to external MSCs for circuit-switched data and voice calls.
- *Serving GPRS support node* – the serving GPRS support node (SGSN) switches the internal packet-switched data traffic.
- *Gateway GPRS support node* – the gateway GPRS support node (GGSN) switches the external packet-switched data traffic.

Visitor Location Register

The VLR is a database, with the location (location area) of all UEs attached to the network. When a UE registers in the network, the VLR will retrieve relevant data about the user (SIM) from the HLR associated with the SIM (IMSI).

Home Location Register

The HLR is a database that contains all the relevant data about the subscriber (SIM). This is the subscription profile, roaming partner and current VLR/SGSN location of the UE.

2.4 Introduction to HSPA

It is not within the scope of this book to describe all the details about HSPA. Only the issues that important for maximizing the HSPA performance inside buildings will be covered. Please refer to Reference [5] for more details about HSPA.

2.4.1 Introduction

Seamless mobility for voice and data service is of paramount importance for the modern enterprise. In particular, service for mobile telephones and data terminals is important, as most professionals leans towards 'one person, one number', also referred to as the mobile office. The service offering and quality should be seamless, no matter whether you are at home, on the road or in the office.

Offices with a heavy concentration of high-profile voice and data users, in particular, pose specific challenges for high-performance in-building data performance using HSPA. HSPA/Super-3G is an alternative to Wi-Fi service. Mobile 3G networks are deploying HSPA on their outdoor networks, and full mobility is now an option for high-speed data. Providing sufficient in-Building coverage on HSPA poses a major challenge for the mobile network operators due to the degradation of the radio service when covering indoor users from the outdoor network. The data users with the highest demand-to-data service rate and speed are typically located indoors, so a special focus on how to solve the challenge is important.

2.4.2 Wi-Fi

Why will the users prefer to use 3G/HSPA rather than Wi-Fi? Wi-Fi is accessible in many places, it is 'free' and the data speed is claimed to be high, in excess of 50 Mbps, so why not just use Wi-Fi to provide mobile data indoors?

The Challenges for the Users, Using Wi-Fi

Lack of Mobility
There are no handovers on Wi-Fi, so users cannot roam around in the building without dropping the service. However, developments are ongoing to make handovers possible between Wi-Fi access points, and even to other systems.

Hassle to get Wi-Fi Service
It is not very user-friendly to obtain Wi-Fi service. It is often expensive and you have to get a specific user access code for every Wi-Fi provider in each building.

Wi-Fi Service is Slow
The data rate on the Wi-Fi–air interface is high, in excess of 50 Mbps. However, the actual user data rate is not limited by the air interface but by the ADSL backhaul to the Wi-Fi access point. Typical user data rates on public-access Wi-Fi services in hotels and airport are often lower than 400 kbps, far below the 54 Mbps on the typical Wi-Fi air interface.

No Guarantee of Service
Wi-Fi uses free open spectrum, so there is no guarantee that other Wi-Fi providers will not deploy Wi-Fi service in the same areas, using the same or adjacent frequencies and degrading the current Wi-Fi service. Therefore there is no quality of service guarantee using Wi-Fi.

Wi-Fi Over Indoor DAS

Why not distribute Wi-Fi over the DAS? To understand why it is not advisable to distribute Wi-Fi over the same distributed antenna system (DAS) that distributes the mobile service for GSM and UMTS, you need to understand the basic characteristics of Wi-Fi, refer to Figure 2.38.

- *Wi-Fi is distributed capacity* – the basic concept of Wi-Fi is to distribute the data capacity out to the location of the data users by the use of distributed base stations, Wi-Fi access

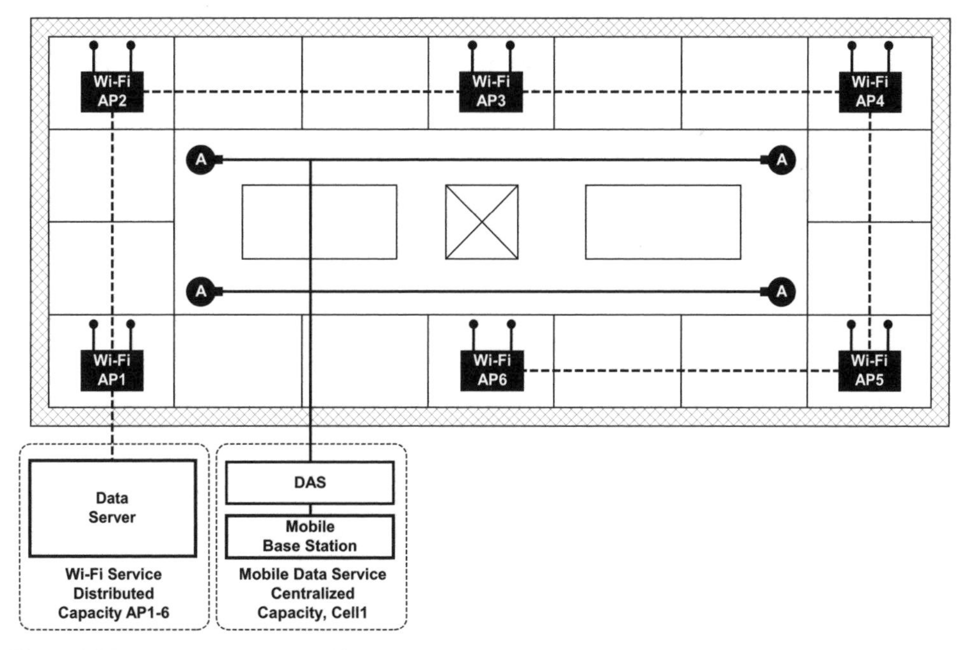

Figure 2.38 There is a generic difference between Wi-Fi and mobile data service. Wi-Fi is distributed capacity via access points. Mobile data service uses centralized capacity from the base station

points. By using the strategy of distribution of the Wi-Fi data capacity and deploying the Wi-Fi access points locally, combined with the limited coverage range from these access points, you will be able to support high data rates in the building. This is exactly the same principle as frequency reuse in a cellular network. If you have limited frequencies on a GSM roll-out, you do not just deploy one tall macro site in the center of a city and blast power on a limited number of frequencies. This is like Wi-Fi only having limited number of frequencies; you need to distribute the GSM base stations to raise capacity. The same principle applies to Wi-Fi: you need to distribute the access points to tender for the capacity need.
- *Hot-spot data coverage* – Wi-Fi is designed to cover the hot-spot data need for specific locations in the building. It is only possible to deploy three or four Wi-Fi carriers in the same area without degrading the service, owing to adjacent channel interference.

Why Not Distribute Wi-Fi Over the DAS?

It does not make sense to combine the RF signal of one to three Wi-Fi access points (APs) and distribute it via the DAS, owing to the inherent difference between GSM/UMTS/HSPA and Wi-Fi: Wi-Fi is distributed capacity via APs, and GSM/UMTS HSPA is centralized capacity, distributed via DAS.

If you distribute the Wi-Fi RF signal over the DAS you face these limitations:

- Limited number of carriers in the same cell, limited capacity.
- Not enough capacity for the data users in most buildings.
- The concept of Wi-Fi is distributed capacity, not centralized in a DAS like the capacity of mobile services such as UMTS/HSPA.
- Wi-Fi is TDD; with more than one AP in the same cell the effect is that the APs are not synchronized and the transmission will collide, degrading the performance.
- Wi-Fi APs use antenna diversity; this is important to boost the performance, which is lost in the DAS.
- The desired coverage area for the Wi-Fi and mobile signal might not correlate.
- The sector plan on the mobile system will dictate the capacity on Wi-Fi.

2.4.3 Introduction to HSDPA

HSDPA High-speed Downlink Packet Access

HSDPA is a high-speed downlink data service that can be deployed on the existing 3G network infrastructure. The operators can deploy the first phase on the existing UMTS R99 network, using the same UMTS CH (as shown in Figure 2.39) and use the power headroom not utilized by the UMTS traffic.

This provides the mobile operator with a deployment strategy that can support the need for high-speed downlink data rates, at a relative low deployment cost. Many operators will prefer to use a separate CH for HSDPA, because loading the existing UMTS R99 carrier with

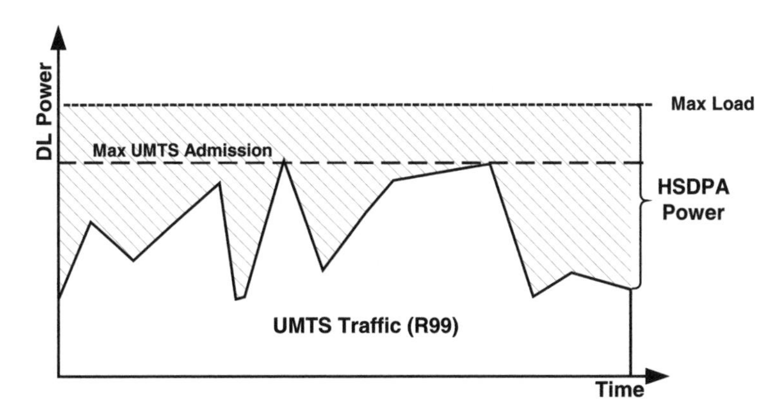

Figure 2.39 HSDPA deployed on the same RF channel as UMTS

HSDPA will cause the network to run constantly on a higher load rate, increasing the noise on the existing UMTS CH, degrading the UMTS capacity.

Why will the operators prefer to use HSDPA?

- Typically the highest data load is on the downlink.
- HSDPA will have a trunking gain of the radio capacity.
- There is no need for distributed APs like with Wi-Fi.
- Operators can use their existing UMTS network to launch HSDPA.
- HSDPA is more spectrum-efficient compared with EDGE and UMTS.
- There is higher spectrum efficiency, more data per MHz.
- There are fewer production costs per Mb.
- There are higher data speeds, up to 14 Mbps, with indoor DAS systems.
- Mobile operators can compete with Wi-Fi using HSPA.

HSDPA Key Features

HSDPA can perform high downlink data rates, provided that the radio link quality is high.

- Downlink data Rates up to 14.4 Mbps.
- No soft handover, so pico cells overlapping in the same building will produce 'self interference' (see Figure 4.38).
- Can use the same channel as UMTS by utilizing the 'power headroom' for HSDPA (as shown in Figure 2.39).
- Typical operators will launch HSDPA on a separate carrier, to minimize noise increase on the UMTS service.
- HSDPA is a downlink data service; users will use UMTS on the uplink until HSUPA is deployed.
- For the uplink data transmission from the mobile, UMTS R99 is used until HSUPA is launched in 2008 to cater for higher uplink data rates.

Indoor HSDPA Coverage

Why do you need indoor active DAS systems to have high quality HSDPA?

- The data speed is related to the SNR. A good quality radio link is needed to produce higher speeds.
- The macro layer will use excessive power to reach indoor users, causing interference and degradation of the macro layer. Only in the buildings few hundred meters from the outdoor base station will it be possible to service indoor users with HSPA coverage. In an urban environment it is not likely to exceed 360 kbps in the major portion of the area.
- Production cost will be lower per Mb for the operators, using indoor systems where the traffic is.

2.4.4 Indoor HSPA Coverage

The best performance on HSPA is achieved by deploying indoor DAS. The data speed is related to the SNR. A good radio link, with a high SNR, is needed to produce higher speeds. Using the outdoor macro layer causes excessive power to reach indoor users, causing interference and degradation of the macro layer. Only in buildings a few hundred meters from the outdoor base station will it be possible to service indoor users with acceptable HSDPA coverage. In urban environments it is not likely to exceed data speeds of 360 kbps indoors, without the use of dedicated in-building systems.

It is possible to lower the production cost per Mb for the operators by deploying indoor DAS to service HSDPA, as dedicated in-building systems help offload the macro network.

The Impact of HSUPA

HSUPA (high-speed uplink packet access) will provide the users the possibility of high-speed data rates on the uplink. HSUPA relies on the mobile's ability to reach the network and will be sensitive to bad macro network quality and to the high loss of passive in-building systems. Therefore, an active DAS system will be essential for providing a high-quality service. The low noise of the active DAS and the high power delivered at the DAS antenna will give the highest possible data rate performance for HSUPA.

The Choice of DAS Distribution for HSPA

HSPA performance is very sensitive to the quality of the radio link, the dominance and the isolation of the indoor signal. How to insure this is explored in Chapters 3 and 4. The DAS that will perform the best is the DAS that will provide the highest downlink radiated power at the antenna, and the lowest possible noise figure. For medium to large buildings it will often be some form of active DAS that will perform the best HSPA service, owing to the attenuation of passive DAS. Be careful and design the DAS solution to maximum isolation; good indoor design strategy is very important for HSPA performance (refer to Chapter 5 for more details). Passive coax solutions do not perform at the higher data rates. Whereas using passive DAS was sometimes acceptable for 2G mobile services for voice only, the future is data, and the high losses on the coax cables will degrade the data speed.

Advantages of Using HSPA

Providing high-speed mobile data via HSPA distributed indoors via a DAS has several advantages for the mobile operator:

- *Better HSPA service* – users on 3G (HSPA) will have coherent coverage; users can roam from the outdoor network to the indoor DAS with full mobile data service.
- *Laptops are HSPA ready* – new laptops come with integrated 3G cards. Like the early days of Wi-Fi, users on HSPA will initially have to use HSPA plug-in cards in their PC, but the new generation of laptops comes with pre-installed HSPA support.
- *Less production cost per Mb* – the production cost on HSPA is significant lower than for EDGE and UMTS data service, boosting the business case for the operators.
- *Seamless billing* – the user does not have to purchase individual Wi-Fi service subscriptions in the individual buildings from individual service providers, often at a high cost. Users on HSPA can get all the data billing via their normal mobile subscription.
- *Easy to connect with HSPA* – users do not need to scan for Wi-Fi and try to connect at various Wi-Fi service providers. Operators will have automatic algorithms enabling the users just to 'click and connect'.
- *Better than Wi-Fi* – HSPA service with indoor DAS can easily outperform Wi-Fi. The DAS will distribute the HSPA service throughout the building, supporting the highest possible data rate. 3G and HSPA will service the mobile data users with higher data rates. The mobile system is not limited by the relative slow ADSL backhaul like the Wi-Fi AP. Typically HSPA will provide DL data rates exceeding 6 Mbps, when using indoor DAS.

Table 2.3 clearly shows how competitive HSDPA can be compared with Wi-Fi service. This example is a shopping mall where a mobile operator outperforms the existing Wi-Fi provider, by deploying HSPA DAS antennas in the Wi-Fi hot spots.

Table 2.3 Typical HDSPA data rate from omni antenna radiating 10 dBm in an open area of a shopping mall

HSDPA speed	Coverage radius
480 kbps	47 m
720 kbps	37 m
1.8 Mbps	33 m
3.6 Mbps	29 m
7.2 Mbps	20 m
10.7 M	15 m

2.4.5 Indoor HSPA Planning for Maximum Performance

The highest mobile data rates are typically needed only in hot-spot areas of the building. Therefore you must apply a strategy to our indoor design that is based on the hot spot areas of the building. Locating DAS antennas in these areas will provide the highest possible data rate; the rest of the building can then be planned around these antenna locations. This principle is described in Section 5.4.4. Hot spot areas will typically be areas where mobile data users are able to sit down and use their PC or mobile data device.

Hot-spot data areas in buildings will typically be:

Office buildings

- Conference areas.
- Meeting rooms.
- Executive floors.
- Wi-Fi zones.

Airports

- Business lounges.
- Sitting areas at gates.
- Restaurants and cafes.
- Meeting rooms.
- Staff meeting rooms.
- Wi-Fi zones.
- Administration areas.

Hotels

- Conference areas.
- Meeting rooms.
- Executive floors.
- Wi-Fi zones.
- Business rooms.
- Lobby.
- Restaurants.

Shopping malls

- Food court.
- Sitting areas.
- Cafes and restaurants.
- Staff meeting rooms.
- Administration areas.

2.4.6 HSDPA Coverage from the Macro Network

It is possible to provide sufficient levels of RF signal inside buildings to provide a strong UMTS signal and to some extent HSPA coverage, provided that the serving macro base stations are within few hundred meters of the building, and have direct line-of-sight. In reality, however, only few a buildings will fall in that category. However, apart from the signal level on UMTS, there are other important factors to consider:

UMTS Soft Handover Load

UMTS service provided from the macro layer might be acceptable seen from the user's perspective. In Figure 2.40, three macro cells are providing good indoor coverage, indicated by the three shades of gray.

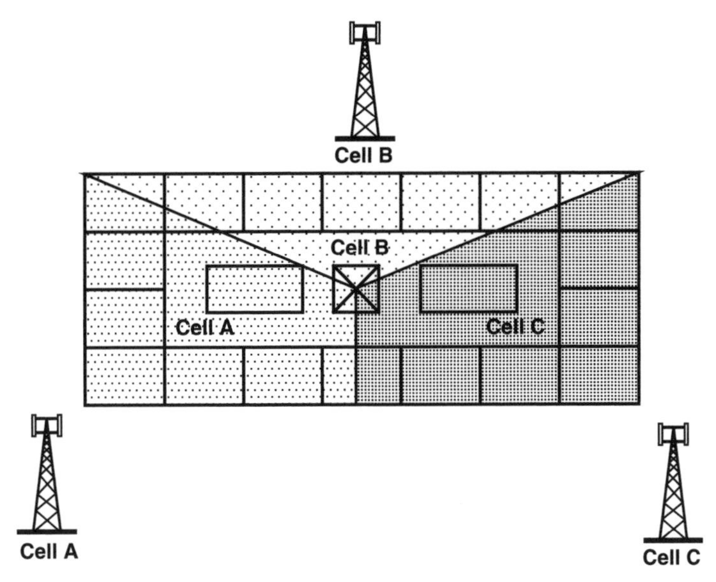

Figure 2.40 Three outdoor macro base stations near the building are providing good UMTS signal level inside the building. Each of the outdoor cells is covering about a third of the floor space

The challenge for the UMTS network is the extensive soft handover area; the white area seen in Figure 2.41. This large soft handover area will increase the load on the network for UMTS traffic, for the users in the part of the building serviced by more than one cell. But for HSPA traffic the data throughput wil be degraded caused by C/I (co-channel interference), because the cells are operating on the same HSDPA frequency.

Degraded HSDPA Performance

The UMTS service will still be acceptable, but the price on the network side is the large soft handover area for about 60% of the users inside the buildings as HDPA does not use soft handover, the three serving all HSDPA cells are on the same frequency and there is a large impact on the *C/I*. In Table 2.4, you can see an example of a HSDPA link budget. In order to provide HSDPA service at high confidence level, you need good isolation between the serving cell and the adjacent cell due to the strict *C/I* requirements.

In the example shown in Figure 2.41, only 20% of the building has a *C/I* better than 5 dB, the gray areas. Only in this area will the DL data rate be better than about 500 kbps. In the major part of the building the data rate will be lower than 480 kbps.

Table 2.4 HSDPA link budget example

Service	RSSI @ Cell Edge [dBm]	90% RSSI @ Cell Edge [dBm]	C/I [dB]	90% C/I confidence [dB]
DL: 120k (QP-1/4-1)	−89.7	**−79.3**	−14.5	**−4.1**
DL: 240k (QP-1/2-1)	−85.2	**−74.8**	−10	**−0.4**
DL: 360k (QP-3/4-1)	−82.2	**−71.8**	−7	**3.4**
DL: 480k (QA-1/2-1)	−80.2	**−69.8**	−5	**5.4**
DL: 720k (QA-3/4-1)	−75.7	**−65.3**	−0.5	**9.9**
DL: 3.6M (QP-1/2-15)	−74.2	**−63.8**	1	**11.4**
DL: 5.3M (QP-3/4-15)	−70.7	**−60.3**	4.5	**14.8**
DL: 7.2M (QP-1/2-15)	−69.2	**−58.8**	6	**16.4**
DL: 10.7M (QP-3/4-15)	−64.7	**−54.3**	10.5	**20.9**

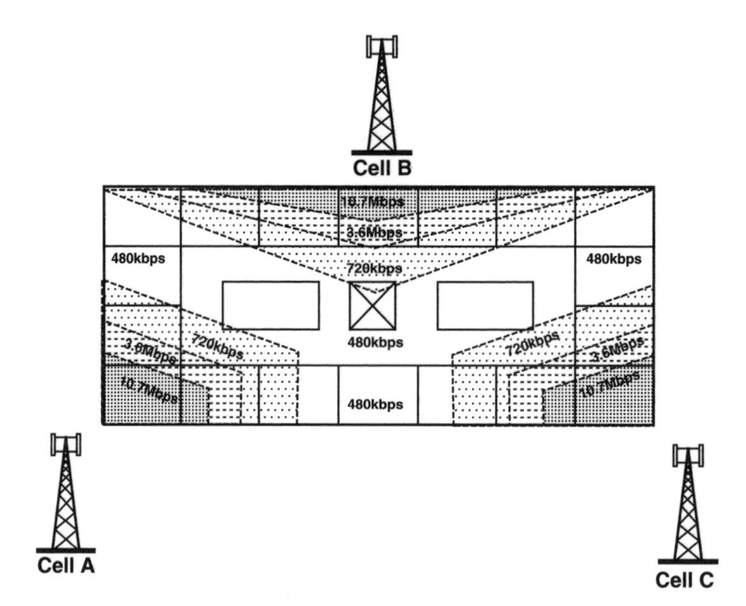

Figure 2.41 The three macro base stations give full coverage, but a larger portion of the indoor floor space is covered by more than one serving cell, the gray area

Solution

The only solution for a consistent and high-performing indoor HSPA service is to provide isolation between the intended serving cell and the interfering HSDPA cells from the outdoor network. In order to be able to provide HSDPA service at the higher data rate, the only solution is an indoor DAS system. The challenge is to dominate the building with the indoor coverage, and at the same time avoid leakage from the indoor DAS system. This can be an issue, especially in buildings with high signal levels from outdoor macro cells.

A viable solution is to deploy directional antennas at the edge and corners of the building and point them towards the center of the building. Using this strategy, the total indoor area is dominated by the indoor cell as long as you dominate at the border. The reason is that the indoor signal radiated by both the corner- and edge-mounted antennas and the outdoor cell will have the same signal path and attenuation from the edge of the building throughout the whole indoor area. Dominating the border of the building will give total dominance, and will at the same time minimize leakage into the macro network.

The solution is described in Section 3.5. In order to archive high HSPA data rates, good isolation between the cells is needed. The challenge with this strategy is the actual installation of the antennas in the corners of the building. Normally there will be limited access to cable ducts and installation trays along the perimeter of the building. This is even more of an issue for traditional passive DAS based on heavy and rigid coax cables. However, using an active DAS that relies on lightweight, flexible IT-type cabling to reach the antenna point, the installation of these antennas is less of a challenge.

2.4.7 Passive DAS and HSPA

Traditional passive DAS systems based on coax cables will in many case have too high an attenuation to provide sufficient HSPA data service inside buildings. Passive DAS can perform on HSPA provided that the attenuation between the base station and the indoor antenna is less than 20 dB. In practice, using passive DAS for indoor HSPA solutions is only possible for smaller buildings that can be covered by six to eight indoor antennas. For larger buildings the HSPA performance will be compromised due to the high losses of the coax cables, splitters, tappers and other passive DAS components.

HSDPA on Passive DAS

In Figure 2.42, the HSDPA service range from the indoor DAS antenna in a typical office environment is shown. The graphs represent benchmarking of the HSDPA performance of passive DAS with 20–30 dB of attenuation between the base station and the DAS antenna. This is compared with the HSDPA performance of the active DAS from Section 4.4, with +20 dBm output power.

The impact from the attenuation of the passive DAS is evident. Only the passive DAS with 20 dB of attenuation performs slightly better than the active DAS owing to the DL power of the +43 dBm base station. This makes passive DAS only applicable for HSDPA solutions in small buildings with less than six to eight antennas. However, the installation challenges associated with the passive DAS remains. Therefore it will be hard to draw up the correct antenna layout as shown in Section 3.5. The typical office building implemented with passive DAS and corner-mounted antennas will have losses exceeding 30–35 dB, limiting the HDSPA service range severely.

HSUPA on Passive DAS

The degrading impact on the HSUPA performance by the attenuation of the passive DAS is even more evident than the HSDPA degradation. The reason for this can be explained by basic cascaded noise theory and practice. Any loss prior to the first amplifier in the receiver will degrade the noise figure of the system, in this case the HSUPA receiver in the base station.

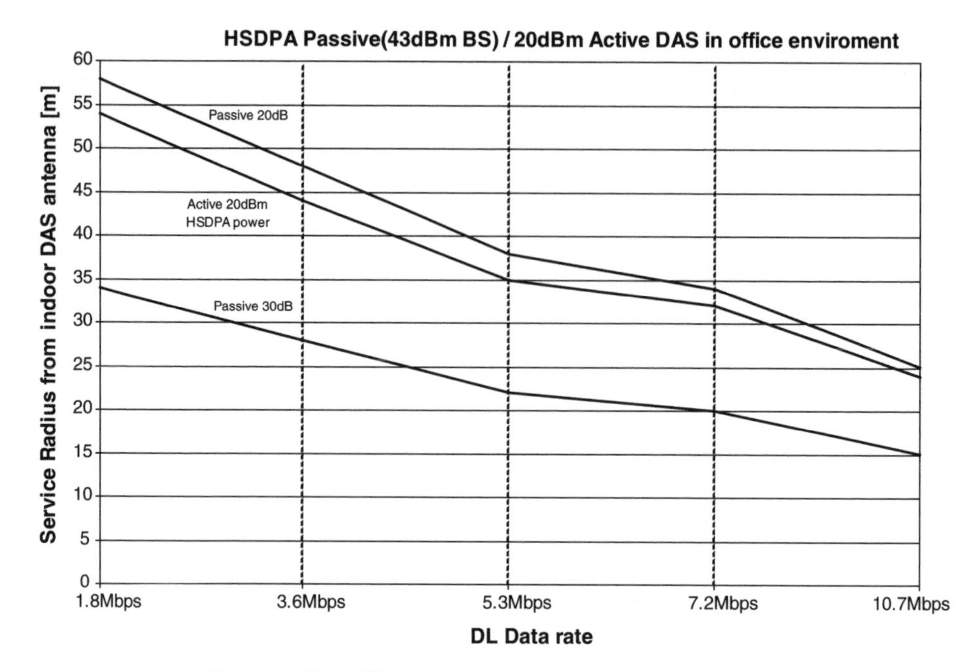

Figure 2.42 HSDPA performance passive and active DAS

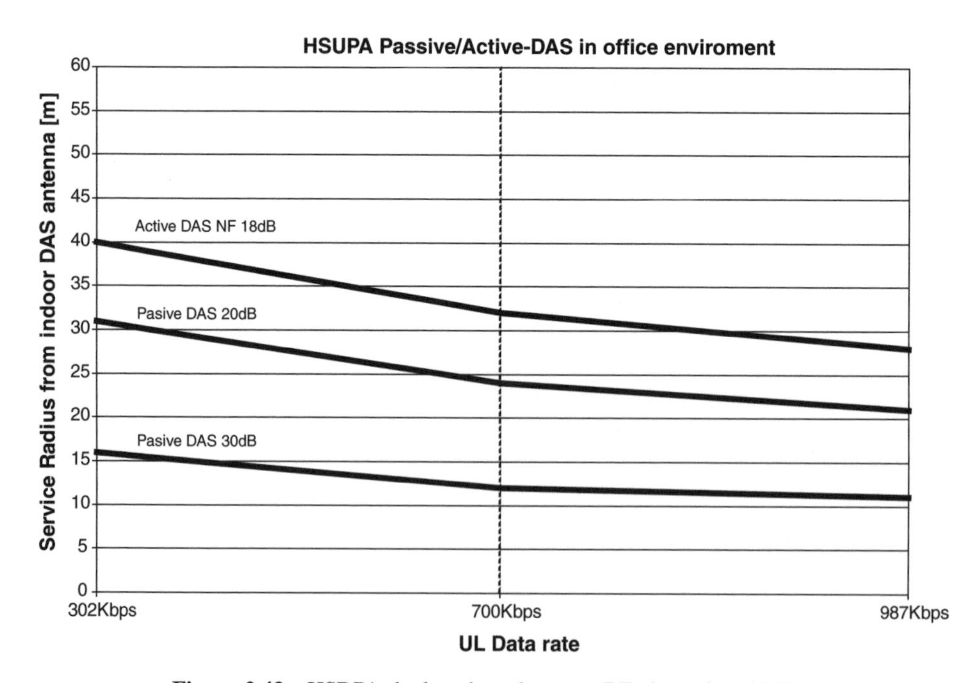

Figure 2.43 HSDPA deployed on the same RF channel as UMTS

Figure 2.43 clearly shows the impact on the HSUPA performance, caused by the passive losses. Even a 'low loss' passive DAS with 'only' 30 dB of attenuation cannot perform at more than 700 kbps just 12 m from the indoor antenna. The active DAS will perform at about 1 Mbps up to 27 ms from the antenna, at the same location in the same environment.

2.4.8 Conclusion

The active DAS concept for distribution of HSPA signals is shown to be superior for the provision of high-speed data services. Both for HSDPA and HSUPA, the conclusion is clear – having the DL amplifier and the UL amplifier as close as possible to the indoor DAS antenna will boost HSDPA and HSUPA performance to a maximum. The concept is to use distributed amplifiers, and locate them at the antenna points, compensating for the losses back to the base station. Delivering the DL power at the antenna and minimizing the UL noise figure is the best solution. Calculations and real-life measurements and implementations back up the theory. Chapter 7 covers the noise calculations for this concept.

3

Indoor Radio Planning

There are numerous challenges, both from a business and a technical perspective, when designing and implementing indoor coverage solutions. Indoor radio planners carry a major responsibility for the overall business case and performance of the network. In many countries 80% of users are inside buildings, and providing high-performance indoor coverage, especially on higher data rates, is a challenge. It is much more than a technical challenge; the business case must also be evaluated, as well as future-proofing of the solutions implemented, among other considerations. As an indoor radio planner, it is important not to focus only on the technical challenge ahead, but to look a few steps ahead.

3.1 Why is In-building Coverage Important?

There are many reasons for the mobile operator, both technical and commercial, for providing sufficient in-building coverage. The technical motivations are typical; lack of coverage, improvement of service quality, need for more capacity, need for higher data rates and to offload the existing macro network. In 3G (UMTS) networks, the need to offload the existing macro network is an especially important parameter. The need for higher-speed data rates inside buildings also plays an important factor. It is evident that you will need dedicated in-building (IB) solutions to provide high-speed data service on UMTS and especially when deploying HSDPA/HSUPA high-speed data services.

This book will focus mainly on the technical part of the evaluation, and design of IB solutions. However, even the most hardcore technical RF design engineer must realize that the main driver for any mobile network operator must be to increase the revenue factor. The purpose is to maximize the revenue of the network and to lower the production cost of the traffic. The cost of producing a call minute (CM) is a crucial factor for the mobile operator, and so is the production cost per Mb of data transmitted in the network.

3.1.1 Commercial and Technical Evaluation

First and foremost, the mobile operator must do a business evaluation, before even considering investing in any in-building coverage solution. The operator must use standard tools and

Indoor Radio Planning: A Practical Guide for GSM, DCS, UMTS and HSPA Morten Tolstrup

metrics to evaluate the business case, in order to be able to calculate the revenue of each individual user of the different user segments. This will enable the operator to compare the business case on all the individual indoor coverage projects, in order to prioritize the projects.

This evaluation must be based on a standardized evaluation flow (see Section 5.1.1), using standard forms templates and metrics in order to secure a valid comparable business case for each indoor coverage project.

3.1.2 The Main Part of the Mobile Traffic is Indoors

Depending on what part of the world you analyze, it is a fact that the bulk of the traffic originates inside buildings. Therefore, special attention to the in-building coverage is needed, in order to fulfill the user's expectations and need for service. This is especially the case in urban environments, and the focus from the mobile users is on higher and higher data rates.

3.1.3 Some 70–80% of Mobile Traffic is Inside Buildings

In most cities it is very interesting to note that typically a few important buildings (hot-spots) will produce the major part of the traffic. In some cities more than 50% of the traffic originates from about 10% of the buildings. These buildings are referred to as the hot-spots. These buildings will typically be shopping malls, airports and large corporate office buildings.

3.1.4 Indoor Solutions Can Make a Great Business Case

Especially for UMTS, the power load per user (PLPU) is an important factor owing to the fact that downlink power on the base station is directly related to the capacity. The higher the PLPU, the higher the capacity drain from the base station per mobile user will be. This results in relative high production costs for indoor traffic on UMTS when trying to service the users inside buildings, from the outdoor Macro network.

Not only will the coverage, quality and data speed be better on UMTS with dedicated indoor coverage solutions, but the PLPU will be much lower due to the fact that with an indoor system the base station will not to have to overcome the high penetration loss of the building (20–50 dB). In addition, when servicing indoor UMTS users from the macro base station, the signal will rely mostly on reflections in order to service the users, degrading the orthogonality.

Implementing indoor coverage solutions is a very efficient use of the capacity (DL power) of the base station, and of the data channel. You can reduce the production cost per call minute or Mb, using IB coverage solutions and also reduce the overall noise increase in the network. *The production cost per call minute or Mb on UMTS can be cut by 50–70% using IB coverage solutions.*

3.1.5 Business Evaluation

Even the most technical dedicated engineer must appreciate that the main reason for providing IB solutions is to make a positive business case. The designer of the IB solution carries a major responsibility, on the one hand, for a well-designed and high-performing technical solution, but also a solution that needs to be future-proof, and will make the investment worthwhile. This is a fine balance between investment and technical parameters. Engineers are often tempted to overdesign the solutions 'just to be sure' – but the cost of doing this is high.

3.1.6 Coverage Levels/Cost Level

Selecting the correct coverage design level for the indoor design is crucial, for the performance of the indoor system, the data throughput performance and the leakage from the building. We will take a closer look at these more technical parameters later on in this book, but there is more to it than 'only' the technical part.

The radio planner must also realize that the design levels come at a cost, and this has a direct impact on the business case. The higher the coverage level, the higher the cost of the system, obviously due to the need for more antennas and equipment for the indoor system, more equipment, more installation work and maintenance costs.

3.1.7 Evaluate the Value of the Proposed Solution

Before considering any indoor coverage solution, you must carefully evaluate the value of the proposed solution. You will need to answer these questions:

1. Will the investment make a positive business case?
2. When will the investment begin to pay back?
3. Is the selected solution optimum for future needs:
   - Higher data rates.
   - New services.
   - More operators.
   - More capacity?
4. Can the selected solution keep up with the future changes in the building:
   - Reconstruction.
   - Extension?
5. Will the solution offload the macro layer, and free needed capacity? This must be part of the business case for the indoor solution; it is added value if you free up power or capacity on the outdoor network that can service other users.
6. Are there strategic reasons for providing the IB Coverage Solution:
   - Competitive edge over other operators.
   - Increased traffic in other parts of the network.
   - International roamer value (airports, harbors, ships, ferries, hotels, convention centers)?
7. Can dedicated corporate buildings be covered, in order to secure the business for the whole account:
   - Better coverage.
   - Better quality.
   - Better capacity.
   - Higher data rates.
   - More loyalty from the users?

3.2 Indoor Coverage from the Macro Layer

Why not just use the macro coverage to provide the needed indoor coverage? When designing a cellular network, especially in the first phase of rollout, many radio designers initially try to cover as many buildings from the macro layer as possible. This is despite knowing that most of the traffic originates from inside buildings.

To some extent, and in certain areas, this strategy makes sense. In many cases you are able to provide reasonably good overall indoor coverage from the macro base stations, but it is a fine balance and a compromise.

In a typical suburban environment you need to rely on a very tight macro grid with an inter-site distance of no more than 1–2 km, depending on the services that are offered. In urban environments the inter-site distance can be down to 300–500 m to provide the deep indoor penetration needed for GSM and UMTS. In many cases even this tight a site grid is not sufficient to provide the higher-data-rate EDGE coverage on GSM, and is not sufficient for providing higher data rates on UMTS (64–384 kps). On UMTS especially, HSPA will be a major concern, when covering from the outside network into the buildings.

As you can see from traffic data from many real-life examples like the one shown in Figure 3.1, even a tight macro grid will in many cases be insufficient to service the indoor users, and certainly not the data users on the higher data rates in particular, but also even voice users on GSM and 3G.

3.2.1 More Revenue with Indoor Solutions

This is an example of the traffic production (Figure 3.1), after implementing solid indoor coverage in a shopping area. This shopping area consists of two major shopping malls, as

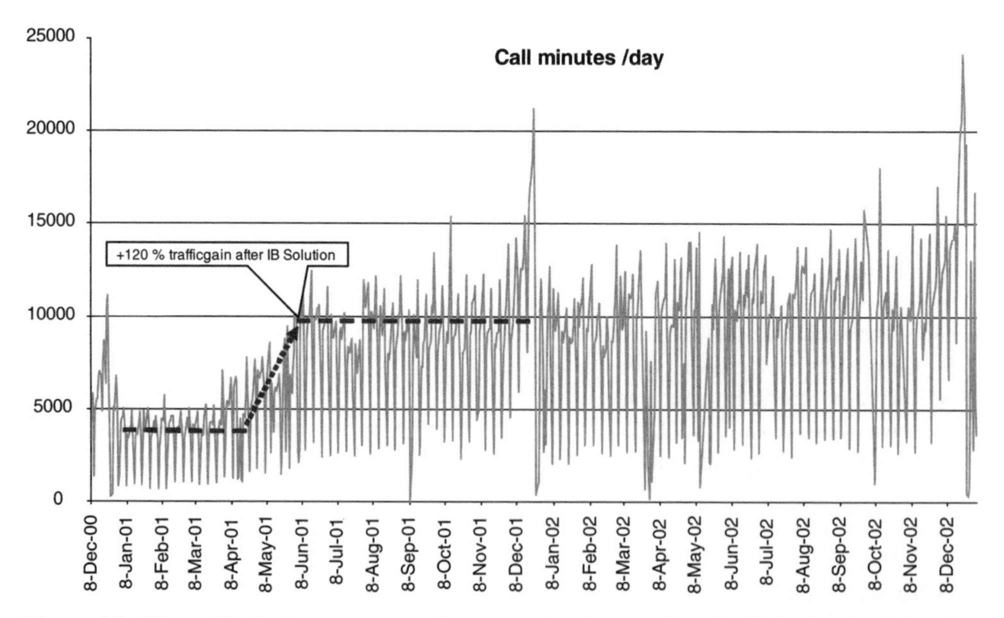

Figure 3.1 The traffic in the area covered by macro sites is more than doubled when implementing indoor coverage

well as many small outside shops. Prior to implementing the indoor coverage in the two shopping malls, the whole area was covered by a tight grid of standard three-sector GSM macro sites. The macro sites are separated by about 450 m, providing an outdoor coverage level measured on-street of minimum −65 dBm (GSM900) in most of the area.

However, even this high a coverage level from the macro sites was not enough to service all the potential traffic in the area, and thus not able to cater for the actual service need. This is clearly documented by the results shown in Figure 3.1.

Traffic Boost

This example in Figure 3.1 clearly shows that, in the traffic statistics for the total traffic production for the whole shopping area, there is an instant 'jump' in the traffic production when the indoor coverage system in the two shopping malls is set into service – the traffic rises 120% instantly.

The total traffic in the area exceeded the existing traffic on the macro (prior to the IB coverage being implemented) plus the new traffic on the IB solutions. Not only did the indoor system pick up more traffic, it also boosted the traffic in the neighboring macro-serviced area, by carrying more traffic into this area due to the increased service level, and reduced the number of dropped calls to less than 0.5% in the area after the implementation.

The growth rate of the traffic in the area with 100% IB coverage also showed an annual gain of traffic of about 30%. This is more than double the growth rate in a similar reference area.

Even High Coverage Level from the Macro was Not Enough

Even with an existing macro coverage level of −65 dBm there had been users that were not covered, that you could now capture with the indoor solutions. This clearly shows that IB coverage solutions are important revenue generators for the operators. This is, of course, provided that the indoor coverage solutions are implemented in the right area, covering the right buildings, and are designed correctly.

3.2.2 The Problem Reaching Indoor Mobile Users

Why is it a problem for macro sites to cover inside buildings? Let us explore some of the challenges for the macro coverage penetrating deep inside buildings and providing sufficient coverage and service level where the users are located.

Urban and Suburban Environments Rely on Reflections

In urban and suburban environments macro coverage will typical reach the users by reflection and diffraction – multipath propagation (see Figure 3.2). The delay profile will typical be 1–2 μs. Only a minor part of the traffic is serviced by the direct signal in line-of-sight to the base station antenna, having only the free space loss plus penetration loss of the building (as shown in Figure 3.2).

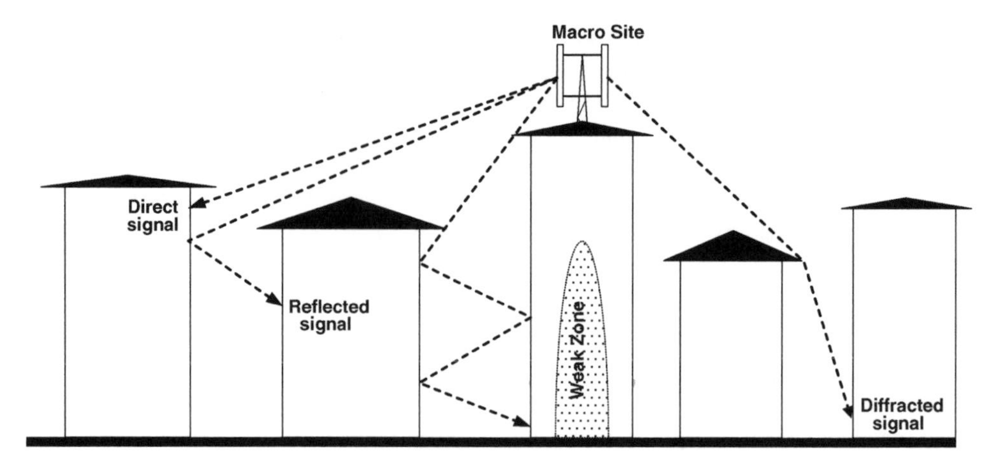

Figure 3.2 Macro sites rely mainly on reflections in urban and suburban environments to provide indoor service

In most or all cases the resulting signal at the receiver will be a result of the multipath radio channel – a 'mix' of different signals with different delays, amplitudes and phases. The result is a multipath fading signal (as shown in Figure 3.3), with a fading pattern that mainly depends on the environment and the speed of the mobile. Surprisingly, very often a building with a macro site on the roof can have coverage problems at the core of the same building, especially on the lower floors of the building and near the core.

This peculiar problem occurs due to the fact that coverage inside the building relies on reflections from adjacent buildings. On the topmost floors the coverage might be perfect, but the further down, problems starts to occur. Starting with lack of coverage in the staircase and in the elevators, in many cases the inner core of the building might have performance problems, especially with data service.

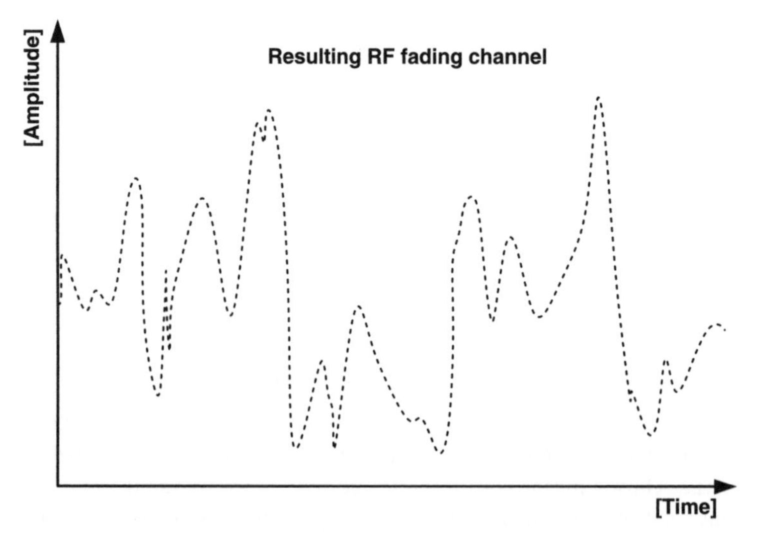

Figure 3.3 Typical multipath radio fading channel

If a building with the rooftop macro has no, or only low, adjacent buildings, this problem can be a major issue, due to the lack of buildings to reflect the signal back into the servicing building. However the problem can be easily solved by deploying a small indoor solution, filling in the black spot. See Section 4.6.3 for details.

The Mobile Will Handle Multipath Signals

If the reflections are shorter than about 16 μs, the equalizer in the GSM mobile will to some extent cancel the reflection (if longer, it will impact the quality as co-channel interference). The rake receiver on UMTS will, if the reflections are offset more than one chip duration (0.26 ms), recover and phase-align the different signal paths and 'reconstruct' the signal to some extent, by the use of the rake receiver and maximum ratio combining; see Section 2.3.6 for more details.

3.3 The Indoor UMTS/HSPA Challenge

When providing radio coverage for mobile users inside buildings, you are facing several radio planning challenges. It is mainly these challenges that motivate the need for indoor coverage solutions. Radio planners with indoor GSM planning experience need to be very careful not to apply all the radio planning strategy gained on GSM when designing indoor UMTS/HSPA solutions. If you are not careful, you will make some expensive mistakes and compromise high-speed data performance, as well as the business case for these indoor solutions.

3.3.1 UMTS Orthogonality Degradation

The UMTS RF channel efficiency is sensitive to degradation of the 'RF environment', typically caused by multipath reflections. Without going into mathematical details, the efficiency of the UMTS RF-channel is expressed using the term 'orthogonality'. The higher the orthogonality of the radio channel is, the higher the efficiency of the radio link is.

The Data Efficiency on the UMTS Channel

The same 5 MHz (3.84 Mcps) UMTS RF channel can in some cases carry high data rates, in excess of 2 Mbps, provided that the users are in line-of-sight to the serving cell, and only if a minor portion of the signal is reflected energy. This is the typical indoor scenario, with a dedicated indoor coverage solution. In that case the orthogonality can be as high as 0.85–0.90, so the channel is very efficient in carrying high-speed data efficiently (as shown in Figure 3.4).

However when covering UMTS indoor user from the macro layer in an urban or suburban environment, the RF channel relies on many reflections, diffractions and phase shifts – a multipath channel. This will degrade the efficiency of the radio channel, and under these circumstances the orthogonality can be as low as 0.55.

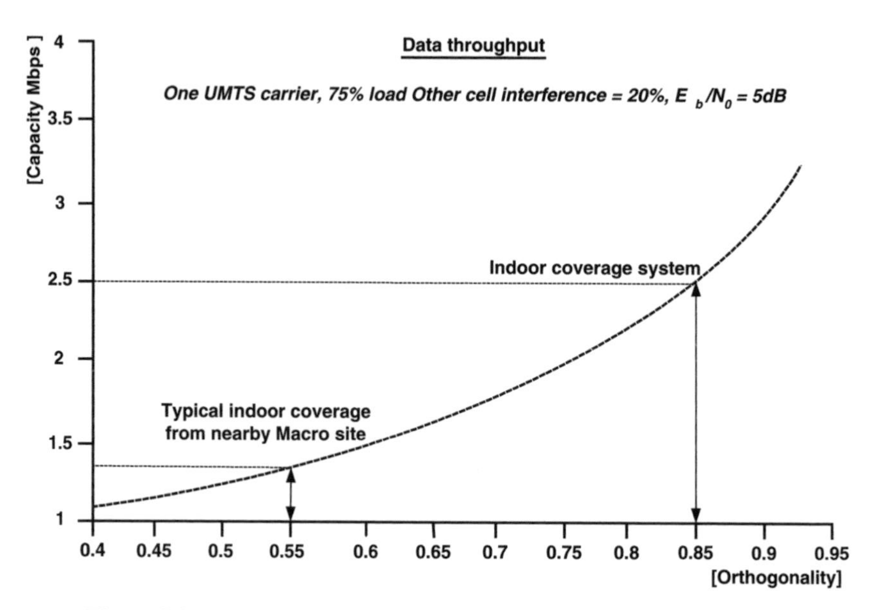

Figure 3.4 Orthogonality affects the efficiency of the UMTS RF channel

Degraded Orthogonality, Higher Costs for the Operator

The degradation of the orthogonality when servicing indoor users from the macro layer is a major concern for the mobile operators. This concern is due to the degraded efficiency of the channel, resulting in a lower data throughput. In practice a particular UMTS base station will be able to serve line-of-sight users with high data throughput, whereas indoor users served by the same base station will only be serviced at lower data rates, due to the degraded orthogonality.

The orthogonality is directly related to the production cost per Mb, and thus directly related to the business case of the operator. In addition to the less efficient radio channel, there are other negative effects, including higher power load per user due to the high penetration loss into the buildings. We will elaborate more on these effects later in this chapter.

Degraded Indoor HSDPA/HSUPA Service from the Macro Layer

The modulation on HSPA service is very sensitive to interference and degradation of the radio channel. HSDPA and HSUPA need a high-performing RF link, in order to support the highest possible data rates. In reality this means that HSDPA/HSUPA will only be served in the buildings in direct line-of-sight to the serving macro, and only in the part of the building facing the nearby macro site. To provide coherent and high-performing indoor HSPA coverage, dedicated indoor coverage solutions are needed.

The Macro Layer on 3G Will Take a Major Impact from Indoor Traffic

As just described, the orthogonality is degraded when servicing indoor users from the outdoor macro base stations. This is mainly due to the reflections and diffractions and phase shifts of the signals from the clutter in the area, buildings, etc.

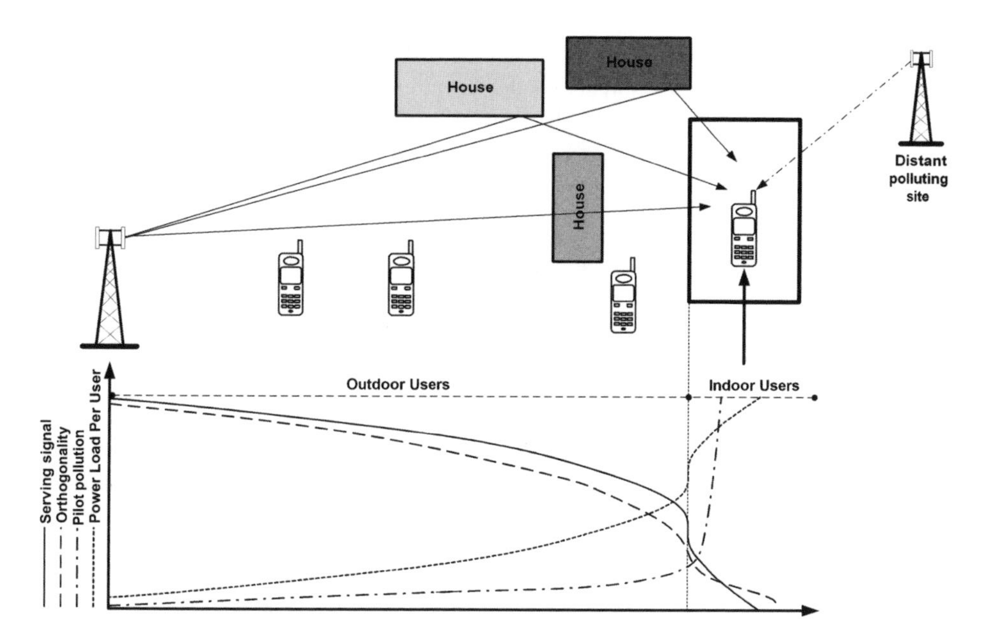

Figure 3.5 The degradation of the UMTS channel, and power load when servicing indoor users from the macro layer

More than Orthogonality Degradation

In addition to degraded orthogonality, there are several additional degrading factors to take into account, in order to evaluate the impact on the macro layer from users inside buildings (as shown in Figure 3.5).

The link loss increases with the distance from the base station, the free space loss and the additional penetration loss into the building. The penetration loss depends on the wall type, thickness and material. Typical penetration losses of the outer wall of a building can vary from 15 to 50 dB or more, depending on the type and the frequency.

In addition to the penetration and free space loss from the base station, the clutter inside the building will also add to the attenuation of the RF link. This is especially true for the users at the core deep inside the building and those on the side of the building opposite the servicing macro base station. These indoor users will have a relative high link loss, and demand a high power resource on the base station.

In order to maintain the downlink, the base station has to assign more power to these high-loss users inside the building. These mobiles will also need to power up their transmit power, in order to maintain the uplink.

3.3.2 Power Load per User

The end result is a high power load per user for users with high link losses inside buildings. It is a fact that the capacity resources of the UMTS base station are directly related to the power resource. The higher the PLPU is, the higher the power drain off the base station and the higher the capacity drain off the base station will be per indoor user. This high power drain per indoor user has a big impact on the remaining power (capacity) pool of the base station, needed to service other users. This directly relates to the business case.

3.3.3 Interference Control in the Building

To add even more complexity to the matter, the higher up a building the users are, the more likely it is that they will receive nonintended distant base stations (pilot pollution). This is especially a concern in high-rise buildings in the topmost floors that rise over the clutter of other buildings in the area (see Section 3.5.3) This is particular problem for those users that are close to windows, opposite to the side of the building of the serving base station. These users will be able to detect pilot signals from distant base stations. This pilot pollution will degrade the quality; the E_b/N_o of the UMTS signal. For GSM the Rx-Qual (bit error rate) will be degraded, and this might result in dropped calls even though the signal level is relatively high. UMTS macro cells close to the building must always be in the neighbor list (monitored set) as soft handover will be active whenever the pilot signals from outside the building are high, in order to prevent pilot pollution and dropped calls. However, soft handover will also pose a potential problem; when the mobile inside the building enters soft handover, more links are used to maintain the same call, so all the degraded orthogonality and power load will in reality hit even more outdoor base stations, draining even more resources from the macro network. HSDPA planning mandates special attention to this problem, as all cells are on the same frequency and interference from other cells will degrade the HSDPA data rate.

3.3.4 The Soft Handover Load

As we know, all serving cells in UMTS are on the same frequency, only separated by codes. Therefore UMTS has to use soft handover (SHO), as shown in Figure 3.6; this is the only way to shift the calls smoothly when the user roams from one cell to another. Soft handover is a main feature of UMTS, securing the traffic transition between the cells, but it comes at a cost. During soft handover the mobile takes up resources on all the cells engaged in soft handover. Typically, with two to three cells, one mobile in soft handover will load the network with a factor of 2–3. However, one must distinguish between softer handover, which occurs within the cells on the same site, and soft handover between cells on different sites. The latter takes more resources due to the need for double backhaul to the RNC/RNCs, see Section 2.3.4 for more details.

3.3.5 UMTS/HSPA Indoor Coverage Conclusion

From the topics just covered, it is clear that it is very costly to cover indoor UMTS users from macro base stations because of the impact of power drain of the macro layer, the degraded RF channel due to low orthogonality, the pilot pollution in high-rise buildings and the load increase due to soft handover (lack of dominance) when more macro cells are servicing users inside a building.

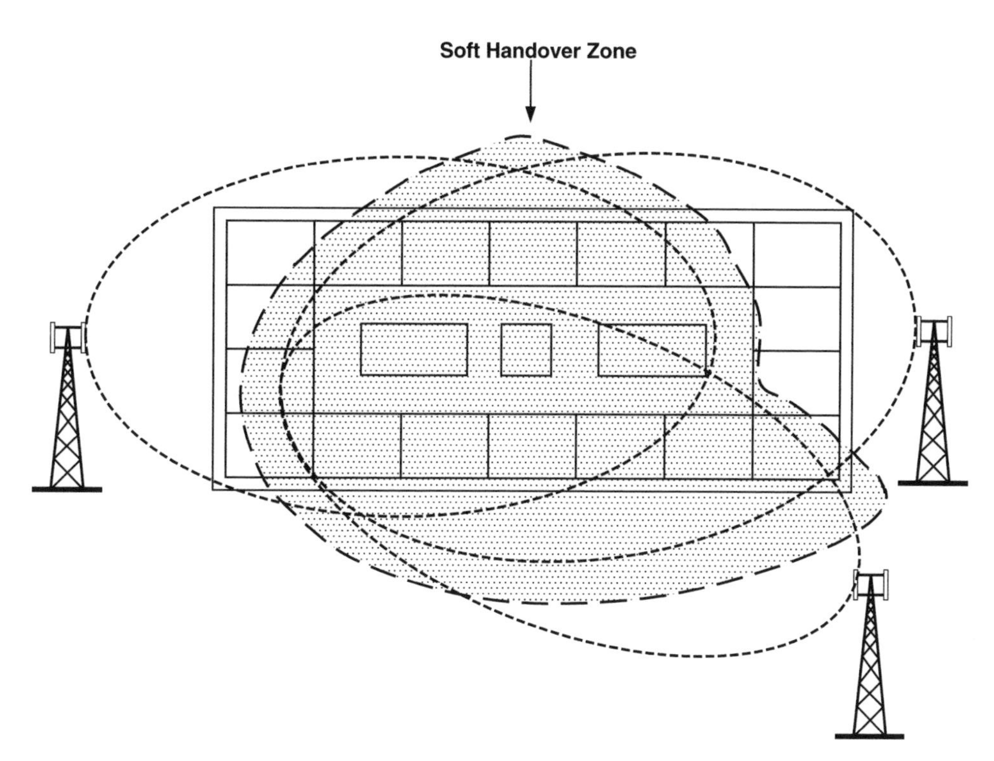

Figure 3.6 Three UMTS cells from the macro layer provide excellent indoor service, but most of the building is in soft handover

Even Perfect Indoor Coverage from UMTS Macro Sites can be a Problem

In the example in Figure 3.6 there are three surrounding macro sites providing good, deep indoor coverage. These sites are close to the building, and the signal level throughout the building is perfect.

Therefore all users inside the building are serviced with high RF quality and high-speed data service. However, the users are covered by three different macro cells with a considerable overlap. This is a big concern; the major part of the traffic inside this building will permanently be in soft handover.

In this example one mobile in the soft handover zone will be in simultaneous communication with all three cells. Therefore the same transmission will load three cells, and also put a factor 3 load on the backhaul network. This has a big impact on the outside network, due to the capacity load, increasing the production cost of the traffic for mobiles in the SHO area.

The probability of soft handover decreases when the margin between the serving cell and the other cells in the monitored set/neighbor list increases. As a guideline, one needs a margin of about 10–15 dB to avoid soft handovers – that gives room for fading.

Traffic inside buildings comes at a high cost when serviced from the macro base stations. This might be a minor problem if this is a building with only few users, then the impact on the surrounding macro network will be marginal. In the case of hot-spot buildings with high traffic density, like a shopping mall, a big corporate building or an airport, the impact on the macro network will be severe. It is important to dedicate an indoor coverage solution for these types of buildings, to secure a dedicated dominant indoor signal, with enough capacity to accommodate the traffic.

Cover the Hot-spot Buildings from the Inside, from Day One

The effects that occur when you try to cover indoor users from the outside are really important to keep in mind; the fact is that only a few hot-spot buildings in a city can overload the macro network. The users inside the building might have perfect service, but the fact is that the macro base stations might not have any resources left to maintain service to other users outside these hot-spots buildings.

This might not be evident in a UMTS macro network with only a minor load of traffic, like in the roll-out phase of the UMTS network. However it might become serious when the traffic increases in the network. Then it might be too late, and if you wait to address the UMTS indoor coverage issue until then, you will compromise the user perception of the UMTS service and the network quality. Users might turn to other operators and the business case for the network will have been compromised. Yes, the radio planner has a direct impact in the business case for the network; it is a big responsibility, one that is not to be underestimated.

3.4 Common UMTS Rollout Mistakes

It is very tempting to place a nearby macro site close to one of these hot-spot buildings, especially when coming from a GSM radio planning background. Often the result is that the side of the building facing the macro site will have really high signal level. However, eventually there might still be some areas on the far side and in the basement that still have an insufficient coverage level.

3.4.1 The Macro Mistake

The classic example is a shopping mall; these are usually important hot-spot buildings with many users, and therefore on the top of the list when prioritizing the roll-out plan. A macro site just across the street from the shopping mall could be a perfect solution, or is it? Typically the operator will realize that sufficient indoor coverage is still lacking in large areas inside the building, especially for HSPA data service.

The capacity needed inside the building might also exceed the resources on the macro site, mainly due to the high power load per user inside the building. Then the operator realizes that the next step is an indoor UMTS solution.

3.4.2 Do Not Apply GSM Strategies

Trying to cover a shopping mall from the outdoor macro mast is tempting, but this is the GSM experience that kicks in; this is how the GSM network has been and is still being planned. However, this mindset is expensive to use for UMTS radio planning, because after implementing the indoor UMTS system in the building, the operator will now realize that about 60–80% of the traffic inside the building is in soft handover, due to the high signal coming into the building from the nearby macro site, causing a lack of dominance. This will also severely degrade the HDSPA performance in the building due to high interference between the cells. This problem can easily be solved on GSM, by using different frequencies, but UMTS uses the same frequency on all cells. You might be tempted to assign a dedicated channel for the indoor UMTS solution, but once again this is GSM planning, because by doing so you have used a huge part of your capacity. Typically you have only two or three UMTS channels in total: one assigned for UMTS, one for HSPA and if you have a third channel assigned to the indoor solution, there is no way to utilize this channel for future capacity needs.

The best solution will then be redirecting or removing the sector of the macro that covers the shopping mall. However, often this sector will also service other smaller buildings and areas that cannot be serviced without this sector. Often the only valid solution is to remove the sector and deploy a smaller site in the new problem area outside the shopping mall.

3.4.3 The Correct Way to Plan UMTS/HSPA Indoor Coverage

The correct way to do UMTS indoor and macro planning, is to plan the UMTS network 'from the inside out' not from the outside in. This is especially important in the high-capacity areas, that is, areas with hot-spot buildings. Operators doing UMTS roll-out need to realize that a portion of the roll-out cost should be reserved for indoor coverage in the most important high-traffic buildings. Deploying indoor systems in these hot spot buildings will save a lot of cost, grief and hassle in the long run.

This will for sure be the most economical strategy, providing better data service with higher data rates and higher quality in the network.

UMTS networks should be planed, from the inside out.

Better Business Case with Indoor UMTS Solutions

It is a fact that the production cost when servicing the users inside the hot-spot buildings from the macro base stations is very high, even if the service inside the hot-spot buildings seems to be perfect. This is due to the less efficient RF link from the macro base station, mainly due to the high power drain per user that uses up the capacity of the UMTS base station.

Depending on the scenario, a cost reduction of up to 65% can be achieved by covering the UMTS users from inside the building with indoor coverage systems, rather than using nearby macro base stations to provide solid indoor coverage.

It is not realistic to cover every building from the inside, but the hot-spot buildings at least must be considered from the first roll-out plan. These buildings will have a big impact on the macro layer. If you cater for them, the macro layer can use their power resources to service all the other buildings and areas, and the revenue in the network will be boosted.

Most operators do a roll-out plan for their macro network three to five years in advance; the most important buildings should be a part of that plan, and this will give the best performance and the best business case.

3.5 The Basics of Indoor RF Planning

No matter what the radio service, whether it is GSM, UMTS, HSPA or other technologies, there are some basic design guidelines one must apply in order to design a high-performing indoor coverage solution.

3.5.1 Isolation is the Key

If you must select one parameter and one parameter only that truly defines the most important success parameter when designing an IB solution, it must clearly be the 'isolation'. Isolation is defined as the difference between the IB signal and the outdoor network, and vice versa.

Users in office buildings are typically close to the windows; therefore the dominance of the indoor system must be maintained throughout the building, even right next to the windows.

3.5.2 Tinted Windows Will Help Isolation

Modern energy-efficient windows with a layer of thin metallic coating will attenuate the macro signal servicing the building. This type of window will attenuate the RF signal from the outdoor network, and create a need for a dedicated indoor solution.

The positive side effect of this 'problem' is that, once the indoor system is installed, these metallic-coated windows will actually help the design, giving us good isolation. These types of windows with a thin layer of metallic coating will typically attenuate the signal by 20–40 dB, depending on the radio frequency, and the incident angle of the radio signal.

Recently new types of windows are being used, or film applied to existing windows in high-profile office buildings. These buildings are being fitted with 'Wi-Fi-proof glass'. This prevents hackers with laptops outside the building camping on the Wi-Fi service from within the building. This type of window or film attenuates the radio signal even more, by up to 50–70 dB.

This will actually help in producing a good indoor radio design: these windows, together with the aluminum-coated facade that is typical for many modern corporate buildings, are relatively easy to plan, with regards to isolation (until someone opens a window...). With 40–70 dB of isolation, even a nearby macro site will be shielded efficiently, and you can rely on omni antenna distribution inside the building, as shown in the example in Figure 3.7, without leakage from the building to the outside network. This makes the 'handover zone' easy to design really close to the building, so that the indoor system does not service outside users or leak interference outside the building.

3.5.3 The 'High-rise Problem'

In some high buildings, typically older buildings with normal windows (no metallic coating), you can experience very high interference levels from the outside macro network; even strong signals from distant macro bases will reach indoor users at surprisingly high signal levels. This is mainly due to the low (or no) attenuation from the windows, providing only limited or no isolation between macro base stations and the area inside the building along the

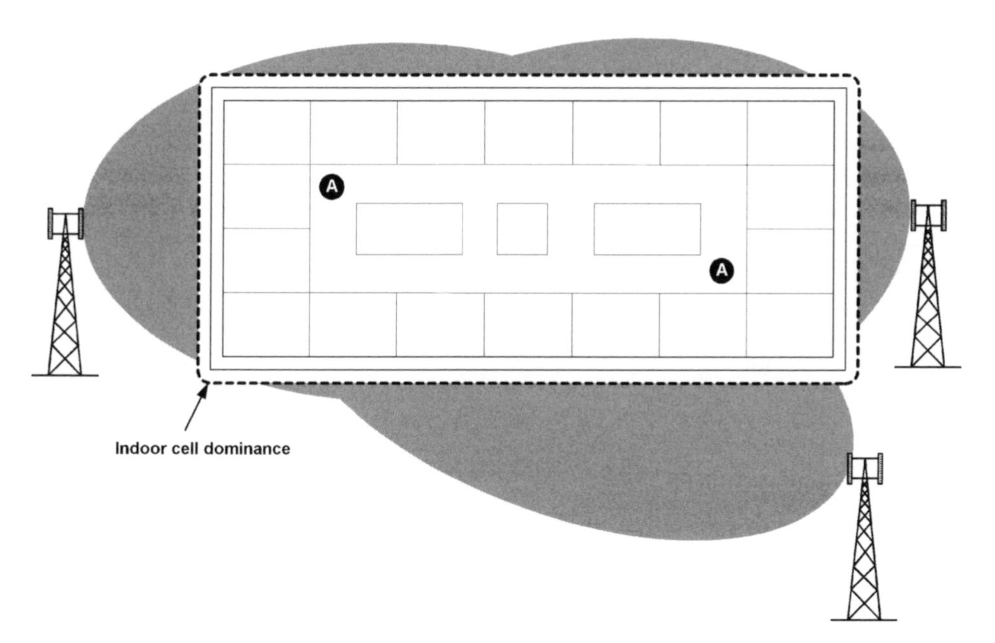

Figure 3.7 A well-designed indoor solution will be dominant throughout the building, but not leak access signal to the surrounding network

windows. Despite these physical factors, you need to insure isolation from the high-level outdoor signal and dominance of the indoor coverage system throughout the building; you need to carefully consider the strategy for designing and implementing the indoor coverage solution.

The traditional approach would be to deploy omni antennas in the walkways near the core of the building (like in Figure 3.7), but this can cause an unwanted side effect. These 'central' omni antennas must radiate high RF levels, in order to overcome the high signal from the nearby macro base stations, and make the indoor cell also dominant along the windows of the building.

The unwanted side effect of this strategy is that the high power from the indoor system will leak high levels of signal from the indoor cell into the macro network, increasing the noise, and degrading the quality and capacity in the outside macro network.

Solution to the 'High-rise Problem'

The solution is to make the indoor cell the dominating cell throughout the building. The indoor cell needs to dominate the total area from along the windows and all the way into the center of the building, without leaking signal out to the outdoor network. You can achieve this by deploying directional antennas along the border of the building, and direct them to the center of the building (Figure 3.8).

This design will insure that the indoor signal is stronger than the outdoor signal, even with very high signals from the nearby macro base station. This is due to the proximity and line-of-sight of the user to the nearby serving indoor antenna. Thus the indoor cell is

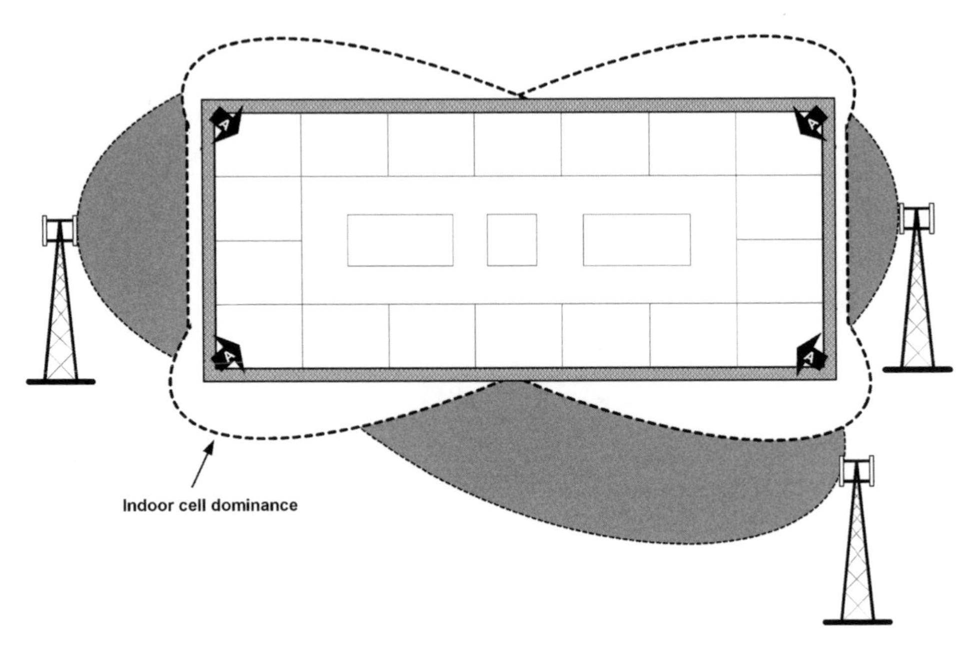

Figure 3.8 The topmost floors in a high-rise building pose specific challenges to isolation due to powerful macro signals. The key is dominate at the border of the building perimeter, in this example using directional antennas mounted in the corners pointing towards the center of the building

dominant at the border of the building, and when the antennas are pointing inwards to the center of the building, the indoor cell will be dominant throughout the indoor area. This due to the fact that both the outdoor signal and the indoor signal are penetrating the same indoor signal path, so the isolation between the outdoor and indoor signal will 'track' into the center of the building.

Plan for Perfect Isolation

Exactly how efficient the 'corner antenna' strategy is, is shown in the simulation in Figure 3.9. The graph shows how efficient the solution can be when you use the approach of directional antennas pointing inward to the center of the building.

This is a simulation, with the indoor directional antennas mounted suspended in free space, having only the front-to-back isolation of the antenna, 8 dB. In a real-life installation, the antennas will be mounted on a corner pylon or on the outer wall, above the window, so the front-to-back isolation ratio will be 5–20 dB better. In either case it is clear that this is a very efficient way to provide indoor dominance, even in buildings with little penetration loss, and even in buildings with high signals from the outdoor network.

Installation is a Challenge

Placing antennas in the corners pointing towards the center of the building as shown in Figure 3.8 might be a perfect radio planning solution to the 'high-rise problem', but one

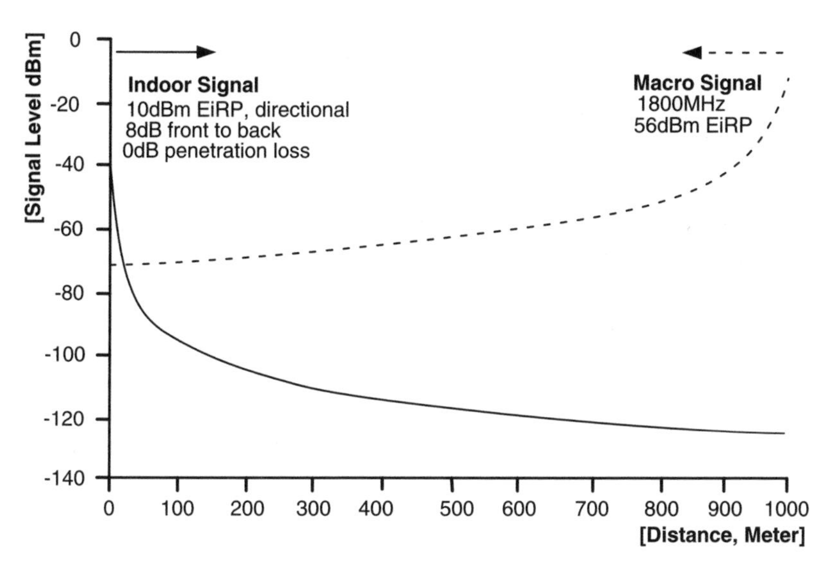

Figure 3.9 The graph shows how efficiently the signal is contained inside the building, with minimum leakage to the outside macro network

practical challenge remains: installing the antennas at this location can be an issue. In particular, the installation of cables reaching these 'perimeter' antennas can be a problem, especially when using passive distributed antenna systems that rely on rigid passive cables. It can be difficult to find an appropriate cable route, and end up with an expensive installation, but using active distributed antenna systems, relying on thin cable infrastructure, is a possible solution to the problem (see Section 4.4.2).

What Level of Isolation is Needed?

You know now that the main parameter when doing indoor radio planning is isolation of the macro layer, but how much isolation do you need in reality?

GSM

The GSM system will be less restricted in terms of the isolation issue, due to the fact that different frequencies are assigned for the indoor and outdoor cells. However, isolation is still important, especially when planning indoor solutions in high-rise buildings in high-capacity macro networks. This is the typical example in cities; in these areas you have to deal with the fact that the indoor frequencies might be in use even by nearby macro sites. When planning GSM voice, the co-channel isolation must be more than 11 dB in order to secure the quality – well, this is the theory, but in reality it depends very much on the actual fading environment, so it is advisable to add a margin of 6–10 dB. When planning for GSM data, EDGE MCS-6, you will need a co-channel isolation of more than 17 dB plus a margin! You must plan for a very good signal quality inside the building; refer to Chapter 8 for more details on EDGE planning levels.

Also for GSM, dominance is a concern in order to ease the optimization, and to avoid 'ping-pong handovers' where users toggle between the outdoor and the indoor cell. Even with perfect Rx-Qual the user perceived quality during the 'ping pong' handover is degraded. This is due to the 'bit stealing' of the traffic channel in order to perform fast signaling for the GSM handover. This often results in bursts of 'clicks' on the audio on the call, degrading the voice quality.

UMTS/HSPA

As you know, on UMTS all users are using the same RF channel, and in order to maintain the link when two or more serving cells are at the same signal level, the mobile enters soft handover, loading more than one link and taking up resources on several network elements and interfaces. If there is no dominant indoor coverage system, then typically two or three base stations will service a user inside an urban building. Even implementing indoor solutions you need to make sure that the mobiles inside the building will only be served by the indoor base station.

You must make sure that this base station is dominant throughout the building; the less dominance, the higher probability there is of soft handover. As a general guideline, you should make the indoor cell 10–15 dB more powerful inside the building than any outside macro signal. However this is a fine balance; you must design the indoor solution in order to make sure that the indoor system does not leak too much signal outside the building, thus 'pushing' the soft handover zone outside the building.

Luckily modern building design will help us, with aluminum-coated exterior walls, metallic-coated windows, etc. In some cases buildings are being installed with 'WLAN-proof windows' in order to protect the IT system within the building from hackers outside, trying to lock in on the WiFi signal. Attenuation of more than 50 dB is in many cases a fact.

However, first of all you need to apply really good radio planning when designing indoor UMTS coverage systems. Use corner-mounted antennas to overcome the high-rise problem and generally distribute more antennas inside the building, radiating lower power, to achieve a good uniform signal level. You must definitely avoid 'hot' antennas, i.e. few antennas radiating very high power (In access of 25–30 dBm) to cover large indoor areas. These 'hot' antennas will often cause more problems than solutions, mainly due to leakage from the building into the nearby macro network, raising the noise load on nearby base stations and degrading the capacity.

3.5.4 Radio Service Quality

The performance of any radio link, GSM, UMTS and HSPA included, is not related to the absolute signal level but rather to the quality of the signal. This is described as the signal-to-noise ratio (SNR), and the bigger the ratio between the desired signal – the signal from the serving indoor cell – and the noise – the unwanted signals from the macro layer – the better performance the link will have, and the higher is the data rates that can be carried on the radio link.

The good radio designer, doing indoor coverage solutions, will make an effort to make sure the design is maximized to have the best possible isolation from the outside network. Leaving some areas of the building in service from the macro sites is tempting, in the areas where the macro coverage is strong. However, this temptation should be avoided if possible; the result on 3G will be large areas of soft handover, large areas with degraded HSPA performance.

3.5.5 Indoor RF Design Levels

Frequently operators use only one design level for indoor coverage and one level only. However, as you know, the quality and data throughput of the radio signal is dependent on the SNR, not the absolute signal level. How does this impact the radio planning design for indoor systems?

As you know, isolation plays a major role in the performance of the indoor system. Therefore the actual design level for the signal from the indoor system must be adapted to the particular building, the existing macro coverage inside the building and the isolation of the building. Using more than one design level in the indoor design procedure will also help assure the business case. It is very expensive to provide too high a coverage level in areas of the building where it is not needed.

3.5.6 The Zone Planning Concept

The advice is not to focus on the absolute signal level, and not to use only one planning level. The planning level must be adapted to the project at hand. In practice you might need to use two or three different planning levels in the same building, depending on the 'baseline' of the existing coverage from the surrounding macro sites. An example to describe the need for adaptation of the design level is evident if you look at the typical high-rise building in Figure 3.10.

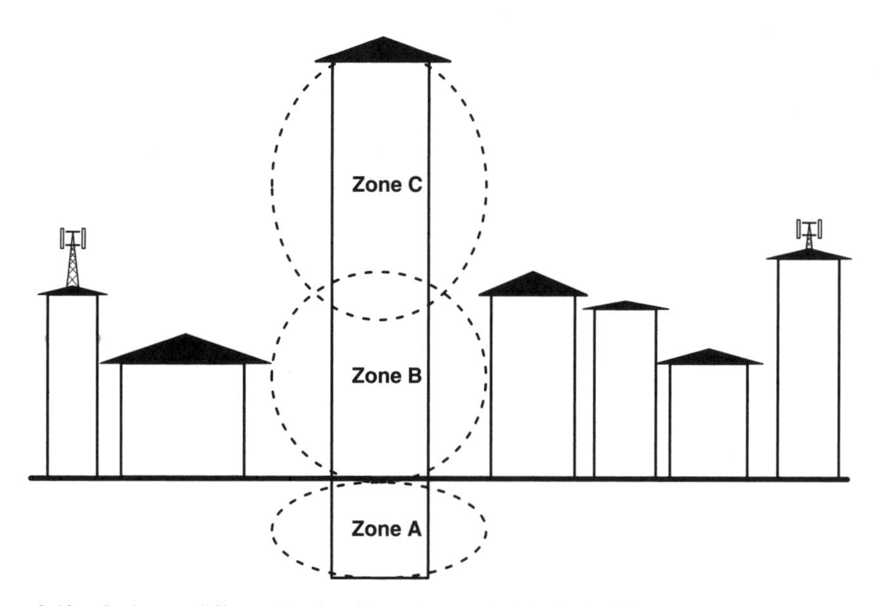

Figure 3.10 Owing to different levels of interference inside the building, it is wise to adjust design levels accordingly, dividing the building into different zones, each zone having individual design levels; this can save cost and maintain RF performance

Divide the Building into Different Planning Zones

For a high-rise building like the example in Figure 3.10 you can typically divide the building into three different planning zones, each zone having special considerations and design requirements.

Zone A: Coverage Limited Area
The isolation to the outside network in this part of the building is really good, which is no surprise this part of the building is sub-ground level. Typical isolation is better than 70–80 dB and the design level for the indoor cell can be relative low, because you do not need to account for macro interference affecting the performance.

The coverage level for the indoor cell and the noise floor of the indoor cell itself are the main driving factors. One part of Zone A that needs careful consideration is the entrance and exit of the building, if Zone A contains an underground parking area. You need to insure sufficient handover level, and allow time for the users entering and exiting the building.

Typical design levels for zone-A are:

- −85 dBm for GSM BCCH level.
- −90 dBm CPICH level for UMTS.

Zone B: Coverage and Interference Limited Area
This midsection of the building is often served by the nearby macro cells. The building is to some extent isolated from the interference coming from distant macro sites, by the neighboring buildings. When planning indoor coverage for zone B, you need to overcome relative low interference coming from distant macro sites, and to insure that the indoor cell will be the dominant server, overpowering the coverage from the macro cells that currently cover this area. Typically a medium signal level is needed, to provide sufficient dominance/isolation in order to avoid 'ping-pong' handovers (GSM) and to limit the soft handover zones (UMTS).

You must be careful with leakage from zone B, and make sure that the indoor DAS system does not service outdoor users near the building, pedestrians or nearby cars. Typical design levels for zone-B are:

- −70 dBm for GSM BCCH level.
- −80 dBm CPICH level for UMTS.

Zone C: Interference Limited Area
The topmost part of the high-rise building is typically above the local clutter of neighboring buildings. Users along the perimeter of the building at the windows will have clear line-of-sight to many distant macro sites. These distant base stations will cause co-channel interference and all the usual high-rise problems. In order to overcome the raised noise floor and pilot pollution due to the unwanted signals from all the distant macro sites, you need a high planning level.

Interference and pilot pollution can come from even very distant macro sites, up to 10–20 km away, and therefore these cells will be impossible to frequency/code plan inside the building. The interference has a side effect; with no indoor system the mobiles will typical run on a relative high uplink power, to overcome uplink interference on the base station, especially when camping on distant, unintended servers. This will generate noise increase on the macro network and on the indoor cell, degrading the capacity and data service on UMTS/HSPA, and degrading quality on the macro base station. The high transmit power from the mobile will also expose the indoor user to higher levels of radiation.

Typically the users will experience high signal level in the display of the mobile, but suffer degraded service, bad voice quality, reduced data service, dropped calls or even totally lack of service, due to the high interference level. Typical design levels for zone C are:

- −60 dBm for GSM BCCH level.
- −70 dBm CPICH level for UMTS.

Be Careful with Leakage from Zone C

However in extreme cases you will need to plan for even higher signal levels from the indoor system. Always be careful when radio planning in zone C and be sure not to leak out excessive power from the building, due to the high indoor design levels. Use corner-mounted antennas facing towards the core of the building, and use many low to medium power antennas uniformly distributed throughout zone C. Do not be tempted to crank up the radiated power in excess of 15–20 dBm in zone C, this might cause excessive leakage into the macro network, degrading performance and capacity.

Beyond Zone C

For very high buildings, there is actually a zone D. In this zone the interference from the outdoor network starts to decrease. However, you still need to be careful about leakage from the building from Zone D.

4

Distributed Antenna Systems

The lesson learned from Section 3.5 is clear; you need to distribute a uniform dominant signal inside the building, from the indoor cell, using indoor antennas in order to provide sufficient coverage and dominance. In order to do so you must split the signal from the indoor base station to several antennas throughout the inside of the building.

Ideally these antenna points should operate roughly at the same power level, and have the same loss/noise figure on the uplink to the serving base station. The motivation for the uniformly distributed coverage level for all antennas in the building is the fact that all the antennas will operate on the same cell, controlled by the same parameter setting. In practice passive DAS will often not provide a uniform design to all antennas; you might have one antenna with 10 dB loss from the base station, and in the same cell an antenna with 45 dB loss back to the base station, and the actual parameter setting for handover control etc. on the base station might not be able to cater for both scenarios. Therefore uniform performance throughout the distributed antenna system is a key parameter in order to optimize the performance of the indoor coverage system.

4.1 What Type of Distributed Antenna System is Best?

There are many different approaches to how you can design an indoor coverage system with uniformly distributed coverage level; passive distribution, active distribution, hybrid solutions, repeaters or even distributed Pico cells in the building. Each of these approaches have their pros and cons, all depending on the project at hand. One design approach could be perfect for one project, but a very bad choice for the next project – it all depends on the building, and the design requirements for the current project, and the future needs in the building.

Seen purely from a radio planning perceptive you should ideally select the system that can give the most downlink power at the antenna points and the least noise load and loss on the uplink of the base station, and at the same time provide uniform coverage and good isolation to the macro network. On top of the radio planning requirement, other parameters like installation time and costs, surveillance and upgradeability play a significant role. In practice the service requirements and the link budget (see Section 8.1.3) will dictate how much loss and noise you can afford and still accommodate the service level inside the building you are designing for.

4.1.1 Passive or Active DAS

Traditionally passive distributed antenna systems have been used extensively for GSM in the past 15 years. Therefore naturally many radio planners will see this as the first choice when designing indoor coverage for 3G systems. However, it is a fact that, for UMTS and especially for HSPA, active distributed antenna systems will often give the best radio link performance and higher data rates. The main degrading effect from the passive systems is the high losses, degrading the power level at the antenna points and increasing the base station noise figure on the higher frequencies used for UMTS/HSPA. UMTS and HSPA can perform really high-speed data transmission, but only if the radio link quality is sufficient, and passive systems will to a large extent compromise the performance; see Section 7.2 for more details on NF increase due to passive loss.

Another big concern with passive distributed antenna systems is the lack of supervision. If a cable is disconnected the base station will not generate any voltage standing wave radio (VSWR) alarm, due to the high return loss through the passive distributed antenna system. Distributed indoor antenna systems are implemented in the most important buildings, serving the most important users, generating revenue in our network. Surely you would prefer to have surveillance of any problems in the DAS system.

On the other hand, passive systems are relatively easy to design; and components and cables are rigid and solid, if installed correctly. Passive distribution systems can be installed in really harsh environments, damp and dusty production facilities, tunnels, etc., places where active components will easily fail if not shielded from the harsh environment. Passive distribution systems can be designed so they perform at high data rates, even for indoor HSPA solutions – but only for relative small buildings, projects where you can design the passive distributions system with a low loss.

4.1.2 Learn to Use all the Indoor Tools

It is important that the radio designer knows the basics of all the various types of indoor coverage distribution solutions. In many projects the best solution will be a combination of the various types of distribution hardware. Good indoor radio planning is all about having a well-equipped toolbox; if you have more tools in your toolbox, it is easier to do the optimum design for the indoor solution. Having only a hammer might solve many problems, but only having a hammer in you toolkit will limit your possibilities. If you only know about passive distribution, learn about the possibilities and limitations of active distribution, repeaters and Pico cells. This will help you design high-performing indoor coverage distribution systems that are future-proof and can make a solid business case. After all that's why you are here – to generate revenue in the network.

4.1.3 Combine the Tools

Indoor radio planning is not about using one approach only. Learn the pros and cons of all the various ways of designing indoor coverage, and then you will know what is the best approach for the design at hand. Often the best approach will be a combination of the different solution types.

4.2 Passive Components

Before you start exploring the design of passive distributed antenna systems, you need to have a good understanding of the function and usage of the most common type of passive components used when designing indoor passive distributed antenna systems.

4.2.1 General

Inside buildings you must fulfill the internal guidelines and codex that apply for the specific building. In general you will be required to use fire-retardant CFC-free cables and components.

Be very aware of how to minimize any PIM (passive intermodulation) problems. Also be sure that the components used fulfill the required specification, especially when designing high-power passive distributed antenna systems. The effect of combining many high-power carriers on the same passive distributed antenna system using high-power base stations will produce a high-power density in the splitters and components close to the base stations. Use only quality passive components that can meet the PIM and power requirements, 130 dBc or better specified components (see Section 5.7.4).

4.2.2 Coax Cable

Obviously coax cable is widely used in all types of distributed antenna systems, especially in passive systems. Therefore it is important to get the basis right with regards to cable types, and losses. Table 4.1 shows the typical losses for the commonly used types of passive coaxial cables. For accurate data refer to the specific datasheet for the specific cable from the supplier you use. Often there will be a distance marker printed on the cable every 50 cm or 1 m, making it easy to check the installed cable distances.

Table 4.1 Typical attenuation of coaxial cable

	Frequency/typical loss per 100 m (dB)		
Cable type	900 MHz	1800 MHz	2100 MHz
$\frac{1}{4}$ inch	13	19	20
$\frac{1}{2}$ inch	7	10	11
$\frac{7}{8}$ inch	4	6	6.5
$1\frac{1}{4}$ inch	3	4.4	4.6
$1\frac{5}{8}$ inch	2.4	3.7	3.8

Calculating the Distance Loss of the Passive Cable

It is very easy to calculate the total loss of the passive coaxial cable at a given frequency.

Example
Calculating the total longitudinal loss of 67 m of $\frac{1}{2}$ inch coax on 1800 MHz, according to Table 4.1:

$$\text{total loss} = \text{distance (m)} \times \text{attenuation per meter}$$
$$\text{total loss} = 67\,\text{m} \times 0.1\,\text{dB/m} = 6.7\,\text{dB}$$

Reduce the Project Cost When Selecting Cable

The main expense implementing passive indoor systems is not the cable cost, but rather the price for installing the cable. Installing heavy rigid passive cable can be a major challenge in a building. In particular, the heavier types of cable from size $\frac{7}{8}$ inch and up are a major challenge. These heavy cables literally take whole teams of installers; after all the cable is heavy, and a challenge to install without dividing the cable into shorter sections.

Carefully consider the price of installing the cable against the performance. You might be alright with 2 dB extra cable loss, if you can save 50% of the installation costs by just selecting a cable size thinner. On the other hand, do be sure that the distribution system will be able to accommodate the higher frequencies and data speeds on 3G and HSPA.

4.2.3 Splitters

Splitters and power dividers are the most commonly used passive components in distributed antenna systems, dividing the signal to or from more antennas. Splitters (as shown in Figure 4.1) are used for splitting one coax line into two or more lines, and vice versa. When splitting the signal, the power is divided among the ports. If splitting to two ports, only half-power minus the insertion loss, typically about 0.1 dB, is available at the two ports. It is very important to terminate all ports on the splitter; do not leave one port open. If it is unused, terminate it with a dummy load.

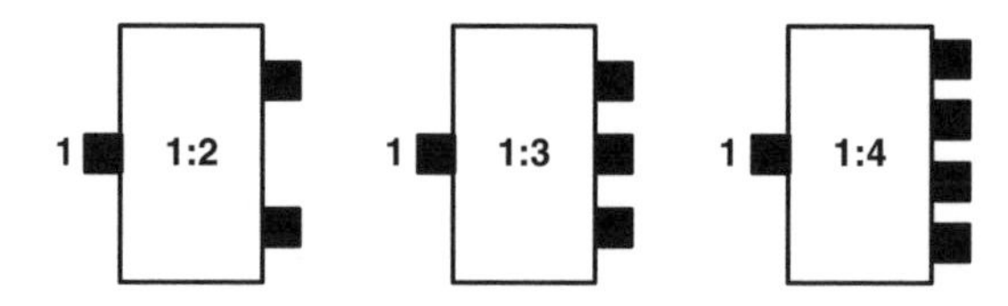

Figure 4.1 Coax power splitters

Example
You can calculate the loss through the splitter:

$$\text{splitter loss} = 10\log(\text{no. of ports}) + \text{insertion loss}$$

For a 1:3 splitter (as shown in Figure 4.2), the attenuation will be:

$$10\log(3) + 0.1\,\text{dB} = 4.87\,\text{dB}$$

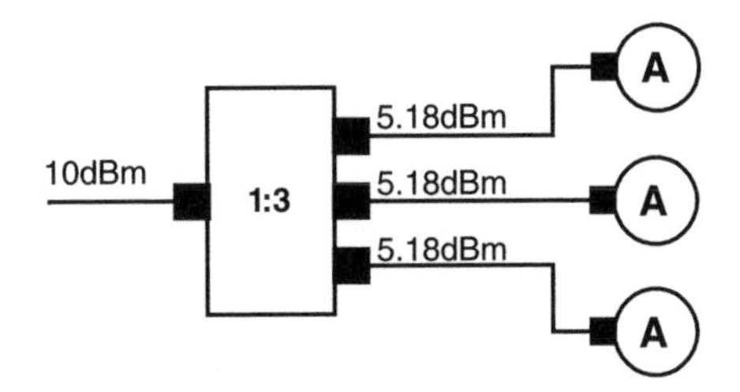

Figure 4.2 Power distribution of a typical 1:3 splitter

In this example, when we feed a 1:3 splitter 10 dBm power on port 1, the output power on ports 2–4 will be $10 - 4.87 = 5.18$ dBm.

4.2.4 Taps/Uneven Splitters

Tap splitters (as shown in Figure 4.3) are used like splitters, used to divide the signal/power from one into two lines. The difference from the standard 1:2 splitter is that the power is not equally divided among the ports.

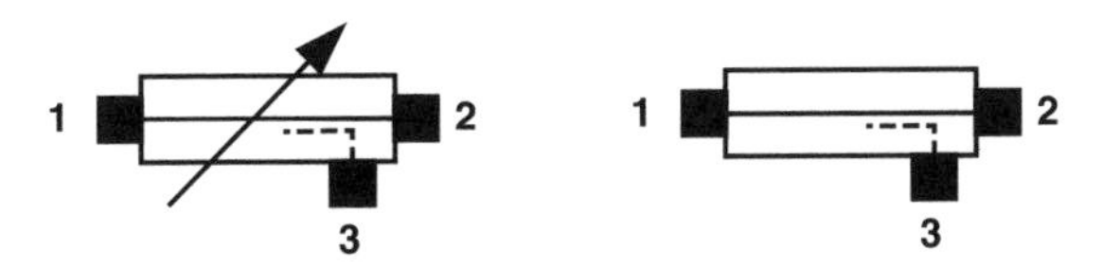

Figure 4.3 Taps, adjustable and fixed

This is very useful for designs where you install one heavy main cable through the building, and then 'tap' small portions of the power to antennas along the main cable. By doing so, you reduce the need to install many parallel heavy cables, and still keep the loss low.

This is an application that is commonly used in high-rise buildings, where you install a heavy 'vertical' cable and tap off power to the individual floors (as shown in Figure 4.14). By adjusting the coupling loss on the different tappers by selecting the appropriate value, you can actually balance out the loss to all the floors in the high-rise building, providing the required uniform coverage level.

Taps come in various types, the principle being that there is a low-loss port (1–2) and then a higher-loss port (3), where you 'tap' power of to a local antenna/cluster of antennas. In a high-rise building, you install one vertical heavy $\frac{7}{8}$ or $1\frac{1}{4}$ inch cable in the vertical cable riser, and then use tappers on each floor to feed a splitter that divides out to two to four antennas fed with $\frac{1}{2}$ inch coax. Standard tappers are available with the values shown in Table 4.2.

Example of Use

In this example (as shown in Figure 4.4), we can see that, even over long distances (200 m at GSM1800), using a $\frac{7}{8}$ inch main cable, and different types of tappers and a splitter, we can

Table 4.2 Typical taps and their coupling losses

Type	Loss port 1–2	Loss port 1–3
1/7 Tap	1 dB	7 dB
0.5/10	0.5 dB	10.5 dB
0.1/15	0.1 dB	15.1 dB
Variable	0.1–1.2 dB	6–15 dB

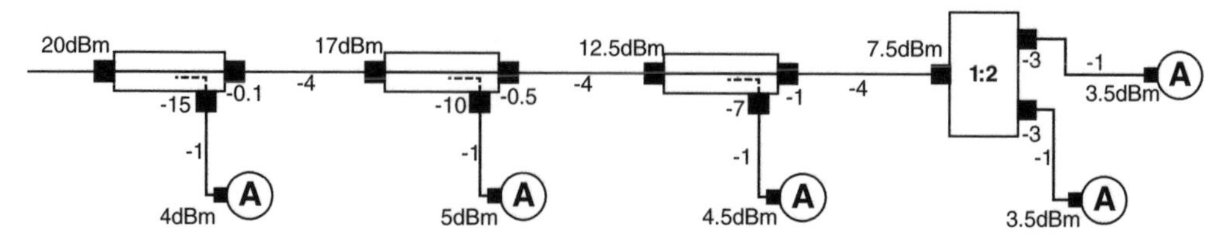

Figure 4.4 Typical configurations of tappers on a distributed antenna system to keep a uniform coverage level for all antennas over a large distance

keep a relative constant attenuation of all of the antennas within a variation of 1.5 dB, even though the longitudinal loss of the main cable varies up to 12 dB. It is evident that it is possible to balance out the loss efficiently when using tappers.

4.2.5 Attenuators

Attenuators (as shown in Figure 4.5), attenuate the signal with the value of the attenuator. For example a 10 dB attenuator will attenuate the signal by 10 dB (port 2 = port 1 − attenuation). Attenuators are used to bring higher power signals down to a desired range of operation, typical to avoid overdriving an amplifier, or to limit the impact of noise power from an active distributed antenna system (see Section 10.2).

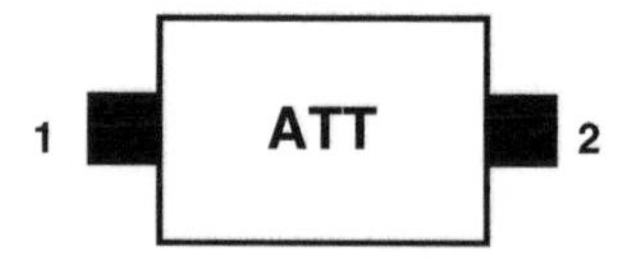

Figure 4.5 RF attenuator

Typical standard attenuator values are 1, 2, 3, 6 10, 12, 18, 20, 30 and 40 dB. When you combine them, you can get the desired value; variable attenuators are also available, but typical only for low power signals. Note that, when attenuating high power signals for many carriers, typically for multioperator applications you should use a special type of attenuator, a 'cable absorber', to avoid PIM problems.

4.2.6 Dummy Loads or Terminators

Terminators (as shown in Figure 4.6), are used as matching loads on the transmission lines, often on one port of a circulator, or any 'open' or unused ports on other components. In applications that are sensitive for PIM, the better option is to use a cable absorber (−160 dBc)

Figure 4.6 Standard 50 Ω dummy load or terminator

4.2.7 Circulators

The circulator splitter (as shown in Figure 4.7) is a nonreciprocal component with low insertion loss in the forward direction (ports 1–2, 2–3 and 3–1) and high insertion loss in the reverse direction (ports 2–1, 3–2 and 1–3).

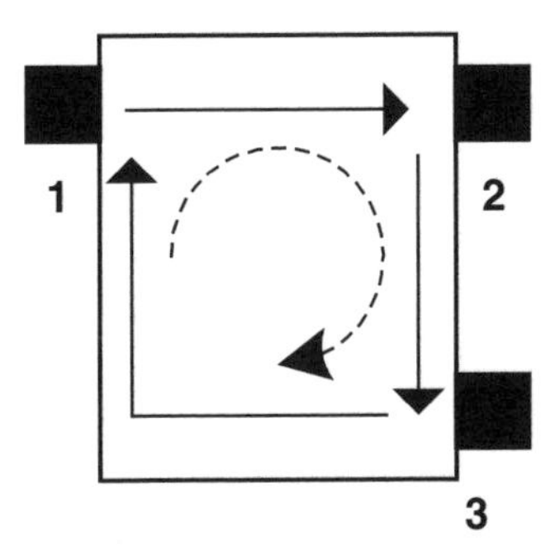

Figure 4.7 RF circulator

The insertion loss in the forward direction is typically less than 0.5 dB and in the reverse direction better than 23 dB. You can get 'double stage' isolators with reverse isolation better than 40 dB if needed.

Examples of Use

The circulator can be used to protect the port of a transmitter (as shown in Figure 4.8) against reverse power from reflections caused by a disconnected antenna or cable in the antenna system.

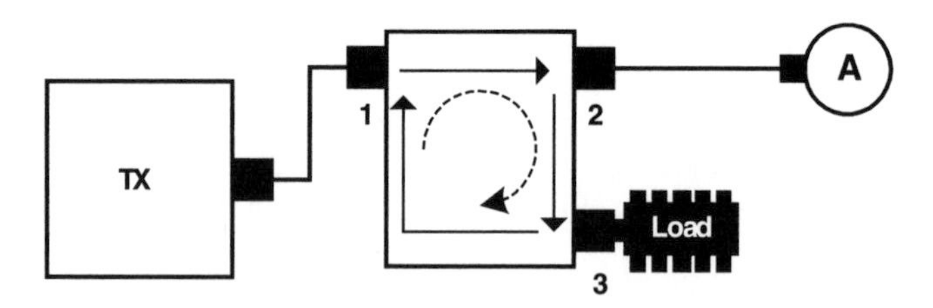

Figure 4.8 Circulator used for protecting a transmitter against reflected power

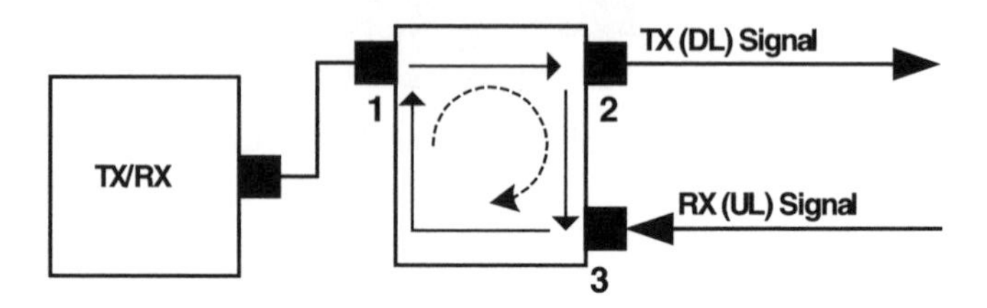

Figure 4.9 Circulator used as a duplexer, separating Rx and Tx from a combined Rx/Tx line

A common application for circulators in mobile systems is to use the circulator to separate the transmit and receive directions from a combined Tx/Rx port (as shown in Figure 4.9). This is mostly used for relatively low power applications due to PIM issues in the circulator. For high-power applications it is recommended to use a cavity duplex filter to separate the two signals.

Circulators can also be used to isolate transmitters in a combined network for a multioperator system (as shown in Figure 5.28).

4.2.8 A 3 dB Coupler (90° Hybrid)

The 3 dB coupler shown in Figure 4.10 are mostly used for combining signals from two signal sources. At the same time the coupler will split the two combined signals into two output ports. This can be very useful when designing passive distributed antenna systems.

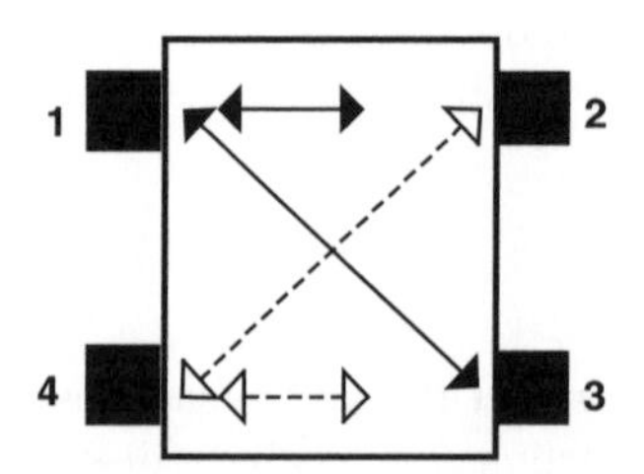

Figure 4.10 A 3 dB coupler

The 3 dB coupler has four ports; two sets of these are isolated from each other (ports 2 and 3/1 and 4). If power is fed to port 1, this power is distributed to ports 2 and 3 (−3 dB). Port 4 will be powerless provided that ports 2 and 3 are ideally matched. Normally a terminator will be connected to port 4.

Example of Use
If you need to combine two transmitters or two transceivers (TRXs/TRUs), you can use a 3 dB coupler (as shown in Figure 4.11). However, if you need to combine the two transmitters and at the same time distribute the power to a passive distributed antenna system with several antennas, you should connect one part of the DAS to port 2 and the other to port 3 (as shown in Figure 4.12). Thus you will increase the power on the DAS by a factor of 2 (3 dB). This method is to be preferred, and will increase the signal level by 3 dB in the building, rather than burning the 3 dB in the dummy load on port 3 (as shown in Figure 4.11) – this will only generate heat!

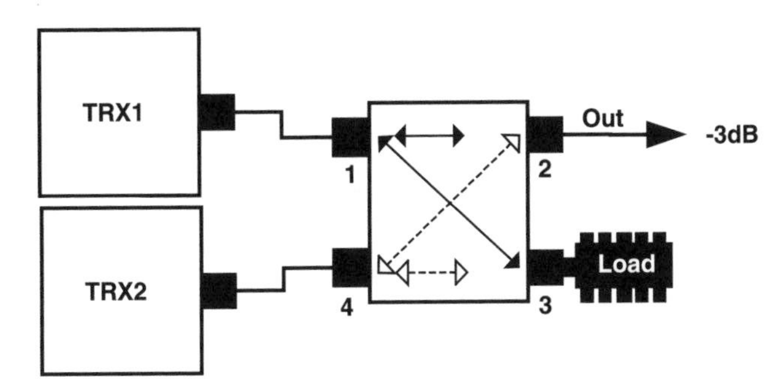

Figure 4.11 A 3 dB coupler used as a two-port combiner

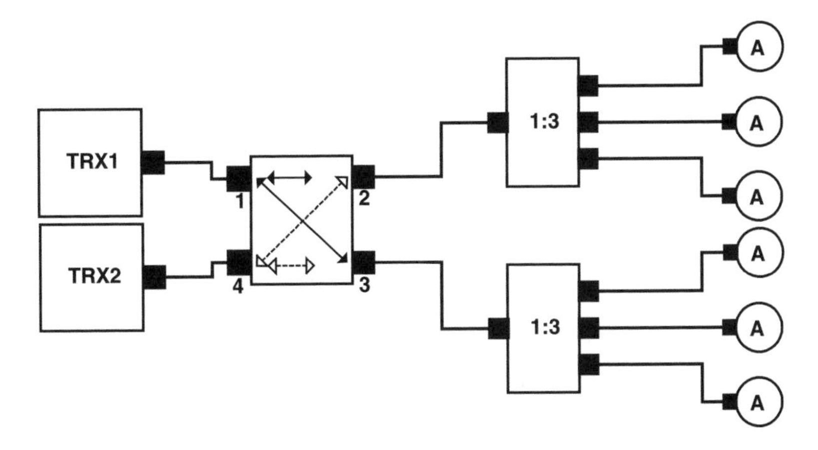

Figure 4.12 Combining two TRX and splitting out to a distributed antenna system

4.2.9 Power Load on Passive Components

One parameter that is very important to keep in mind when designing with passive components is not exceed the maximum power rating that the passive component can handle. This is a particular problem for high capacity or multioperator passive DAS solutions, where you combine many carriers and base stations at high power levels into the same passive distribution system.

Calculating the Total Power from the Base Stations

How do we calculate the total power?

Example, calculating composite power on a passive DAS
We have a multioperator system in an airport; there are three GSM operators connected at the same point, and each operator has six TRX. The base stations output 46 dBm into the distributed antenna system. In total we have $3 \times 6 = 18$ TRX.

Worst case is that all carriers are loaded 100%; therefore each GSM radio transceiver transmits a full 46 dBm constantly on all time slots. We need to sum all the power, but first we have to convert from dBm to Watt:

$$P\ (\text{mW}) = 10^{\frac{\text{dBm}}{10}} = 40\,000\ (\text{mW}) + = 40\,\text{W}$$

We have 16 signals of 40 W each; the total composite power is:

$$\text{Total power} = 16 \times 40\,\text{W} = 640\,\text{W}$$

Then back to dBm = 10log (640 000 [mW]) = 58 dBm

Therefore we need to insure that the passive components can handle 640 [Watt]/58 dBm constantly in order to make sure the system is stable.

The PIM Power

PIM is covered in Section 5.7.4. However, let us calculate the level of the PIM signals. If we take a passive component, with a PIM specification of −120 dBc, the maximum PIM product will be 120 dB below the highest carrier power. For example, continued from the previous example, we can expect the worst case PIM product to have a signal level of 46 dBm − 120 dB = −74 dBm.

This is a major concern; −74 dBm in unwanted signal, especially if the inter-modulation product falls in the UL band of one of the systems, it will become a very big problem that will degrade the performance of that system/channel.

Exceeding the maximum the maximum power rating on a passive component will make it even worse. Be sure to use passive components with very good specifications when designing high-power, high-capacity solutions, and be absolutely sure to keep within the power specifications of all the components.

4.2.10 *Filters*

When designing indoor solutions there are basically two types of filters that you will encounter, the duplexer and the diplexer or triplexer, as shown in Figure 4.13.

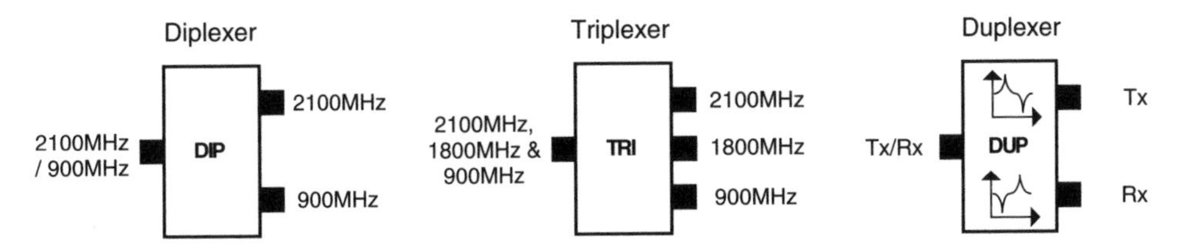

Figure 4.13 The typical filters used to separate frequency bands: diplexer and triplexer. Also, the duplexer used to separate uplink and downlink

Duplexer

The duplexer is used to separate a combined TX/RX signal into separate TX and RX lines. Bear in mind the isolation between the bands as well as the insertion loss and the PIM specifications.

Diplexer/Triplexer

The diplexer will separate or combine whole bands from or with each other, for example, input combined 2100 and 1800 MHz and output separate 2100 and 1800 MHz bands. Bear in mind the isolation between the bands as well as the insertion loss and the PIM specifications.

A three band version that can separate or combine 900, 1800 and 2100 MHz is also available, called a triplexer. Some manufacturers even do combined components that contain both a diplexer or triplexer and a duplexer.

4.3 The Passive DAS

Now that we know the function of all the passive components, we are able to make a design of a passive distributed antenna system. Passive DAS systems are the most used approach when providing indoor solutions, especially to small buildings.

4.3.1 *Planning the Passive DAS*

The passive DAS is relatively easy to plan; the main thing you need to do is to calculate the maximum loss to each antenna in the system, and do the link budget accordingly for the particular areas that each antenna covers. You will need to adapt the design of the passive DAS to the limitations of the building with regards to restrictions to where and how the heavy coax can be installed. Often the RF planner will make a draft design based on floor plans before the initial site survey, and afterwards adapt this design to meet the installation requirements of the building. In fact, the role of the RF planner is often limited to installation planning, not RF planning, when designing passive DAS.

It is very important that you know all the cable distances and types so that you can calculate the loss from the base station to each individual antenna. Therefore you must do a detailed site survey of the building, making sure that there are cable routes to all of the planned antennas. When doing passive DAS design, you will often be limited and restricted as to where you can install the rigid passive cables. Frequently, the limitations of installation possibilities will dictate the actual passive DAS design, and because of this the final passive solution will often be a compromise between radio performance and the reality of the installation restrictions. You need the exact loss of each coax section in the system, in order to verify the link budget (see Chapter 8), and place the antennas.

A typical passive DAS design can be seen in Figure 4.14, showing a small office building. This design relies on a main vertical $\frac{7}{8}$ inch cable, using tappers on each floor that tap off power to the horizontal via $\frac{1}{2}$ inch coax cables to 1:3 splitters on each floor.

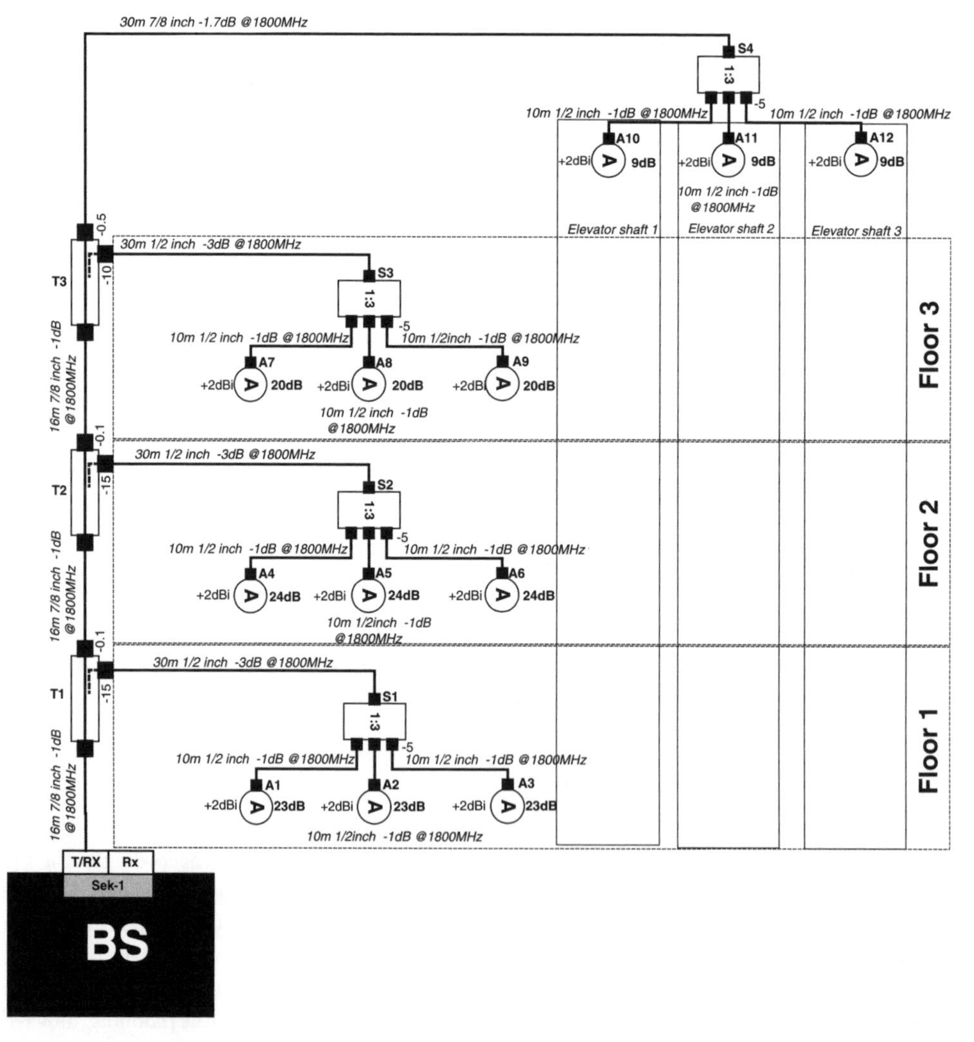

Figure 4.14 Typical passive DAS diagram with the basic information and data

The advantage is that the heavy $\frac{7}{8}$ inch cable can be installed in the vertical cable raiser where there is easy access and the installation of the rigid coax will be relative simple. On the office floors it is more of a challenge to installing coaxial cable as no cable trays are available. By using 'thin' $\frac{1}{2}$ inch coax for these horizontal cable runs, we can simply strap the cable onto the frame of the suspended ceiling with cable ties, making the installation relatively fast and inexpensive.

The coverage in this example is to some extent uniform; there is acceptable balance between the antennas serving the three office floors. It was decided to radiate more power into the top of the three elevator shafts, in order to penetrate the lift-car, and provide sufficient coverage inside the lift for voice service.

When doing the diagram (Figure 4.14), all information must be documented, including losses of components and cables, type numbers, component numbers and total loss to each antenna. This design documentation must contain all relevant information, and must be available on-site in case of trouble-shooting.

Trouble-shooting is an issue with passive systems. You will need to use a power meter connected to selected points throughout the passive DAS in order to disclose any faults, in case of a fault on the system. You could use the 'one meter test' from Section 5.2.9 for the worst faults, but you need to connect a power meter to the DAS to be absolutely sure of the power level.

This is the main downside of using passive DAS: trouble-shooting is painstaking. Also, even realizing that there is a fault in the system is pretty much a question of customer complaints from the building; even severe faults on the passive DAS will give no alarm at the base station.

4.3.2 Main Points About Passive DAS

There are many arguments for and against the use of passive DAS. Remember that passive DAS is just one of the tools in the indoor radio planning toolkit, sometimes passive DAS will be the best choice, sometimes not. The clever indoor radio planner will know when to use this approach, and when not to.

The advantages of passive DAS are:

- It is straightforward but time-consuming to design.
- Components from different manufacturers are compatible.
- It can be installed in harsh environment.

The disadvantages of passive DAS are:

- There is no surveillance of errors in the system – the base station will not give VSWR alarm even with errors close to the base station due to high return loss.
- It is not flexible for upgrades.
- High losses will degrade data performance.
- It is hard to balance out the link budget for all antennas, and to get a uniform coverage level.
- It requires a high-power base station and dedicated equipment room for site support equipment, power supply, etc.

The fact remains that passive DAS is the most implemented type of DAS on a global basis. However the need for 3G/UMTS and HSPA service and even higher speed data service in the future will affect the preference for selecting DAS types.

The attenuation of the passive DAS is the main issue in this context: frequencies used for future mobile services will most likely get higher and higher, and the modulation schemes applied for high-speed data services are very sensitive to the impact of the passive cable loss. This will degrade the downlink power at the antenna, and on the uplink the high noise figure of the system caused by the passive losses will limit the uplink data speeds. Surely passive systems will continue to be used in the future, but only for small buildings with a few antennas, and the losses must be kept to a minimum.

4.3.3 Applications for Passive DAS

Passive DAS is the most widely used distribution system for indoor coverage systems for mobile service. Passive DAS can be used for very small buildings with a low-power base station and a few antennas, all the way up in size to large airports, campuses, etc.

The main challenge in using passive DAS is the installation of the rigid cables, which have a high impact on the installation cost, and might limit the possibilities as to where antennas can be installed. This could be an issue in solving the high-rise problem covered in Section 3.5.3. The building will more or less dictate the design, because of the installation challenge.

Degraded data service can be an issue if the attenuation of the system gets too high, epically on 3G/HSPA. This problem can be solved by dividing the passive DAS into small sections or sectors, each serviced by a local base station. However, this is costly and often ineffective use of the capacity resources; refer to Section 6.1.9 regarding trunking gains. The extra backhaul costs, interface loads on the core network and software licenses to the equipment supplier also add to the cost of distributing the base stations.

High user RF exposure with passive DAS. Mobiles inside the building will radiate on relatively high power, due to the fact that the mobile has to overcome the passive losses by transmitting at a higher power level (see Section 4.13). Thus the mobile will expose the users to higher levels of electromagnetic radiation.

Maintenance and trouble-shooting are challenging in passive DAS systems. Be sure to use a certified installer, and do not underestimate the importance of proper installation code and discipline when installing coax solutions and connectors.

'Passive systems never fail' – this is not true. Most likely, you just do not know that there is a fault! Trust me, when an installer is on top of a ladder at four o'clock in the morning, mounting the 65th coax connector that night, he might be a little sloppy. Even small installation faults can result in severe problems, inter-modulation issues, etc. Do not underestimate the composite power in a passive system when combining many carriers and services into the same system at high power.

4.4 Active DAS

The principle function of an active distributed antenna system is that, like a passive distributed antenna system, it distributes the signal to a number of indoor antennas. However, there are some big differences. The active distributed antenna system normally relies on thin cabling, optical fibers and IT type cables, making the installation work very easy compared with the rigid cables used for passive systems.

The active distributed antenna system consists of several components, the exact configuration depending on the specific manufacturer. All active distributed antenna systems will to some extent be able to compensate for the distance and attenuation of the cables.

4.4.1 Easy to Plan

The ability to compensate for the losses of the cables interconnecting the units in an active distributed antenna system makes the system very easy and fast to plan, and easy to implement in the building.

Whereas, when designing a passive distributed antenna system, you need to know the exact cable route and distance for each cable in order to calculate the loss and link budget, when designing active distributed antenna systems it does not matter if the antenna is located 20 m from the base station or even 5 km. The performance will be the same for all antennas in the system; the active DAS system is transparent. This 'transparency' is obtained automatically because the active system will compensate for all cable losses by the use of internal calibrating signals and amplifiers. This is typically done automatically when you connect the units and commission the system. Therefore the radio planner will not need to perform a detailed site survey. It does not matter where the cables are installed, and the system will calibrate any imbalance of the cables. Nor does the radio planner need to do link budget calculations for all the antennas in the building; all antennas will have the same noise figure and the same downlink power, giving truly uniform coverage throughout the building. These active DAS systems are very fast and easy to plan, implement and optimize.

It is a fact that modern buildings are very dynamic in terms of their usage. Having a distributed antenna system that can easily be upgraded and adapted to the need of the building is important. It is important for the users of the building, the building owner and the mobile operator. The active DAS can accommodate that concern, being easy and flexible to adapt and to upgrade. There is no need to rework the whole design and installation if there are changes and additions in the system; there is always the same antenna power, whatever the number of or distance to the antennas.

4.4.2 Pure Active DAS for Large Buildings

Ideally in an active DAS there will be no passive components that are not compensated by the system. Therefore the active DAS is able to monitor the end-to-end performance of the total DAS and give alarms in case of malfunction or disconnection of cables and antennas. These active DAS systems can support one band–one operator, or large multioperator solutions.

No Need for High Power

The philosophy behind the purely active DAS architecture is to have the last DL amplifier and the first UL amplifier as close to the antenna as possible. Co-located with the antenna is the remote unit (RU), avoiding any unnecessary degrading losses of passive coax cables.

When using this philosophy, having the RU located close to the antenna, there is no need to use excessive downlink transmit power from the base station to compensate for losses in passive coax cables; therefore the system can be based on low to medium transmission power from the RU, because all the RU downlink power will be delivered to the antenna with no losses.

Better Data Performance on the Uplink

Purely active DAS has big advantages for the uplink data performance. Having the first uplink in the RU, with no losses back to the base station, will boost the data performance. This is very important for the performance of high-speed data, the higher EDGE coding schemes on GSM, high-speed data on UMTS and in particular, HSUPA performance.

The main difference between the passive and active DAS on the uplink performance is that, even though the active DAS will have a certain noise figure, it will be far lower compared with the high system noise figure on high-loss passive DAS systems. The effective NF performance is basic radio design; refer to Chapter 7 for more details on how the loss and noise figure affect the system performance.

Medium to Large Solutions

By the use of transmission via low-loss optical fibers, a typical active DAS can reach distances of more than 5 km. The cable between EU and RU up to 250 m makes these types of solutions applicable in medium to large buildings, typically large office buildings, shopping malls, hospitals, campus environments and tunnels.

Save on Installation and Project Cost

The active DAS will typically only require about +10 dBm input power from the base station; there is no need for a large, high-power base station installation, with heavy power supply, air-conditioning, etc. A mini-base station can be used to feed the system, and the system components are so small that an equipment room can be avoided; everything can simply be installed in a shaft.

Less power consumption due to the need for less power from the base station with no ventilation saves on operational costs, and makes the system more green and more CO_2 friendly. The fact that the installation work uses the thin cabling infrastructure of an active DAS can also cut project cost and implementation time.

Time to deployment is also short compared with the traditional passive design. You are able to react faster to the users' need for indoor coverage, and thus revenue will be generated faster, and the users will be more loyal.

The Components of the Active DAS

In order to understand how you can use active DAS for indoor coverage planning, you need to understand the elements of the active DAS. Some active DAS systems use pure analog signals; other systems convert the RF to digital and might also apply IP transmission internally.

The names of the units, numbers of ports, distances and cable types will be slightly different, but the principle is the same (Figure 4.15). Typically these types of DAS systems will be able to support both GSM and UMTS, so you will need only one DAS system in the building to cater for all mobile services and operators.

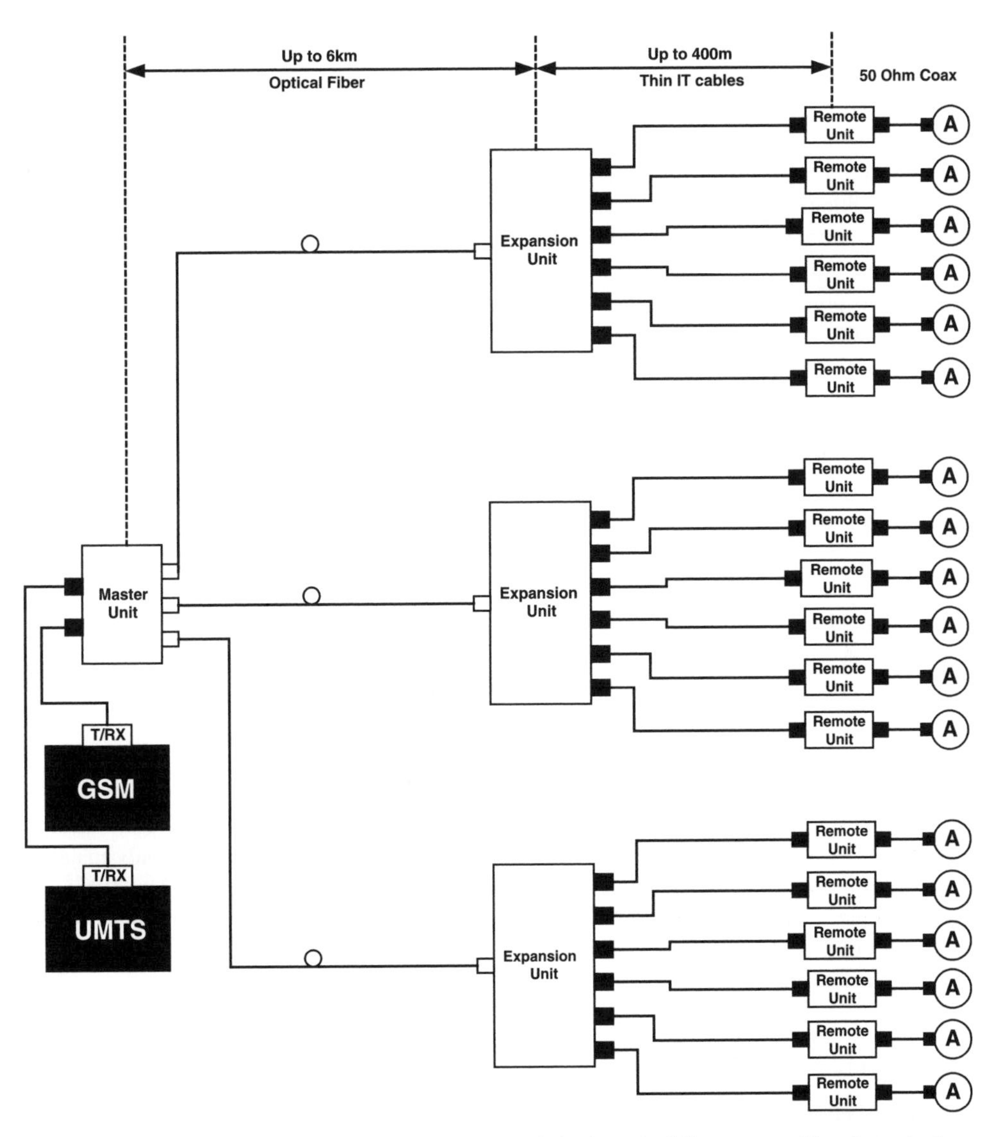

Figure 4.15 Example of a pure active dual band DAS for large buildings; up to 6 km from the base station and antennas with no loss

Main Unit

The main unit (MU) connects to the low-power base station or repeater; the MU distributes the signals to the rest of the system via expansion units (EU). The MU will typically be connected to the EU by optical fibers. The MU is the 'brain' of the system and also generates and controls internal calibration signals in the system together with internal amplifiers, and converters adjust gains and levels to the different ports in order to compensate for the variance of internal cable loss between all the units.

The MU will also monitor the performance of the DAS system, communicating data to all units in the DAS. In the event of a malfunction or a warning it is able to send an alarm signal to the base station that enables the operator to exactly pinpoint the root of the problem and resolve the problem fast.

Detail about specific alarms is typically good; the system will normally pinpoint the exact cable, antenna or component that is the root of the problem. Thus the downtime of the DAS can be limited, and the performance of the system re-established quickly.

It is possible to see the status of the whole system and the individual units at the MU, using LEDs, an internal LCD display or via a connected PC – it all depends on the manufacture of the system.

It is also possible to access the MU remotely via modem or via the internet using IP, perform status investigation, reconfiguration and retrieve alarms in order to ease trouble-shooting and support.

Expansion Unit

The EUs are typically distributed throughout the building or campus and are placed in central cable raisers or IT X-connect rooms. The EU is connected to the MU using optical fibers, typically separate fibers for the UL and the DL. The EU converts the optical signal from the MU to an electrical signal and distributes this to the RU.

Ideally the EU will also feed the DC power supply to the RUs via the existing signal cable in order to avoid the need for local power supply at each antenna point (RU). In many cases LEDs will provide a status of the local EU and subsystem (the RUs).

Optical Fiber Installations

Some systems can use both single mode fiber (SMF) and multimode fiber (MMF), and some systems only SMF. It is important to consider this when planning to reuse already installed fiber in the building, since old fiber installations are typically only MMF. Installation of fibers and fiber connectors takes both education and discipline, so be sure that your installer has been certified for this work, and always refer to the installation guidelines from the manufacturer of the DAS.

Remote Unit

The RU is installed close to the antenna, to keep the passive losses to a minimum and improve the radio link performance. The RU converts the signal from the EU back to normal DL radio signals and the radio signal from the mobiles on the UL is converted and transmitted back to the EU. The RU is located close to the antenna, typically only connected to a short RF jumper. This will insure the best RF performance and the possibility of active DAS to detect when the antenna is disconnected from the system.

The RU should be DC fed with power from the EU in order to avoid expensive local power supply at each antenna point. In addition, the RU should be designed with no fans or other noisy internal parts, in order to enable the system to be installed in quiet office environments.

The RU is connected to the EU with a thin coax, or CAT5 cables or similar 'thin cabling', making it very easy and quick to install compared with the rigid passive coax cables used for passive DAS.

4.4.3 Pure Active DAS for Small to Medium-size Buildings

Even though that active DAS is normally considered applicable for large buildings, small buildings can be designed with pure active DAS (Figure 4.16), thereby utilizing the benefits of having a high-performance DAS, with light cabling infrastructure, a small low-power base station and full surveillance of the DAS.

Main Unit

The system consists of one MU connected directly to RUs with thin coax, CAT5 or other 'IT-type' cables; there is no use of optical fibers. This small system has all the same functions and advantages with regards to auto calibration, uniform performance and improved radio link as the large systems. Even the detailed alarming is the same; full monitoring of the system, including the antennas, is possible.

For medium-sized buildings, you can feed more of these systems in parallel to obtain more antenna points. This makes this medium system a very cost-effective and high-performing system.

Cost-effective Installation

Both of these pure active systems are very cost-effective and easy to install. Often the IT team of the specific building can carry out the installation for the operator, thus sharing the cost load with the operator. In this way the building owner or user pays for the installation and the operator pays for the equipment and infrastructure. This will boost the business case, and create a more loyal customer, who is now a part of the project. This also give an advantages for system performance, as the end user can provide more detailed help with inputs as to where in the building it is most important to plan and implement perfect coverage, and also in understanding why certain areas have a lower signal. However, it is important to note that the units in active DAS cannot always be installed in tough environments, as moisture, vibrations and dust might cause damage. Therefore it is very important to check the manufacturer's guidelines and installation instructions. Many systems can be installed in harsh environments, if shielded inside an appropriate IP-certified box.

Applications for the Pure Active DAS

The large version of the pure active DAS as shown in Figure 4.15 is a very versatile tool for large buildings or campuses. The low-impact installation of the pure active DAS, using and reusing thin 'IT-type' cabling, makes this solution ideal in corporate buildings, hotels and hospitals. The installation is fast and easy, making it possible to react to requests for indoor coverage swiftly. The concept of having the RU close to the antenna boosts the data performance in these high-profile buildings, where 3G/HSPA service is a must.

Radiation from mobiles and DAS antenna systems are a concern for the users of the building. The pure active DAS has the RUs installed lose to the antenna; this boosts the uplink data performance. In addition there is a side effect: because the system calibrates the cable losses, there is no attenuation of the signal from the antenna to the base station. Hence the mobile can operate using very low transmit power, because it does not have to compensate for any passive cable loss back to the base station in order to reach the uplink target level used for power control. This makes this approach ideal for installations in hospitals, for example. See Section 4.13 for more details on electromagnetic radiation.

The small version of the pure active DAS shown in Figure 4.16 has all the advantages of the large system, but it does not rely on fiber installation. This takes some of the complexity out of the installation process. The system is ideal and cost-effective for small to medium-size buildings. Given that the system can reach more than 200 m from the MU to the RU, if installed in the center of a high-rise tower it could cover a 300 m-high building without any need for fiber (reserving some distance for horizontal cables). The systems are fully supervised all the way to the antenna; there is full visibility of the performance of the DAS.

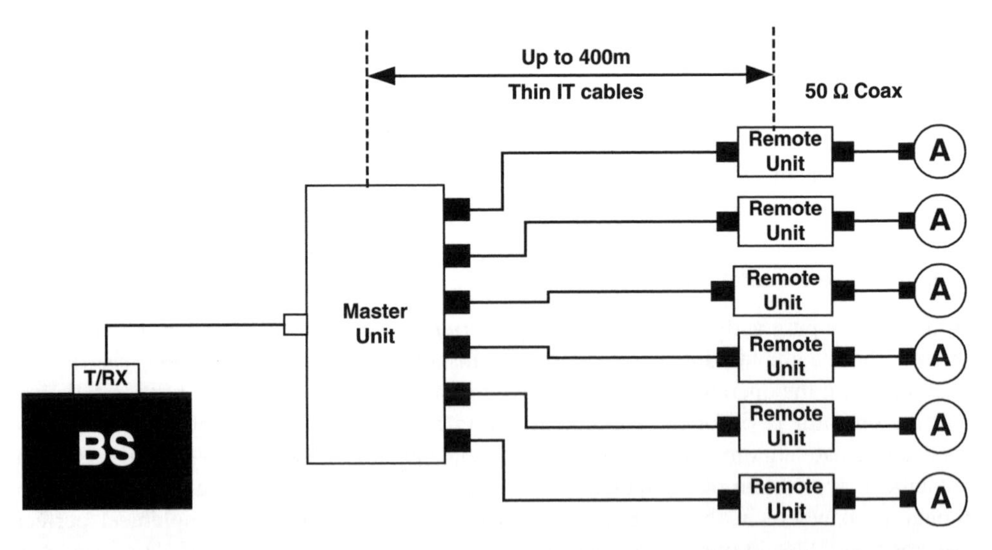

Figure 4.16 Example of a pure active DAS for small buildings; up to 400 m distance between the base station and the antennas with no loss

Reliability is a concern, due to the number of units distributed throughout the building. You must select a supplier that can document good reliability and mean time between failures (MTBF) statistics. You also need to make sure that the location of all the installed units is documented, so they can be accessed for maintenance.

You must be careful when installing these types of systems in moist, damp, dusty environments and shield them accordingly.

4.4.4 Active Fiber DAS

The increasing need for more and more bandwidth over the DAS to support multiple radio services, GSM, DCS, UMTS, Wi-Fi, WiMAX, Tetra, etc. has motivated a need for the indoor fiber DAS system to support a wider bandwidth to accommodate all the radio service, as shown in Figure 4.17.

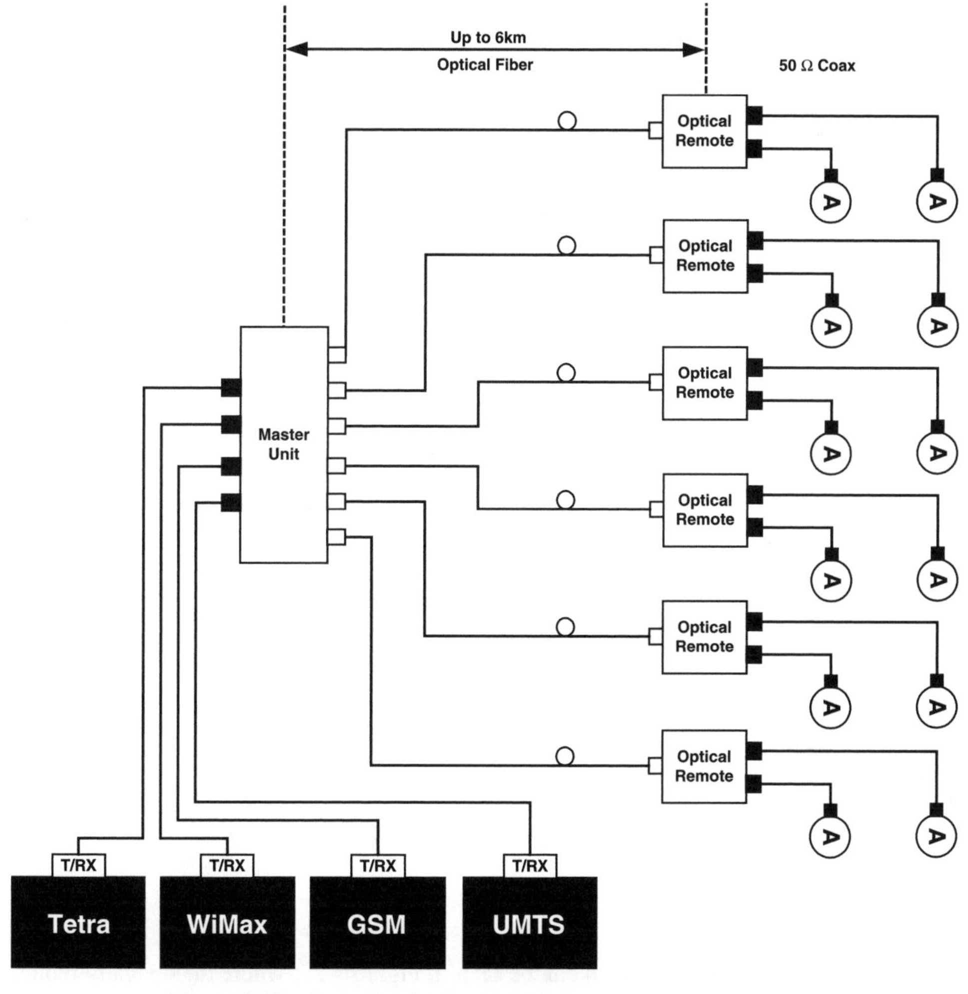

Figure 4.17 Optically distributed DAS for multiservice solutions

Although it is debatable if it is feasible to combine many radio services with quite different design requirement into the same DAS, it is a fact that it is an installation advantage to have only one system and one set of antennas. The compromise will often be the performance on one or more of the services, and a costly DAS.

If all services require 'full' indoor coverage, the consequence is that you will have to design according to the lowest dominator. For example, if the DAS design turns out to show a coverage radius for GSM = 24 m, UMTS = 21 m, but WiMAX = 10 m, you have to place the antennas according to the WiMAX requirement, thus overdesigning the GSM and UMTS system. In fact you often have to adjust the gain of the various services accordingly, to prevent signal leak from the building.

Another concern is the sector plan. Be sure that the DAS allows you to granulate all the different radio systems into independent sector plans.

The Components of the Fiber DAS

Main Unit

The MU interfaces the fiber DAS to the different base stations. The MH distributes the signals directly to the optical remote units by the use of optical fibers. In many cases this fiber is a composite cable, containing both the fiber and the copper cable for power supply to the remote unit. Alternatively the power is fed locally at the optical remote unit.

The MU is the 'brain' of the system and also generates and controls internal calibration signals in the system together with internal amplifiers, and converters adjust gains and levels to the different ports in order to compensate for the variance of internal optical cable loss between the MU and remote units.

The MU will also monitor the performance of the DAS system. In the event of a malfunction or a warning, it must be able to send an alarm signal to the base station, which enables the operators to exactly pinpoint the root of the problem and resolve the problem fast. Details about specific alarms is typically good; the system will normally pinpoint the exact cable, antenna or component that is the root of the problem, thus the downtime of the DAS can be limited, and the performance of the system quickly re-established.

It is possible to see the status of the whole system and the individual units at the MU using LEDs, an internal LCD display or via a connected PC – it all depends of the manufacture of the system. It is also possible to access the MU remotely via a modem or the internet using IP, and perform status investigation or reconfiguration.

Optical Remote Unit

The ORU is installed throughout the building; it converts the optical signal from the MU back to normal RF and the RF UL signal from the mobiles is converted and transmitted to the MU. The ORU will typically need to operate on medium to high-output RF power and will often have two or more antenna connections, for the DAS antennas.

Daisy-chained Systems

A variant of the optical DAS system can be seen in Figure 4.27, where the optically remote units can be daisy-chained, thereby avoiding the star configuration of Figure 4.17, where all optical remote units need an individual fiber or set of fibers back to MU. This makes the optical system very applicable in tunnels, street DAS and high-rise buildings.

The downside with the daisy-chained system is that, if the fiber is cut, or the optical by-pass in one optical remote fails, then the rest of the DAS on the chain will be out of service.

Applications of the Optically Active DAS

This system is ideal for systems where there is a need for multiple services, other than 'just' GSM and UMTS. However, combining all these radio services into the same DAS is a challenge when it comes to inter-modulation and composite power resources.

Power supply to the ORU can also be a challenge; when using the composite cable that has both the copper cable for the DC power and the fiber cable, you need to be careful about galvanic isolation between the buildings and grounding. The concept of having the ORU close to the antenna boosts the data performance in these high-profile buildings, where 3G/HSPA service is a must.

Radiation from mobiles and DAS antenna systems are a concern for the users of the building. The optically active DAS has the ORU installed close to the antenna; this will boost the uplink data performance. In addition there is a side effect; because the system calibrates the cable losses, there is no attenuation of the signal from the antenna to the base station. Hence the mobile can operate using very low transmission power, because it does not have to compensate for any passive cable loss back to the base station in order to reach the uplink target level used for power control. This makes this approach ideal for installations in hospitals, for example. See Section 4.13 for more details on electromagnetic radiation.

The systems are fully supervised all the way to the antenna; there is full visibility of the performance of the DAS. Reliability is a concern, due to the number of distributed unit throughout the building. It is essential to select a supplier that can document good reliability and MTBF statistics. You also need to make sure that the location of all installed units is documented, so that they can be accessed for maintenance.

Care must be taken when installing these types of systems in moist, damp or dusty environments and they must be shielded accordingly. This is a particular challenge in tunnel installations, where the daisy-chained version of this type of DAS is often used.

4.5 Hybrid Active DAS Solutions

It is important to distinguish between the pure active DAS that we covered in the previous section and hybrid DAS solutions. As the name suggests, a 'hybrid' DAS is a mix of an active DAS and a passive DAS.

The passive part of the hybrid DAS will, as for the pure passive DAS, limit the installation possibilities, impact data performance on 3G/HSPA and to some extent dictate the design, due to installation limitations.

4.5.1 Data Performance on the Uplink

The basic of RF design is to limit any loss prior to the first amplifier in the receive chain. The fact is that the passive portion of the hybrid solution between the antenna and the hybrid remote unit (HRU) will degrade the UL data performance. This is explained in more detail in Chapter 7. The impact of the passive portion of the hybrid DAS becomes a concern for UMTS and HSUPA performance.

4.5.2 DL Antenna Power

Even though the typical hybrid DAS produces medium to high power levels out of the HRU, the power at the antenna points will typically be significantly lower. The reason for this is that the power is attenuated by the passive DAS between the HRU and the antenna, but if you only service a few antennas and keep the losses low, you can obtain relatively high radiated power from the antennas.

4.5.3 Antenna Supervision

The small passive DAS after the HRU will give a high return loss back to the HRU; therefore it is often a problem for the HRU to be able to detect any VSWR problems due to the attenuation of the reflection from a disconnected antenna. This, in practice, makes the antenna supervision nonexistent.

4.5.4 Installation Challenges

In order for the HRU to provide high output power levels, the HRU power consumption is quite high. Therefore you will typically need to connect a local power supply to the HRU; alternatively you can use a special hybrid cable from the MU to the HRU that contains both the fiber and copper wires feeding the DC power to the HRU. The need to use a local power supply for each HRU will add cost and complexity to the system, and more points of failure. Most likely, you will not be allowed to use any local power group, but are requested to install a new power group for all the HRUs in the DAS. Therefore, using the hybrid DAS type with composite fiber cables that also accommodate copper wires for power supply of the HRU is often to be preferred; however, there might be a concern in campus installations where you need to make sure that there is galvanic isolation between the buildings. Therefore copper cables might not be allowed to be installed between the buildings, due to grounding issues. Disregarding this might cause severe damage to the DAS and the buildings, and in the worst case start a fire.

Owing to the high power consumption, the HRU normally has active cooling, and often a fan to help with ventilation. This might limit the installation possibilities because of the acoustic noise, which is often restricted to the vertical cable raiser.

4.5.5 The Elements of the Hybrid Active DAS

Refer to Figure 4.18.

Main Unit

The MU connects to the low-power base station or repeater, and distributes the signals to the HRUs in the system. Typically the MU is connected to the HRU using optical fibers. The

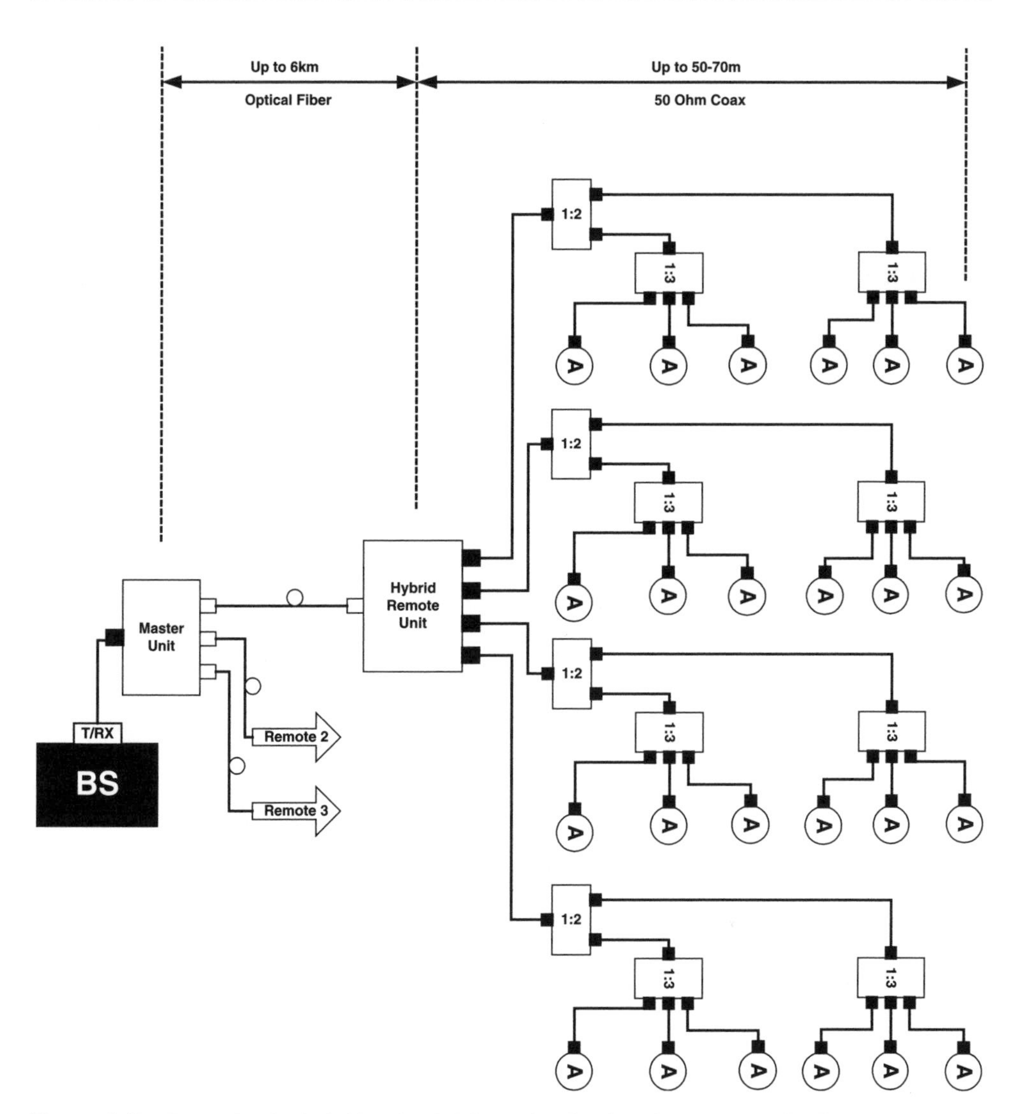

Figure 4.18 Example of a hybrid active DAS, a mix of active elements and distribution, combined with a passive DAS

MU is the controlling element, the 'brain' of the system and also generates and controls internal calibration signals in the system, and then adjust gain and calibration levels to the different ports in order to compensate for the internal cable loss between the MH and HRU. However the system cannot include the passive DAS after the HRU; this part still relies on manual calculation and calibration.

The MU will also monitor the performance of the active part of the DAS system, and in the event of a malfunction is able to send an alarm signal to the base station so the operator can resolve the problem quickly. The system will not be able to detect any problems on the passive DAS, as there will normally be no antenna surveillance. It is possible to see the status of the active part of the system at the MU, on LEDs, an internal LCD display or via a connected PC.

It will often be possible to access the MU remotely via modem or the internet using IP, perform status investigation or reconfiguration, and retrieve alarms in order to ease trouble-shooting and remote support.

Hybrid Remote Unit

The HRU is installed throughout the building. It converts the optical signal from the MU back to normal RF, and the RF UL signal from the mobiles is converted and transmitted to the MU. The HRU will typically need to operate on medium- to high-output RF power, in order to compensate for the losses in the passive DAS that feed the signals to the distributed antennas connected to the HRU.

Applications for Hybrid DAS

Hybrid DAS solutions are ideal solutions where you need high output power at the remote unit. This could be in tunnels; for 'T-Feed systems' see Section 4.8. It could also be sports arenas and multioperator systems where the composite RF power of the remote unit must be shared by many channels.

However, you must be careful with the high output power, and make sure that the uplink can track the coverage area of the downlink, or else the DAS will be out of balance. You must also be careful not to use the high power to only power one 'hot' antenna in the building; this could cause electromagnetic radiation (EMR) concerns (see Section 4.13) and cause interference in the surrounding network.

The relative high cost of the hybrid DAS system normally makes it applicable only for only large structures with high traffic, and high revenue basis.

4.6 Other Hybrid DAS Solutions

Often it can be effective to combine different type of DAS designs in a project, using passive DAS in one part of the building close to the base station, and active DAS in other more distant areas.

It is also possible to combine DAS solutions with macro sectors; for example an outdoor macro site might also be connected to an indoor DAS in the same building, for example where the outdoor sites are located on the roof.

The combination of different concepts will often enable the radio planner to design the DAS as economically as possible, and at the same time maximize performance. Once again it is all about using more tools in the toolbox; do not always rely on only one type.

4.6.1 In-line BDA Solution

It is possible to add an in-line bidirectional amplifier (BDA) to a passive DAS in order to boost the performance of both the uplink and downlink on distant parts of the DAS, as shown in Figure 4.19. However it is a fact that all passive attenuation prior to the BDA, between the BDA and antenna, will seriously impact the noise figure on the system, limiting the uplink performance, which is especially a concern for 3G/UMTS/HSUPA indoor designs. Preferably the BDA should be installed as close as possible to the antenna. Refer to Section 7.2 for more details on how to optimize the BDA design.

As the name suggests, the BDA is a two-way amplifier with two lines, two amplifier systems, one for the DL one for the UL, and a filter system at both input and output.

In some applications the DC power to the BDA can be fed via the coax. This is very useful for tunnel solutions (see Section 4.9.2) and will save costs and complexity for power distribution. In that case it is important to use passive components that are able to handle the DC power that are feed over the coax cable.

Owing to the remote location of the repeater, the only way to get an alarm back to the operations center is to use a RF modem located at the BDA location. Obviously this concept only works if the RF modem does not have to rely on the coverage signal from the BDA itself!

It is possible to cascade BDAs, i.e. have more than one BDA in the DAS system. This could be done in a daisy-chain structure or even in parallel. However, it is important to be careful about noise control for this type of application. The noise increase in cascaded BDA systems can cause major problems with uplink degradation and have a serious impact on the performance of the base station and the network if you are not careful in the design phase. Refer to Chapter 7 with regards to noise calculation, and how to optimize the BDA design.

4.6.2 Combining Passive and Active Indoor DAS

Often the most ideal solution for an indoor project would be to combine the best parts of passive and the best part of active DAS design (Figure 4.19). Passive DAS is cost-effective in

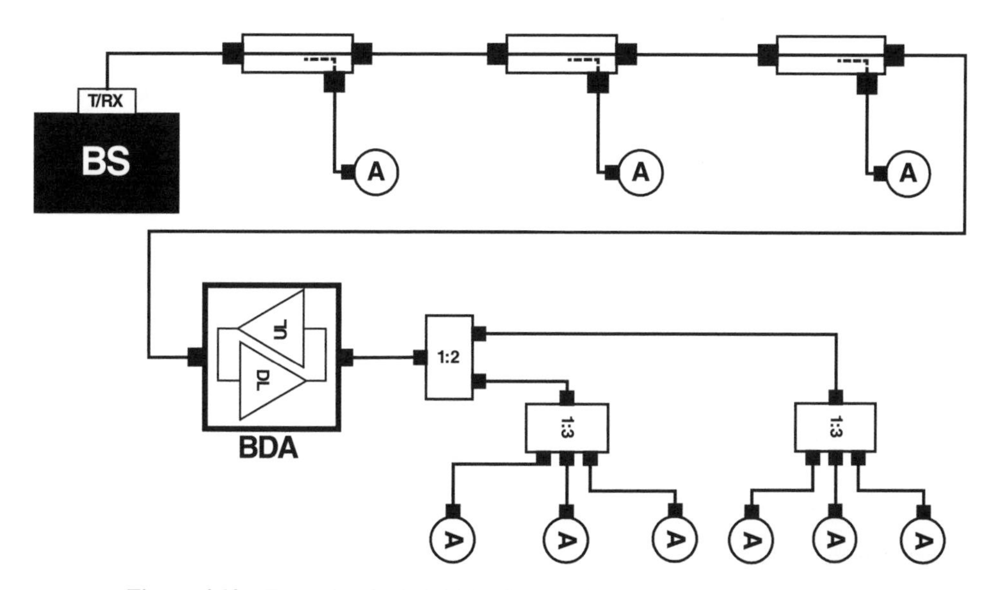

Figure 4.19 Example of a hybrid DAS, a passive system with a BDA added

basements, easy to install in the open cable trays available in basements and parking areas, and with low distances and only servicing a few antennas, the RF performance can be good. Active DAS is often more expensive, but it has the edge on performance at longer distances, and is easier to install in the more challenging parts of the building.

In the typical indoor project you will find that the equipment room you are assigned for the base station is located in the basement. Therefore it would be natural to cover the areas near the base station, parking areas, basement, etc., using a passive DAS with few antennas. This is 'zone A' (see Section 3.5.6), so you can get by with a relatively low coverage level.

In the areas far from the equipment room, however, the office floors in the topmost part of the building, the high loss on a passive DAS might degrade the performance. Also, the installation challenge with rigid heavy passive cables might be an issue, and too expensive. Then the natural choice would be to use passive DAS close to the base station, but active DAS in the more challenging areas. Then you can use the best of both applications, and avoid the downsides of any of them – now that is good use of the 'radio planning toolbox'.

Mind the Noise Power

There is one issue you need to be careful about when combining active DAS with a passive DAS, as shown in Figure 4.20 that is the noise power that the active DAS will inject into the

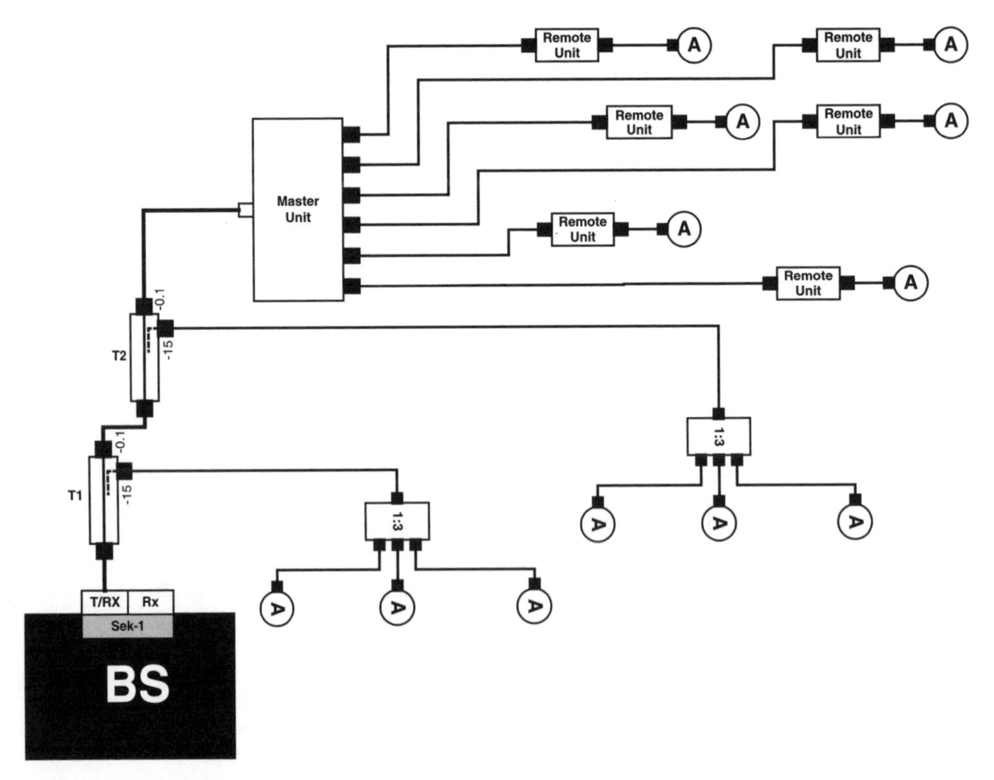

Figure 4.20 Example of a hybrid DAS, a passive system with active DAS in the remote part of the building

passive system (base station). If you have a scenario where the uplink in the passive part of the DAS is the limiting factor, then the noise rise caused by the active DAS could degrade the uplink coverage in the passive part of the DAS.

However the problem can be solved by carefully choosing the correct value of UL attenuator on the main hub of the active system. Refer to Section 7.5 for more detail on how to control the noise power.

4.6.3 Combining Indoor and Outdoor Coverage

Often you will find that a building where a macro base station is located on the rooftop has surprisingly poor indoor coverage near the central core, especially on the lower floors. This is mainly due to the fact that the radio power is beamed away from the building by the high-gain antennas; the coverage inside the building has to rely on reflections from the adjacent buildings and structures. This is a problem especially in high-rise buildings that are not surrounded by high adjacent structures that can reflect the RF signal back into the building (as shown in Figure 4.21). In these cases it might make sense to use the rooftop macro base station as a donor for an indoor DAS. This saves the costs of an indoor base station, backhaul, etc. The capacity is trunked between the indoor and outdoor areas. If the traffic profiles between the two areas are offset, it can be a very efficient use of resources; see Section 6.1.9 for load sharing of the traffic profiles.

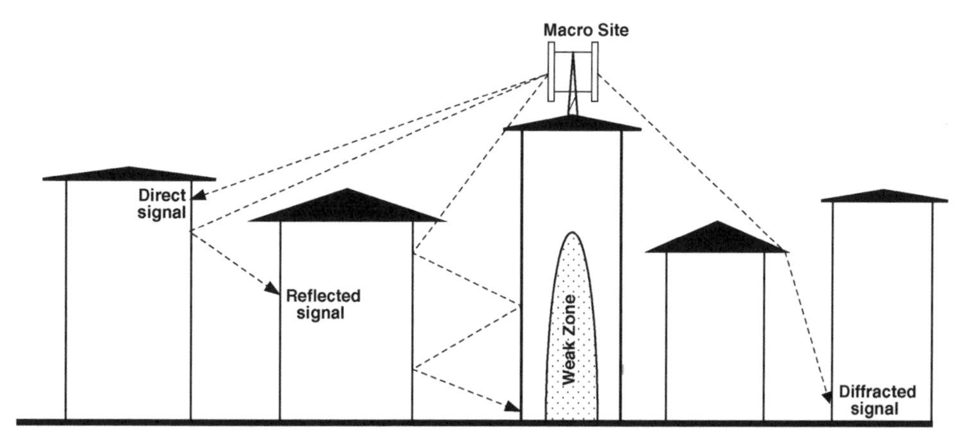

Figure 4.21 RF coverage in the building with the rooftop macro is low due to the need for reflections to provide indoor coverage

Minimize the Impact on the Donor Macro Sector

When you split a rooftop macro cell into also serving an indoor DAS there are concerns that need to be addressed. One option is to split the power to one of the outdoor sectors to a passive DAS. However, this approach costs power and coverage area of the outdoor cell, especially if you need to feed a large indoor DAS.

The solution could be to tap off a fraction of the power (0.1 dB) to the macro cell to an active DAS (as shown in Figure 4.22). The advantage of this approach is that the outdoor coverage for the donor cell is maintained. The active DAS typically needs only +5 dBm input power, so only very little power needs to be tapped off from the outdoor sector.

Mind the Noise Power from the DAS

There is one issue you need to be careful about, which is the noise power that the active DAS will inject in the UL of the macro donor sector. This can desensitize the receiver in the base station, limiting the uplink performance and impacting the UL coverage area, and on UMTS the noise load will offset admission control. The problem can be solved by installing an attenuator on the uplink port of the active DAS and very carefully choosing the correct value of the attenuator. Refer to Section 7.5 for more detail on how to control the noise power and to Chapter 10 on how to optimize the performance.

Applications for Tapping of a Macro Base Station

Only a fraction of the power from the macro base station is needed, which will be more than enough to feed a large active indoor system. This makes this solution shown in Figure 4.22 applicable to many cases, like shopping malls and sports arenas.

However it is a fine balance; you must remember that all the downlink power and capacity used inside the building will also be radiated from the rooftop macro sector, causing interference load and noise increase in the macro sector. By configuring the settings correctly and carefully planning the indoor system to provide high levels of indoor coverage, it should be possible to minimize this effect, and make sure that the power control on the base station will power down the signals to the indoor users to a minimum.

4.7 Radiating Cable Solutions

This book is focused on indoor coverage solutions for buildings, and traditionally radiating cable is only used for tunnel coverage solutions. However there are cases where a building has one or several long tunnels interconnecting buildings in a campus, long vertical shafts, emergency staircases or elevator shafts where a radiating cable could be considered as a possible solution. It is not the intention to cover all the aspects of using radiating cable in tunnel environments – covering this topic would take a whole book on its own. However, in order to understand the basics with regards to radiating cable design, the most important parameters will be addressed here, in order to enable the RF designer with the most basic level of understanding of the issues to design these types of solutions.

4.7.1 The Radiating Cable

The radiating cable or 'leaky feeder'(as shown in Figure 4.23) is typically based on a traditional coaxial cable with (1) an inner conductor, (2) a dielectric, (3) an outer shield, (4) tuned slots in the outer shield and finally (5) the jacket. The size, shape, orientation and placement of the slots are optimized to tune the coverage from the cable. The cable can be

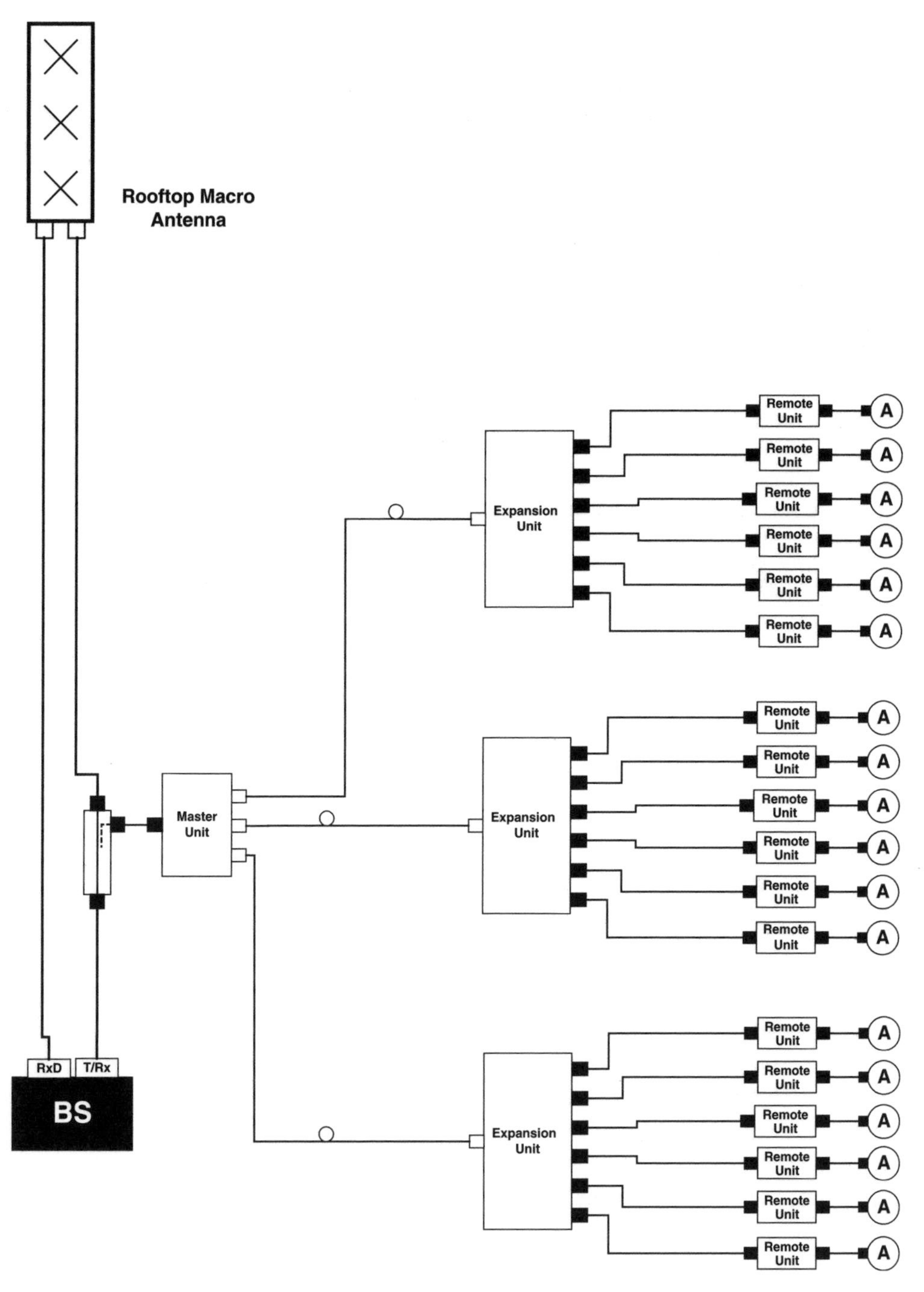

Figure 4.22 Tapping off a fraction (0.1 dB) of the power to the outdoor sector is enough to feed an active indoor DAS, improving the utilization of the base station

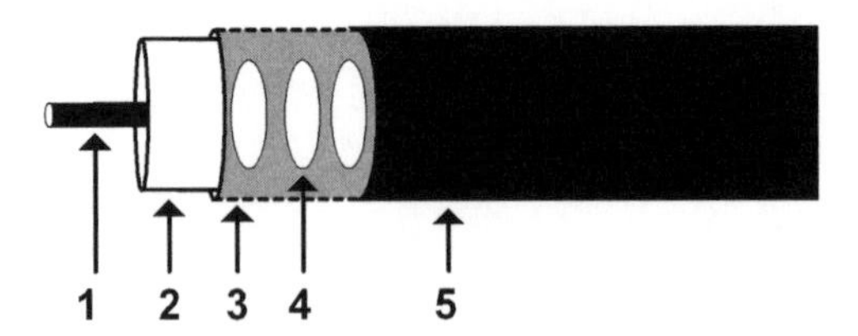

Figure 4.23 Radiating cable, principle

tuned with regards to coupling loss at certain distances, frequency bandwidths, insertion (longitudinal) and loss.

The cable actually acts as a long antenna, or actually many small antennas. The slots will radiate and pick up signal along the length of the cable. Typically the cable will be optimized to serve a specific frequency range of RF spectrum, and optimized to maximum coverage within a certain distance, within the space of the tunnel. Normally a radiating cable is not applicable for servicing users who are not in the proximity of the cable; it is typically ideal for distances of 2-10 m perpendicular to the cable.

The Technical Data of the Radiating Cable

Radiating cable might initially look very simple and easy to understand, but designing and installing radiating cable coverage solutions is a challenge. To understand how to utilize the cable, you first of all need to understand the main parameters of the cable. The most important parameters to consider when selecting, designing and installing radiating cables are listed below.

Frequency Range

The radiating cable will typically be designed and optimized to support specific frequency ranges or bands. For use in designing systems for GSM/DCS/UMTS you can find standard radiating cable optimized to support 900, 1800 and 2100 MHz. Be careful if you plan to apply other radio services on the same cable system, and make sure that the cable will be able to support it.

Longitudinal Loss

The longitudinal loss increases with the distance and will vary with frequency as with standard coaxial cable. Note that the longitudinal loss is also related to the coupling loss of the cable, so the lower the coupling loss the cable is designed for, the more signal is coupled out of the cable, and hence there will be a higher longitudinal loss.

Coupling Loss

The coupling loss is the loss between the cable and the mobile terminal, specified at a given distance and probability, typically 2 or 6 m. The specification of coupling loss usually has a relatively high variance margin, typically ±10 dB. Be sure to include this in the design overhead margin when calculating the link budget.

Note that the coupling losses stated in the datasheet are related to a certain probability. A coupling loss of 80 dB at 50% probability could mean a coupling loss of 95 dB at 95% probability.

System Loss
The system loss is the sum of the longitudinal and coupling loss. This information is sometimes used in data sheets.

DC Resistance
In many tunnel applications DC is injected onto the radiating cable at a central point, and distributed via the inner and outer copper conductor of the coax. This is used for distributing power supply to distributed active elements, amplifiers, etc., connected in the system. It is important that the DC characteristics of the cable will be able to accommodate the power load requirements of the amplifiers, BDAs (bi-directional amplifiers) and other equipment on the line.

Mechanical Specifications
As for other types of cables, mechanical specifications, size, bend radius and fire rating are important. Be sure to fulfill the installation speciation and local guidelines for the specific site where the cable is to be installed, as well as the specifications from the manufacturer of the cable.

Delays in the System
Owing to long cables, optical links, etc., delay and timing can be an issue in tunnel solutions. Remember that the velocity of the coax cable and fiber as well as delays in active elements will cause the timing to shift. This will offset the timing advance in GSM (and for very long tunnels you might even reach the limit for maximum timing advance in the GSM system). For UMTS you might need to widen your 'search window'. Refer to Chapter 10.3 for more details.

4.7.2 Calculating the Coverage Level

Realize that fading in the tunnel and coupling variance from the cable as shown in Figure 4.24 has to be accounted for in the link budget. It is fairly easy to calculate the service level in the tunnel. It is important to include fading margins; these margins will depend on several 'tunnel factors' such as tunnel size, shape, material, speed of the users and obstructions in the tunnel. Most manufacturers of radiating cable will provide you with design support and help with tools that can help you do the radio planning of the tunnel sections.

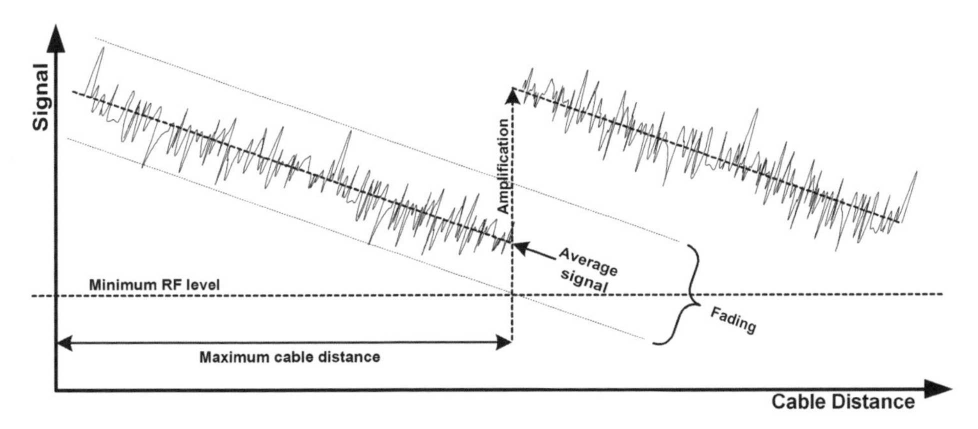

Figure 4.24 Fading and variance of the signal in a tunnel can be high, so good design margins are recommended

Ideally, especially on a large tunnel project, is recommended to verify RF calculations and the link budget with a trial installation in a section of the tunnel. It can be very expensive if mistakes are made in tunnel projects; be absolutely sure to check, re-check and double-check everything, twice. Believe me, I learned that the hard way!

4.7.3 Installation Challenges Using Radiating Cable

In tunnel projects, installation challenges, limited space and lack of access play a major role in the project, especially with regards to time consumption and installation cost.

- *Grounding*: installation of radiating cable inside tunnels, especially metro tunnels with operational train services and high-power DC lines to feed power to the trains, is a major challenge. The power supply to the trains via the typical third rail or overhead power line is so powerful that it is likely to induct power on the radiating cable in the tunnel, therefore grounding is very important.
- *Follow the instructions*: it is very important to install the radiating cable exactly as specified by the specific vendor of the cable. It is equally important to carefully follow the local on-site guidelines 100%. Terminate all points in the system, do not leave the cable 'open' at one end, terminate with an antenna or use a terminator (dummy load).
- *Use cable clamps*: special clamps that stand off the wall have to be used to install the cable at a distance from the wall; if you do not do this, then the coupling loss will increase drastically. Normally you will need to use metallic fire protection clamps every 3 m or so, to insure that the cable does not fall off the wall in case of a fire and block rescue access. In tunnels with a high speed train service, special attention must be paid to the mechanical stress on the cable installation. Often a special type of heavy duty cable clamp has to be used to make sure the cable is secured to the wall.
- *Align the cable*: normally, especially on higher frequencies, such as 1800 or 2100 MHz, the cable has to be aligned perfectly so that the slots can leak to the environment with minimum coupling loss; a guide is typically printed on the jacket of the cable to assist with the alignment.
- *Mechanical stress*: do not underestimate the mechanical stress on a tunnel installation; this applies to all equipment installed, and therefore also the cables. Normally you could get by with letting a coax cable lean against the corner of a concrete wall, but vibration on the coax caused by the vehicles or trains will cause the concrete to wear off the plastic jacket and into the copper of the cable. Over time, moist and corrosion will degrade the performance. It is very important to observe that, if you have to pass the radiating cable through a heavy wall, you need to terminate the radiating cable, using a normal coax jumper through the wall, and continue with a new section of radiating cable on the other side of the wall. If this is not done correctly, the system loss will be increased.
- *Clean the cable frequently*: another practical issue with a major degrading impact on performance is dirt on the radiating cable. Dust from diesel exhaust residue, brake dust from trains, etc., will dramatically increase the system loss. Therefore routine cleaning of the cables will need to be scheduled to preserve performance.

The biggest challenge in using radiating cables is the installation. It is often very expensive and time-consuming. This is especially true if the system is to be installed in an operational tunnel. In this case you might only get access to the tunnel for a few hours every night, while it is closed for scheduled maintenance. Only rarely will you be allowed to close the tunnel, only for the purpose of installing the mobile radio system. Installing radiating cable is a logistical challenge. The cables have to be transported on a special trolley, cable stands pre-mounted and connectors fitted. Everything needs to fall into place in order to be able to install just a few hundred meters of cable during a night shift. Remember, you need to evacuate and clean the site in good time prior to the traffic opening in the tunnel. Tracks need to be inspected in order to approve the tunnel for rail traffic again; it takes just a small cable stand on the track to derail a fast-moving train.

4.8 Tunnel Solutions, Cascaded BDAs

One way to design a tunnel system with radiating cable is to use daisy-chained BDAs to 'ramp up' the signal at certain intervals (Figure 4.25) in order to compensate for the longitudinal attenuation and to keep the coverage level above the required design level.

In this type of application the gain of the BDAs will typically be adjusted in order to compensate for the longitudinal loss of the preceding cable, but not higher – that would just cause noise build-up and degraded dynamic range (see Chapter 7 for noise calculations). This design methodology is sometime referred to as 'zero loss systems'.

4.8.1 Cascaded Noise Build-up

One concern about these types of systems with cascaded amplifiers is the noise increase and limitations in dynamic range. You will need to be careful when configuring the gain of the amplifiers in the system. This is because each amplifier will amplify the signal and the noise of the preceding link and also add its own noise figure to the performance. If you are not careful there will be a severe cascaded noise build-up in the system. Noise calculations and guidelines on how to design amplifier systems and control the noise are covered in Chapter 7. In practice, you should at a maximum cascade three or four BDAs, but this has to be verified in the link budget of the actual project.

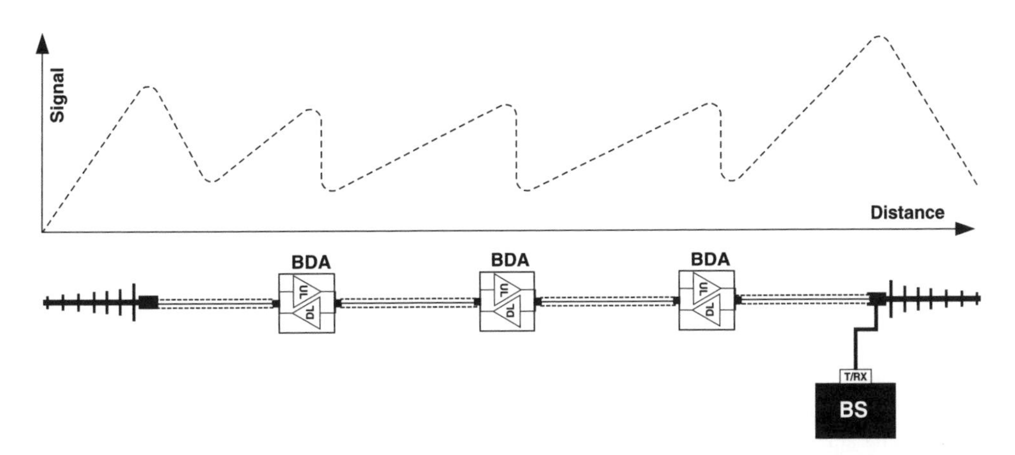

Figure 4.25 Radiating cable fed using in-line BDAs (cascaded system), and the corresponding RF signal

Cascaded System, Principle

The principle of the 'zero loss system' is shown in Figure 4.25, where a base station drives the first section of radiating cable and the handover antenna, before the longitudinal loss attenuates the service level too much. The first BDA amplifies the signal and feeds the next section of radiating cable. This section then drives another section of cable, and so on. The two handover antennas will to some extent overpower the signal from the radiating cable when the mobile is close to these antennas. Note that the BDA works only in one direction (one direction for the downlink, the opposite direction for the uplink), unlike the 'T-feed system' (Section 4.9), that radiates out in both directions of the radiating cable.

Power Supply to the BDAs

To save on project cost, limit failure points and ease maintenance, centralized DC power is injected into the radiating cable near the base station. This DC will power all the BDAs on the line. Make sure that all the components on the line are able to handle the DC power, and are able to pass the DC when needed.

Supervision and Stability

One major week point in this type of system is that, in the case of an amplifier malfunction or if a cable is cut, all signals could be lost in the remaining part of the tunnel system. Special BDAs for tunnels will often have a 'bypass' function, to make sure that signals are passed on 1:1 with no amplification in case of an error. In most cases this will make only a small difference; the signal to the next BDA will be too low for the system to perform due to the lack of signal.

Specialized inline BDA systems are available that are able to communicate internally via the coax. Each active element will be assigned its own address, and the system will be able to communicate internally, perform auto calibration and have a central point where the status of the system, performance and alarming can be monitored and controlled locally or remotely via an RF modem or IP connection.

4.8.2 Example of a Real-life Cascaded BDA System

The measurement shown in Figure 4.26 is a real-life measurement of a cascaded BDA or leaky feeder system in a tunnel, showing a section of the tunnel with three BDAs.

You can clearly see the HO antenna that is placed just before the rightmost BDA, the 'peak' on the signal level, and on the left side of the graph is a similar peak from an antenna. The measurement shown is over an 800 m section of the tunnel, from one metro station to the next station. The radiating cable was placed in a 5 m wide circular concrete tunnel, carefully aligned with the placement of the windows of the train. Measurements were made at 2 m from the radiating cable, in the center of the track during a walk test by foot in the empty tunnel. The distance between the BDAs was about 260 m, and the radiating cable was a heavy $1\frac{1}{4}$ inch.

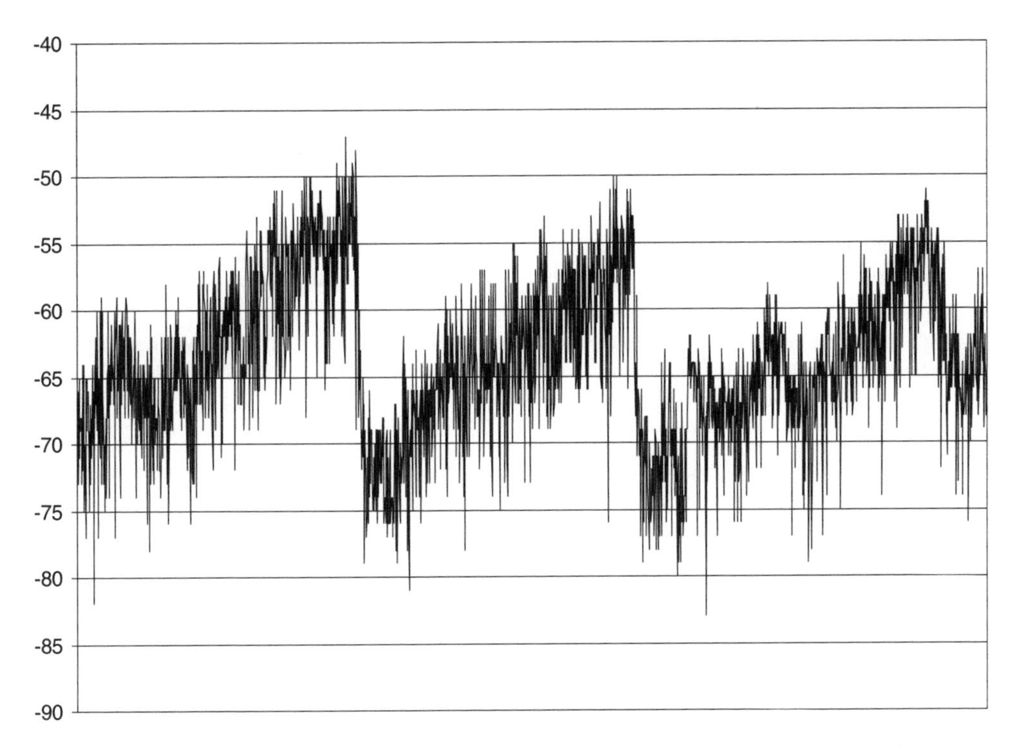

Figure 4.26 Real-life measurement of a cascaded BDA system in a metro tunnel

The design level was −75 dBm at 98% GSM1800 in the empty tunnel (penetration loss into the train was verified by measurements to be only 5 dB). The longitudinal loss per 100 m was 6 dB according to the specification of the cable. From the measurement, the longitudinal loss can be estimated to be about 15 dB. This confirms the information on the datasheet with regards to the longitudinal loss performance.

Note that the measurement was conducted during a walk test, at quite a slow speed compared with the speed of the final users (inside a metro train). This slow-moving measurement tends to emphasize the fading, as can be seen in Figure 4.26.

4.9 Tunnel Solutions, T-Systems

One of the most commonly used designs is to use a 'T-feed' system. This system has several advantages over the cascaded system from Section 4.8. Signal distribution to the BDAs is done in parallel with the radiating cable system. Each BDA is fed an input, and provides output in both directions of the radiating cable. In the case of BDA malfunction, or a broken radiating cable, most of the system will actually still work.

Centralized control and monitoring are done at the 'fiber unit' co-located with the base station. This main unit of the tunnel system also feeds the base station with external alarms in case of malfunction. Remote configuration can be done via an RF modem or IP for easing trouble-shooting and maintenance.

Even though the optical transmission and converters add noise to the system, T-systems will often be superior in performance on the uplink compared with cascaded systems.

4.9.1 T-Systems, Principle

The principle for the 'T-feed' system (as shown in Figure 4.27) is that the base station connects to an optical converter that converts the RF into optical signals. Using optical transmission the signal can be transmitted over relative long distances with only limited degradation compared with RF transmission over coax.

Each BDA has an optical interface and converts the signal back to RF, and then a power amplifier that transmits the radio power in both directions on the radiating cable. Naturally the BDA has a reverse amplifier and RF-to-optical converter for the uplink signal.

The example in Figure 4.27 shows the principle of a commonly used "daisy-chain" distribution of the optical signal, from BDA to BDA, thereby reducing the need for optical fiber to only one link. This is a big cost saver if you need to rely on renting pre-installed 'dark fiber' from a third party.

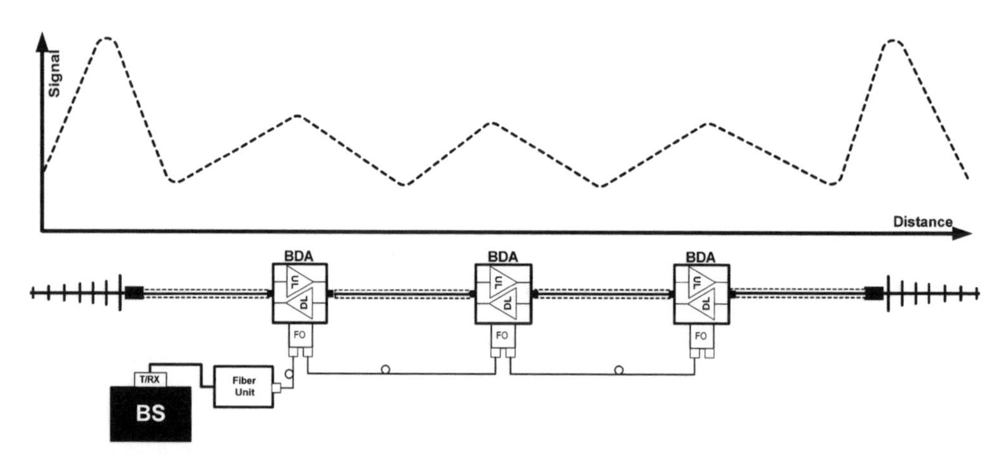

Figure 4.27 Radiating cable fed by a 'T-system'; the signal is distributed via optical fiber and fed via a "daisy-chained" optical fiber system to each repeater

Better Noise Performance than Cascaded BDA Systems

Even with the optical links, converters and amplifiers, the 'T-feed' system is often to be preferred to the cascaded BDA solution (Section 4.8). The reason is better noise control, mainly due to the fact that the cascaded noise build-up is avoided because the BDAs do not feed each other, thus ramping up the noise figure.

4.9.2 Example of a Real-life T-System with BDAs

An example of the signal in a tunnel using a T-system for signal distribution can be seen in Figure 4.28. The measurement shows about 2.6 km of a 4 km-long tunnel. As you can see

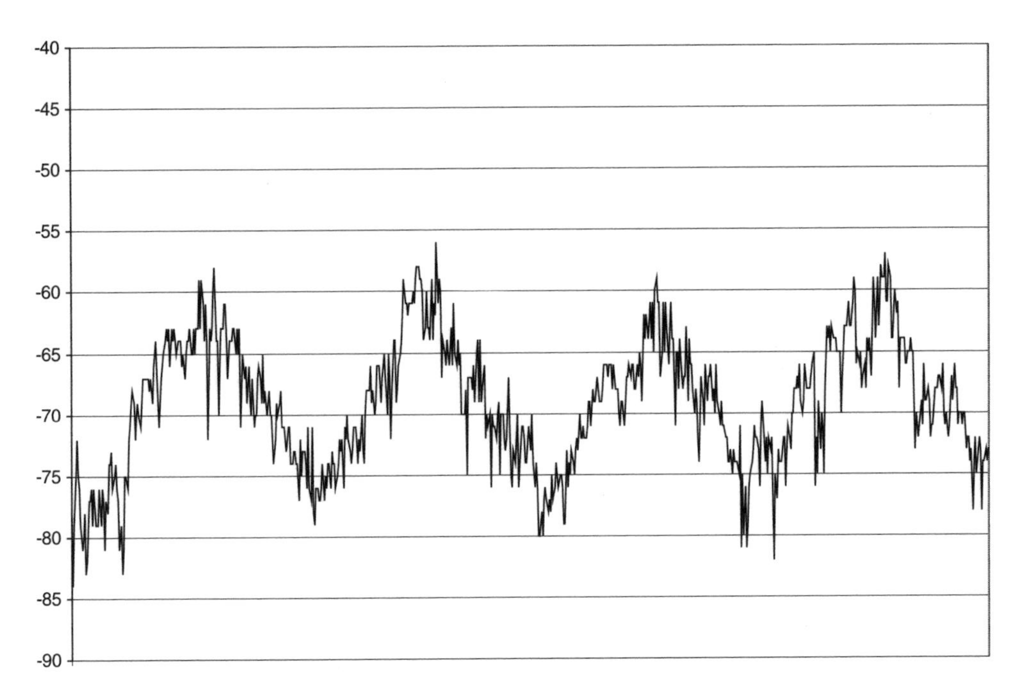

Figure 4.28 Measurement of the GSM900 signal on radiating cable fed by a 'T-system', optically distributed BDAs

there are four BDAs servicing this section of the tunnel, each supporting 660 m of tunnel, 330 m in each direction.

The system was designed using a $1\frac{5}{8}$ inch radiating cable installed in this large square-cut (about 4×6 m) concrete rail tunnel. The measurement in Figure 4.28 was done sitting on top of an open diesel-operated rail trolley at 80 km/h (while I tried to cling on to my PC, measuring receiver and a faint hope of continuing my life!).

The design level for this tunnel was -75 dBm at 98% area on GSM900 for the downlink, measured in the open tunnel. The measurement clearly shows that there are areas that are about 5 dB lower than the design level. In particular, the handover zone, shown in the leftmost part of the measurement, caused problems. The main problem turned out to be caused by the installer having used $\frac{1}{2}$ inch coax instead of the specified $\frac{7}{8}$ inch coax cables from the BDA equipment rooms to the actual tunnel (no equipment was allowed in the tunnel itself; this was restricted to emergency rooms placed along the tunnel), and the cable route was longer than expected. According to the installer it was much easier to install the thin cable! It probably was, but this resulted in about 10 dB extra loss, not accounted for in the link budget. This is a perfect example on how the practicalities of tunnel radio planning can hit you like a ton of bricks!

Do include a level of safety margin, but on the other hand, overdesigning will be expensive. However, the reality is that in a tunnel most of the cost is related to installation work. Therefore, 10% extra on equipment expenses will have close to no impact on the final installed project price.

By the way, it also turned out that the penetration loss into the train was about 8 dB more than expected! So do remember to measure the penetration loss on the actual trains that are supposed to operate in the tunnel, at the service frequency, and using a radiating cable outside the train, preferably in the actual tunnel, prior to committing to any design.

The radiating cable part of the system in the tunnel was actually performing as specified. As can be estimate from the measurement, the longitudinal loss is about 15 dB per 330 m. This is spot on the data from the manufacture of the cable. However, unfortunately the extra loss from the coax cable caused some concern and reinstallation of new $\frac{7}{8}$ inch cable (at the expense of the installer).

4.9.3 T-Systems with Antenna Distribution

Owing to the installation challenges using radiating cable and time to deployment, radio distribution using yagi antennas can be a viable option. If the distance between the tunnel wall and the vehicles allows, an alternative to the radiating cable could be to use distribution via yagi antennas. This still uses the 'T-feed' principle, but instead of a leaky coax you connect the BDAs to yagi antennas installed 'back-to-back', as shown in Figure 4.29.

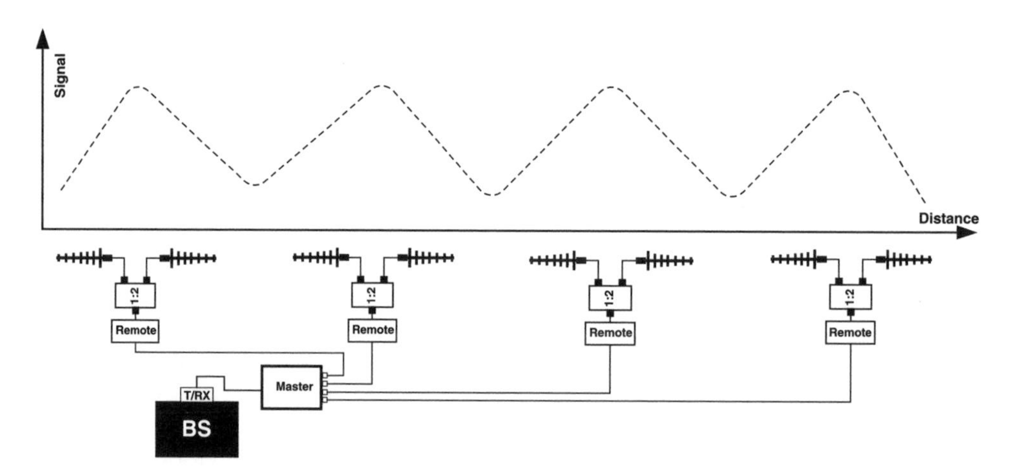

Figure 4.29 Yagi antenna distributed tunnel system

You need to be careful though; vehicles, especially trains, will to some extent act as a moving 'RF-blocker' for the signal. Make sure that there is only one train in each section of the tunnel at one time.

For the link budget calculation, do realize that the longitudinal loss of a train can be very high. Using yagi antennas, clearance between the train and tunnel is important in order to get the signal into the users inside the train, by reflections through the windows. Luckily many metro trains are quite 'open' from an RF perspective.

If the clearance is very tight and you have to rely on the radio signals to penetrate along the longitudinal axis of the train, the penetration loss can easily exceed 35 dB to the far end of the train opposite the yagi antenna! This will depend on the RF frequency and how many passengers there are, i.e. how much 'body loss' there is, inside the train. In these cases you need to make sure that the train constantly is service by two yagi antennas, one from either direction.

Installation Challenges

Special yagi antennas are available, mechanically designed for installation inside tunnels. These antennas will often be totally enclosed in a plastic tube (random) so no dust can degrade the performance over time.

Mechanical stress and reliability is a real concern, especially in rail tunnels, where there is mechanical stress due to air pressure and vibration caused by the moving train inside the tunnel. Special 'tunnel yagi antennas' will often have mechanical fixtures in both ends, making installation more stable.

Be careful with any hardware that you install inside a tunnel. If the equipment detaches from the wall, it can be a hazard for traffic and the users of the tunnel.

Handover Design with Yagi Antennas in Tunnels

One major challenge with yagi-distributed systems is the handover zone design, if the system uses multiple cells. Refer to the following section for more details on this issue.

4.10 Handover Design in Tunnels

Large tunnel systems often need to be divided into several cells, especially if it is a rail tunnel system over a large area. Therefore you need to plan an adequate handover zone between the overlap of the adjacent cells that will insure a successful handover of the traffic. There are some basic parameters and tricks that are essential for a successful design handover design.

4.10.1 General Considerations

HO Zone Placement

In a metro rail application it is worthwhile considering having the HO zone planned at the station area, preferably in a well-defined zone, with clear definition of the level difference of the two cells at the center of the platform. This approach will have several benefits:

- *Slow moving traffic*: a train fully loaded with mobile users will severely load the signaling for the two cells handing over all the calls. The network will have to hand over all the users in the traffic within a limited time window if the train is moving at high speed (relative of course to the planned handover zone). Therefore it is preferred not to do the handover at high speed; you need a margin for 'retries' (GSM) if the first handover fails.
- *Capacity overlap*: having the HO zone at the station actually means that there will be capacity overlaps from both cells at platform level where most passengers are placed. This helps in peak load situations as both cells will be able to carry the traffic.
- *Redundancy of the radio service*: if one cell is out of service, there will to some extent still be coverage from one cell at least at the platform level, even if the preceding section of tunnel coverage is missing.
- *UMTS*: for UMTS you need to be careful not to create extensive soft HO zones; ideally only the platform should be served by both cells, not the whole metro station.

- *HO Zone Size*: remember that the HO zone size will depend on the speed of the train, the HO evaluation and the processing time, and an additional safety margin should be added to make further HO attempts, should the first fail.

4.10.2 Using Antennas for the HO Zone in Tunnels

As tempting as it might be, it is not recommended to use antennas as the primary source of signal in the HO zone. Antennas should clearly be used in the station hall area and pedestrian tunnels, but in the HO zone for the train the signal from two antennas is much too unpredictable and too easily affected by the physical presence of the trains or cars occupying the tunnel. It is very difficult to install antennas in the tunnel, in such a way that they cover a well-defined handover zone with clear signal margin difference between the cells in order to make a well-defined handover zone. If needed, the solution in Figure 4.31 could be considered.

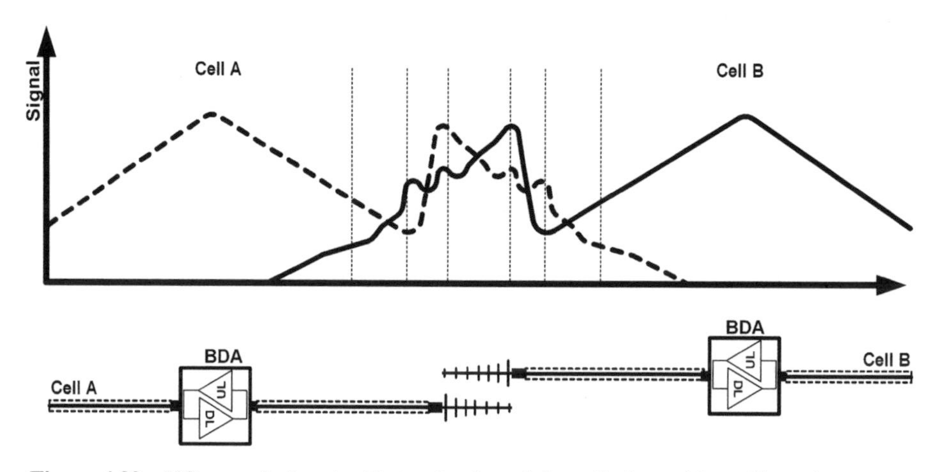

Figure 4.30 HO zone designed with termination of the radiating cables with yagi antennas

The example in Figure 4.30 shows the potential problem: there is no clear definition of the HO zone due to the fading pattern of each antenna. Actually there is a potential 'ping-pong' HO, where the mobile will HO actually do five HOs in each direction, before finally settling on the new cell.

This will increase the signal loading by a factor of 5, and increase the potential for dropped calls and quality degradation ('bit stealing' on GSM for HO signaling can be heard on the audio of the call during HO).

Example, Problems in an Antenna-controlled Handover Zone in a Tunnel

A real-life GSM example showed the extreme case of the problem (Figure 4.30), when it turned out that there was an average of seven HOs in one direction, and nine in the other. This multiplied the signaling load of the two cells when a train full of mobile users passed through the handover zone. After many nights of failed optimization sessions of the handover parameters, the problem was finally clarified and solved by resorting to the design that can be seen in Figure 4.33.

Controlling the HO Zone with Antennas

In some installations it is not possible to install radiating cable in the tunnel, so you simply must have the HO zone work, even with antennas. The trick is to have a clear-cut HO margin as a specific point, preferably symmetrical in both directions (as shown in Figure 4.31). The clear-cut handover is assured if both signals shift their relative signal strengths a certain point, and if both signals use the same signal path, in this case the same antennas.

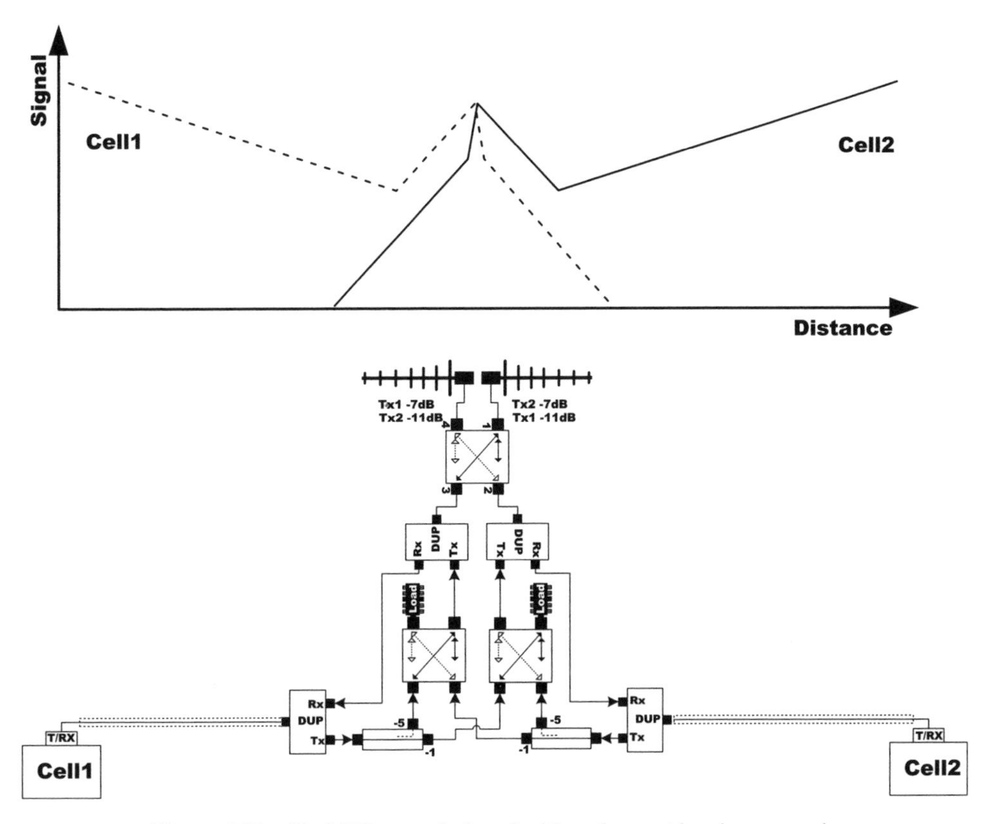

Figure 4.31 Yagi HO zone designed with a clear-cut handover margin

The HO zone in Figure 4.31 works by offsetting the two transmit signals with a specific margin, in this case 4 dB. By choosing other values for the tappers, you can design the HO zone to the exact margin you want.

The signals from the cells are separated in DL and UL; the DL is offset in level, and fed to both antennas. The receive signal path is the same from both directions or cells. Therefore the HO trigger must be the DL signal level. This design of HO zone will have a good margin for handover retries, should the first attempt fail.

In a tunnel where there is traffic in both directions, the HO zone is designed to be symmetrical. However, you could do an asymmetrical HO zone design if that is preferred in a one-way tunnel.

4.10.3 Using Parallel Radiating Cable for the HO Zone

A commonly used method for designing the HO-zone on a radiating cable system is to overlap the cable from the two cells (as shown in Figure 4.32). This is a valid design strategy, but it will take some distance (time) before the difference between the two signals is large enough to trigger the HO margin. In addition, you need to apply the time for the HO evaluation and execution, and time to retry the handover if the first attempt fails (GSM).

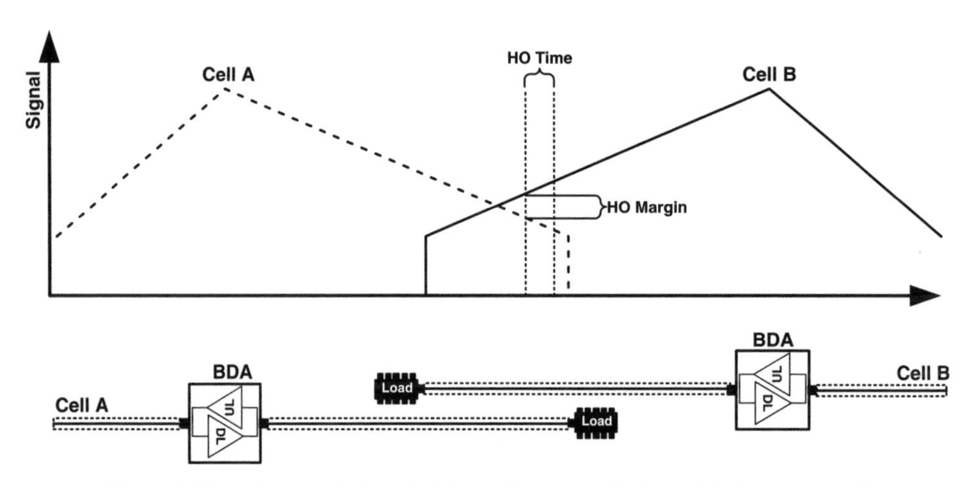

Figure 4.32 HO zone designed with overlapping radiating cable from each cell

Remember that the serving signal level still needs to fulfill the design criteria, so in practice you do not want to get below this level before, during or after the HO, or the following attempts should the first HO fail. Note that the signal from the one cell will be gone once the terminator has been passed.

In reality this means that you will need to start the HO procedure 6–10 dB before you reach the minimum design level, so the two adjacent cells in the handover zone will not be able to service the same distance of radiating cable as the other cells in the system.

In particular, if you need to place the HO zone in between stations in a metro, where the train is moving at full speed, you will need a long (expensive) overlap to the two cells per cable in the tunnel, in reality doubling the installation cost in this part of the system.

The overlapping cable handover zone is a possible solution, but it can be tricky to design and optimize due to the long overlap of the two cells. The two end amplifiers will be located relatively close to each other in order to be able to power the long handover zone. One big advantage, though, is the galvanic isolation between the two systems.

4.10.4 Using a Coupler for the HO Zone

The design that will give the best defined HO zone is to use a RF coupler to connect the two cells on the same leaky coax cable (as shown in Figure 4.33). Thus the two cells will have the exact same fading pattern and coupling loss to the mobile. However, the system loss will be different; this is the result of the coupler 'offsetting' the longitudinal loss at the point where the two cells are joined.

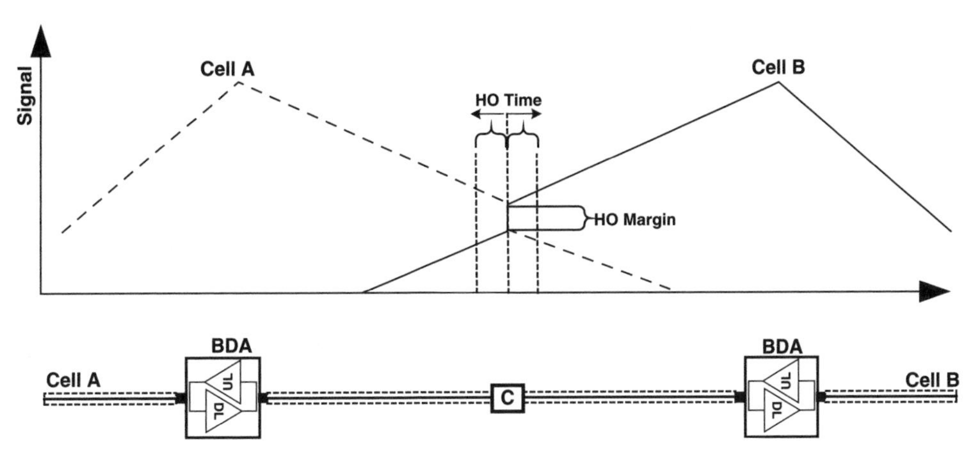

Figure 4.33 HO zone designed connecting the two cells via a coupler

Normally a 3–6 dB coupler is used, giving a 6–12 dB signal offset between the cells. The HO offset is instantly applied to the serving and adjacent cell once the location of the coupler is passed in the tunnel. The result is a well-defined HO point. You still need to make sure that the HO is performed before reaching minimum required signal level.

The roll-off of the signal is not as steep as for the overlapping cable example in Section 4.10.2, but you need to incorporate a safety margin that allows for handover to be retried without the serving cell falling below the design level. There is only one problem worse than a handover problem in a tunnel, which is a handover problem in a tunnel at low signal levels!

This approach has been proven during many practical installations, and is highly recommended. But you have to bear in mind the DC/grounding issue when connecting the two cells, and you might be required to install a 'DC separator' that isolates the DC but passes the RF signals

4.11 Designing with Pico and Femto Cells

4.11.1 The Pico/Femto Cell Principle

The pico base station (as shown in Figure 4.34) is a very effective tool for deploying in-building coverage at low cost on a standalone basis. It is a small device with integrated antenna, typically designed for visible wall mounting in an office environment. Most pico cells will use IP for transmission backhaul, making the pico a very fast solution to implement.

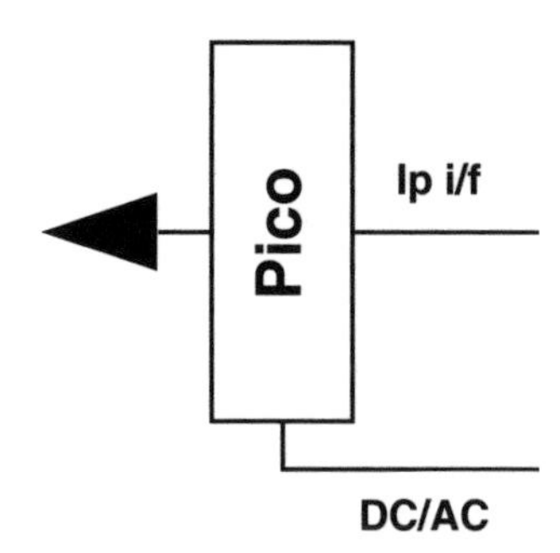

Figure 4.34 Pico/femto, low-power base station with internal antenna

The output power from the pico is sufficient to cover about 20–40 m distance depending on the service requirements, radio design level and the environment where the pico is installed. In most cases the Pico cell will need a local power supply, although some types can be powered over the Ethernet cable which also supplies the IP connection. A local power supply can be expensive, and often you are not allowed to plug it into the existing power group in the building, but will have to install a dedicated AC group for all of the pico cells in the building.

Femto Cells

Smaller versions of the pico cells are starting to emerge in the market, such as the femto cell (for UMTS). The femto has very limited output and capacity and processing power, and is typically only applicable for residential use. The low radiating RF power and autotuning features in the femto make it to some extent 'self-configuring' in the network. However careful tests should be done before deploying femto cells on a large scale.

Many mobile users are likely to compare the femto cell with their WiFi access point, and might be tempted to 'improve' the coverage around their house by installing the femto close to the window, beaming out coverage into the surrounding area. Consider this scenario: in a building with 50 rental flats each having a UMTS femto cell beaming out coverage on the same frequency to cover the common outdoor facility even with the low power of the femto cell it might cause a considerable noise load in the surrounding network.

However, even with these potential issues, the concept of using Femto cells in the network for covering some segments of the market is often attractive from a network revenue point of view, but it is a fine balance.

The radio planning principles are the same as for the pico cells, just radiating with a lower power.

4.11.2 Typical Pico Cell Design

The typical pico application is a small to medium-sized office building (as shown in Figure 4.35), where two pico cells are covering an office floor. This could be repeated throughout several floors. In the case of a building with several floors, be sure to place the pico cells 'symmetrically', that is, in the same place on each floor – so that the one pico

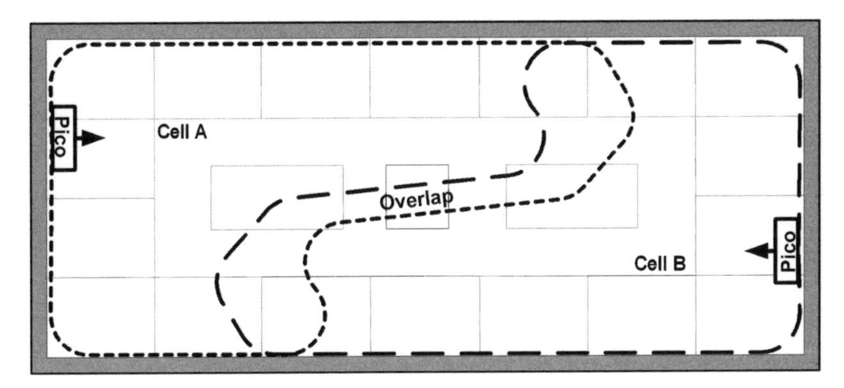

Figure 4.35 Two pico cells covering an office floor

cell will not leak coverage to adjacent floors, creating soft handover zones or degraded HSDPA capacity. Remember that pico cells are individual cells, therefore the 'interleave' concept (see Section 5.3.9) cannot be applied.

Coverage Overlap

As for any other multicell environment, you need to design for some overlap of the cells to insure coherent coverage and sufficient HO zone, but try to minimize the HO zone, by using heavy structures, firewalls, elevator shafts, etc., to separate the cells.

Capacity Overlap (GSM)

The coverage overlap of the Pico cells will, for GSM, give capacity overlap and trunking gain in the overlapping area (as shown in Figure 4.36). In the area where only one pico is covering, the capacity will be 2.15 Erlang (GSM, 7TCH 0.5% GOS, see Chapter 6 for more details), and in the area where both cells are overlapping the users will have access to both cells, by the means of traffic controlled handovers; if one cell is loaded, they can set up calls on the other cell.

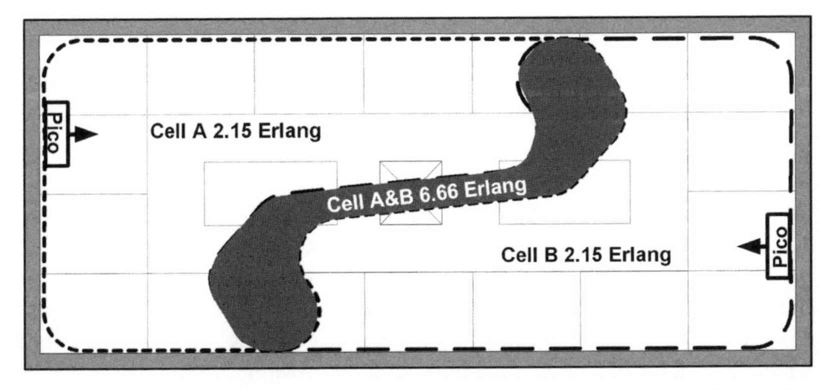

Figure 4.36 Two GSM pico cells with overlapping coverage and capacity trunking

This capacity overlap of the cells yields 6.66 Erlang in trunked capacity in the overlapping area; refer to Chapter 6 for traffic calculations.

Hot-spots

The capacity overlap makes it possible to cater for GSM traffic hot-spots in the building by deploying overlapping pico cells in these hot-spot areas. However, in reality these hot-spots in the building tend to be dynamic and do not always occur in the same place in the building. This makes GSM hot-spot design with pico cells a challenge.

Coverage Overlap UMTS

In a UMTS Pico cell application (as shown in Figure 4.37) with the same two Pico cells overlapping with a margin of 3–6 dB, there will be a large soft-handover zone. If you are not careful and do not limit the overlap of the Pico cells using internal heavy structures in the building when planning the pico cell deployment, the overlap might cannibalize the capacity of the system; remember that this is two different cells so you need both radio processing and transmission capacity on both cells for one call in soft-handover.

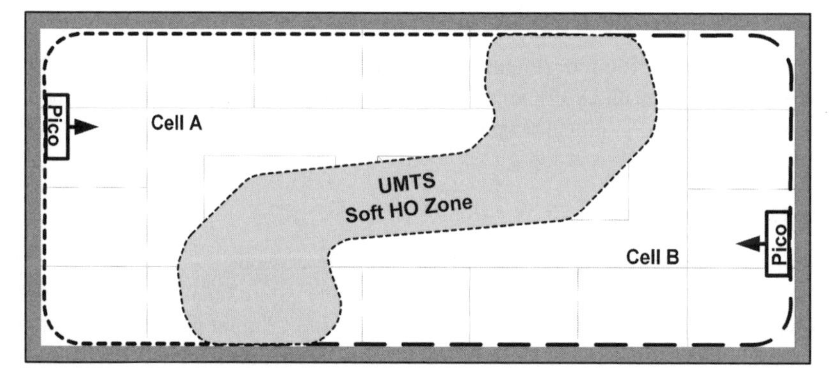

Figure 4.37 Pico cell overlap on UMTS creates a soft handover zone

Coverage Overlap HSPA

The two indoor pico cells (as shown in Figure 4.38) will typically operate on the same frequency. Unlike one carrier assigned to UMTS R99 traffic, HSDPA does not use soft-HO. The problem is that the cells will cause co-channel interference to each other ('inter-cell interference').

This will severely degrade the HSPA performance in the building. There will be HSPA throughput in the overlapping area, but the speed will be limited as the two signals become more and more equal. When relying on pico cells for indoor HSPA applications, it is of the utmost importance to use internal structures and firewalls to separate the cells and to avoid coverage overlap in areas of high traffic. With careful planning, pico cells could be a very strong and viable tool for indoor applications, but try to avoid using more than one pico in an 'open' environment, with high traffic density.

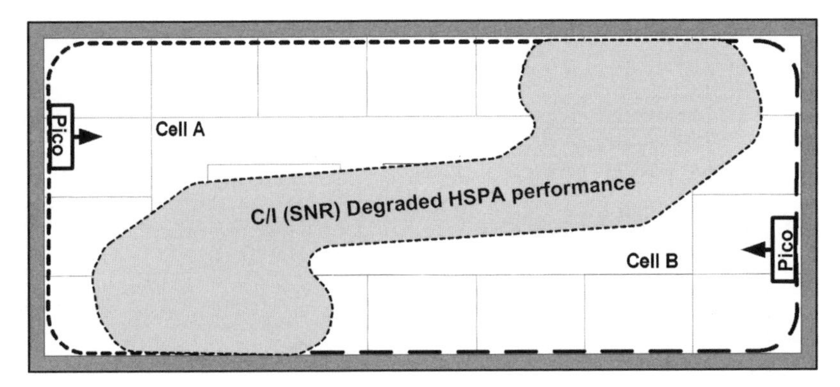

Figure 4.38 Pico cells causing 'inter-cell interference', degrading the HSPA performance in the building

4.11.3 Extending Pico Cell Coverage with Active DAS

The pico base station is a very effective tool for deploying in-building coverage at low cost on a standalone basis. However, there are circumstances in which the coverage area of the pico base station needs to be extended without increasing the capacity of the system, in warehouses (as shown in Figure 4.39) and residential building towers for example.

If you combine the small active DAS system from Section 4.4.3, you can extend the coverage footprint of a single pico cell. The reason for selecting the active DAS is the 'no loss concept' of the active system. The attenuation of passive distribution will degrade the performance of both downlink and uplink.

In GSM applications you can also increase the capacity by connecting two pico cells or more to the same active DAS, and letting traffic controlled handovers take care of the traffic flow between the two cells. You *do not* want to combine two or more pico cells for UMTS or HSPA into the same DAS; this will turn the whole cell into a soft-HO zone for UMTS or give co-channel interference in 100% of the area and thereby degraded HSPA performance.

4.11.4 Combining Pico Cells into the Same DAS, Only GSM/DCS

In some pico cell applications, there are scenarios where a highly mobile user community within a building that has several pico base stations can overwhelm the capacity of a single pico base station system (hot-spot blocking). An example is when employees congregate in the lunch-room. This situation can be improved by taking the same number of GSM pico base station systems and feeding them into an active DAS system to cover the same area. The results is all the radio channels being available in all of the building, leading to higher trunking efficiency, more capacity available over a larger area (as shown in Figure 4.40) and hot-spot capacity throughout the building.

Some manufacturers of GSM pico cells let you combine the picos into one logical cell with only one BCCH, thus freeing up capacity on the individual Pico cells and boosting the trunking efficiency even more.

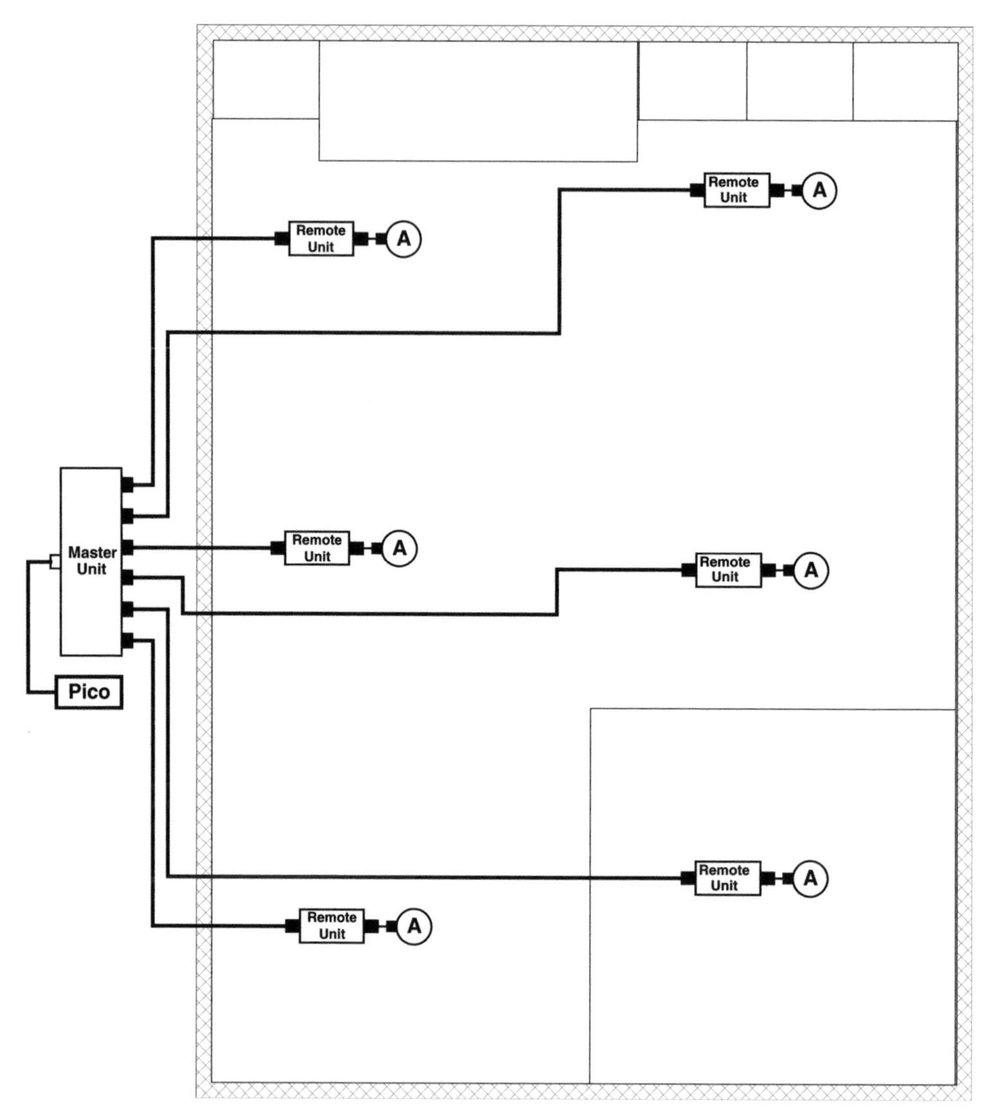

Figure 4.39 The coverage of a single pico cell can be extended, using active DAS to distribute the signal over a wider area

Only for GSM!

Do not combine two or more pico cells for UMTS or HSPA into the same DAS; this will turn the whole cell into a soft-HO zone for UMTS or give co-channel interference in 100% of the area, degrading HSPA performance on the whole coverage area.

4.11.5 Cost Savings When Combining Capacity of GSM Pico Cells

There is a big potential for cost savings by combining the capacity for two or three pico cells rather than deploying several more pico cells to cover the same area. This saves on the

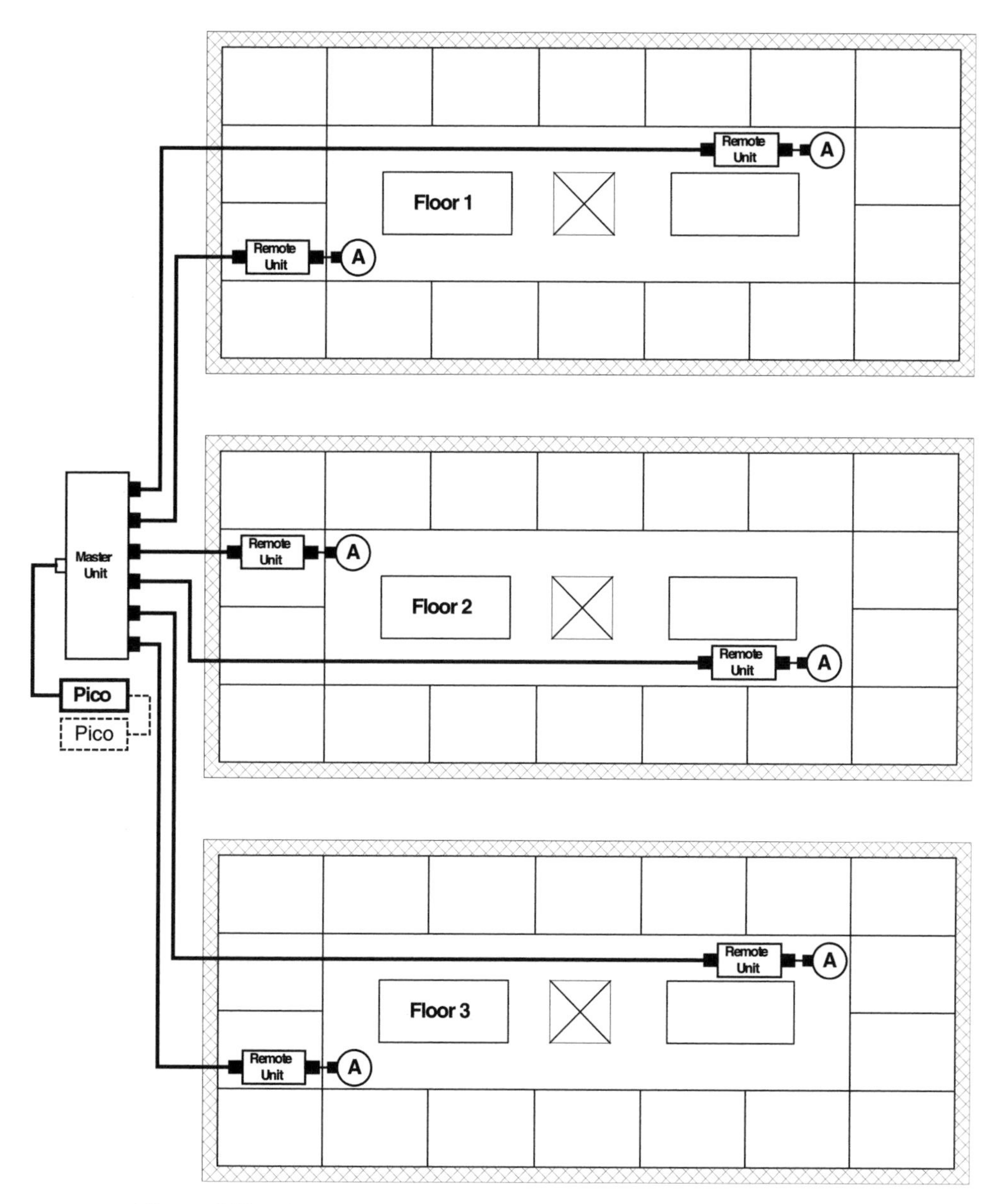

Figure 4.40 Pico cells distributed via a small active DAS over three floors

number of pico base stations needed to cover a building and provides cost savings on backhaul costs due to:

- Better trunking efficiency of the pico BS (more efficient use of the deployed radios, see Section 6.1.9).

- Saving on BSC cost (need to support fewer BTSs in the same building) – each BTS is an 'element' on the BSC, and subject to annual license fee.
- Saving on transcoder and interface costs.
- Saving on MSC costs.
- Saving on ASDL costs.

Fast deployment will provide savings because of:

- The concept of distribution pico cells via a small active DAS providing the mobile operator with the ability to react quickly to coverage requests from customers.
- Less churn.
- More revenue.
- Easy and fast extension of the coverage when new areas in the building need to be added to the system.

Operational savings will come from:

- The uniform coverage from the DAS system, which is easy to plan, implement and optimize.
- The fewer GSM RF channels used in the building, with less interference leaking out.
- Detailed alarming on the DAS, which will limit down-time.

4.12 Active DAS Data

Active DAS systems consist to a large extent of amplifiers and repeater/BDA systems; therefore it is important for the RF planner to understand the basic data of these system components. Many of these standard metrics are used to benchmark the radio performance of different manufacturers and systems. Make sure that the data you are comparing all use the same standard benchmark reference.

The amplifiers used in active DAS, repeaters and BDAs have to be very linear, in order not to distort the signal and degrade the modulation. The more complex the modulation used, the higher demands are on linearity and performance. This is very important for the higher coding schemes on EDGE, UMTS and especially HSPA. Be careful when selecting equipment used for indoor DAS systems, since performance of the system is often directly related to the price. The most basic parameters and merits are described in this section.

4.12.1 Gain and Delay

Gain

Gain is the amplification of the system (as shown in Figure 4.41), the difference between input signal and output signal power. The power of the output signal is:

$$\text{output signal} = \text{input signal} + \text{system gain (dB)}, \text{ or input} \times \text{gain factor (linear value)}$$

Gain is typically stated in dB. For a system with a factor 2 power gain, for example, 1 W input (+30 dBm) will lead to 2 W output.

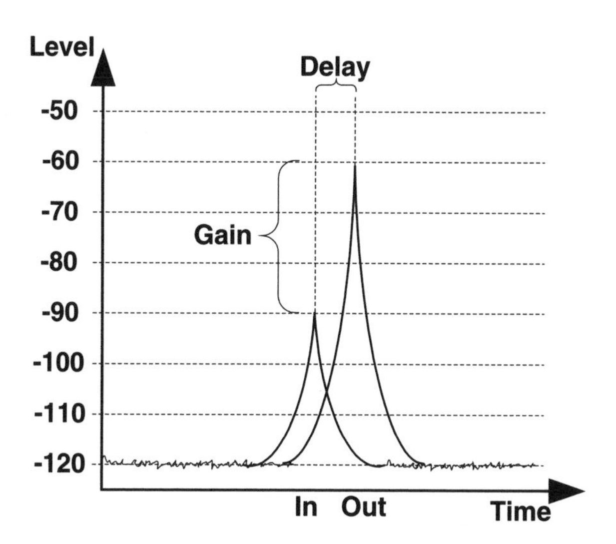

Figure 4.41 Input/output signal of an amplifier vs time

Power gain in dB can be calculated as:

$$\text{Gain (W)} = 10\log(\text{gain factor}) = 3\,\text{dB}, \text{ thus the output power will be} + 33\,\text{dBm}$$

For voltage, gain factors can be converted to dB using:

$$\text{Gain (voltage)} = 20\log\ (\text{gain factor})$$

Gain can be negative; often it will be referred to as attenuation or loss.

Delay

Delay is the time difference between the input and output signal (as shown in Figure 4.41). In practice this will offset the timing for mobile systems. In GSM the timing advance will be offset (increased), and the synchronization window on the UMTS base station network may have to be adjusted wider to accommodate the delay introduced.

Note that you have to include both the delay of the active elements, and the delay of cables (also optical) due to the decreased velocity of the propagation on the cable, for large systems.

dBm

The signal level in RF design is described as absolute power related to 1 mW (in 50 Ω) and expressed in dBm.

$$P\,(\text{dBm}) = 10\log\left(\frac{P(W)}{1\text{mW}}\right)$$

$$P\,(\text{mW}) = 10^{\text{dBm}/10}$$

4.12.2 Power Per Carrier

Amplifiers in active DAS systems, repeaters and BDAs are normally composite amplifiers. This means that the same amplifier amplifies all carriers throughout the bandwidth. All carriers share the same amplifier resource; the result of this is that the more carriers the amplifier must support, the less power can be used for each carrier. The sum of all the powers will remain the same, hence the name composite power.

Every time you double the number of carriers, the power decreases about 3 dB per carrier, depending on the efficiency of the amplifier. For an example, see Table 4.3

Table 4.3 Power per carrier from an active DAS

Number of carriers	Power per carrier
1	20 dBm
2	17 dBm
4	14 dBm
8	10 dBm

4.12.3 Bandwidth, Ripple

Bandwidth

Normally, when defining the bandwidth of an amplifier, it is the 3 dB bandwidth (as shown in Figure 4.42) that is referred to. The 3 dB bandwidth is the band that supports the amplification with a gain decrease of maximum 3 dB.

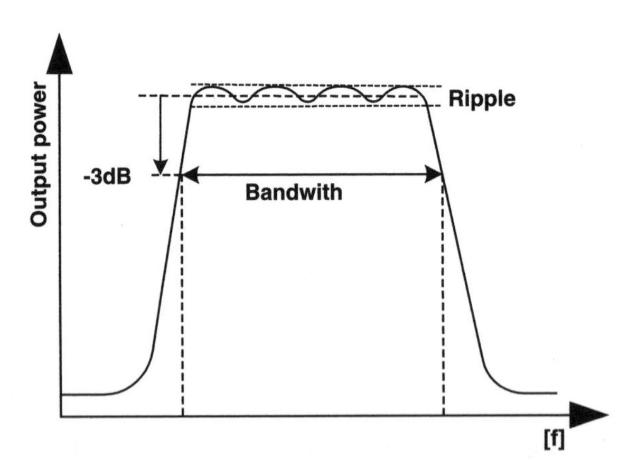

Figure 4.42 Output signal of an amplifier, over the whole operating bandwidth

Gain Ripple

Gain ripple describes the variance in gain over the bandwidth (as shown in Figure 4.42).

4.12.4 The 1 dB Compression Point

The 1 dB compression point (P1dB) is a measure of amplitude linearity. This figure is used for defining output power capabilities (as shown in Figure 4.43). The gain of an amplifier falls as the output of the amplifier reaches saturation; a higher compression point means higher output power. P1dB is at an input (or output) power where the gain of the amplifier is 1dB below the ideal linear gain. P1dB is a convenient point at which to specify the output power rating of an amplifier.

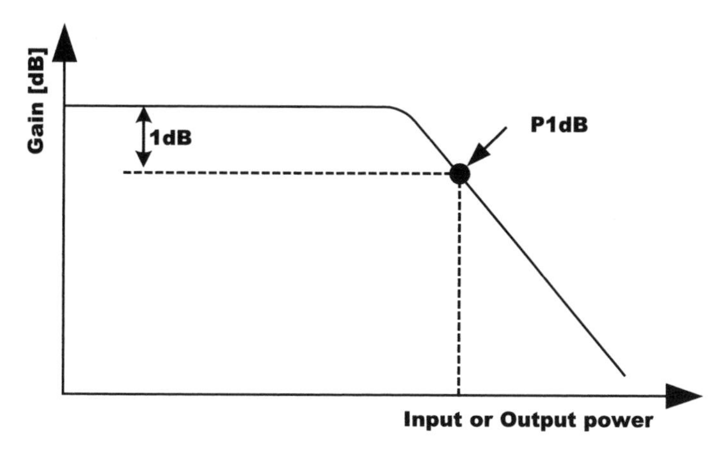

Figure 4.43 The P1dB compression point

Example
If the output P1dB is +20 dBm, the output power from this amplifier is rated at +20 dBm maximum.

Avoid Intermodulation Problems

Reducing the output power below the p1dB reduces distortion. Normally manufacturers back off about 10 dB from the P1dB point: an amplifier with 20 dBm P1dB is normally used up to +10 dBm.

4.12.5 IP3 Third-order Intercept Point

IP3

IP3 is a mathematical term (as shown in Figure 4.44). It is a theoretical input point at which the fundamental (wanted) signal and the third-order distorted (unwanted) signal are equal in level to the ideal linear signal (the lines A and B).

The hypothetical input point is the input IP3 and the output power is the output IP3. IM3 'slope' (B) is three times as steep (in dB) as is the desired fundamental gain slope A.

Unlike the P1dB, the IP3 involves two input signals. The P1dB and IP3 are closely related: roughly IP3 = P1dB + 10 dB.

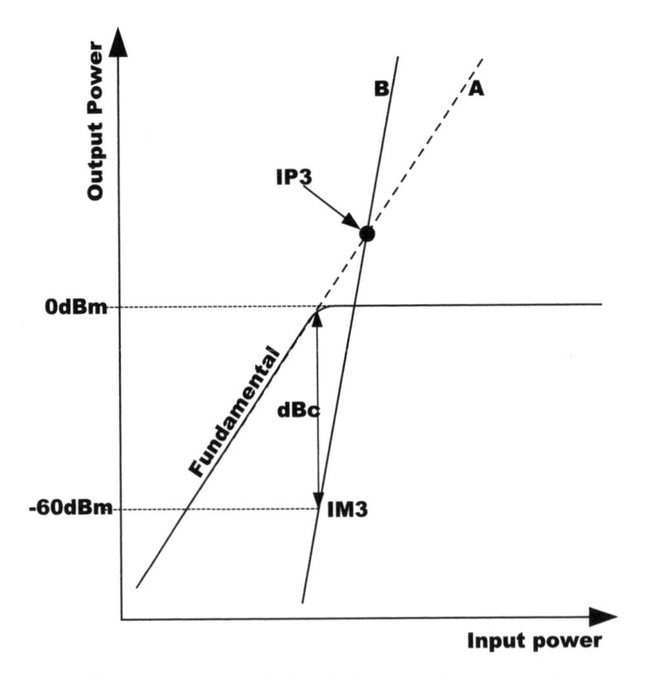

Figure 4.44 The IP3, third-order intercept point

Testing IP3

IP3 is used as a merit of linearity or distortion. Higher IP3 means better linearity and less distortion. The third-order inter-modulation products are the result of inter-mixing the inputs by the nonlinearities in the amplifier:

$$\text{fIM3 } 1 = 2 \times \text{f1} - \text{f2}$$
$$\text{fIM3 } 2 = 2 \times \text{f2} - \text{f1}$$

The two-tone test (as shown in Figure 4.45) is often used to test IP3. Third-order inter-modulation products are important since their frequencies fall close to the wanted signal, making filtering of IM3 an issue.

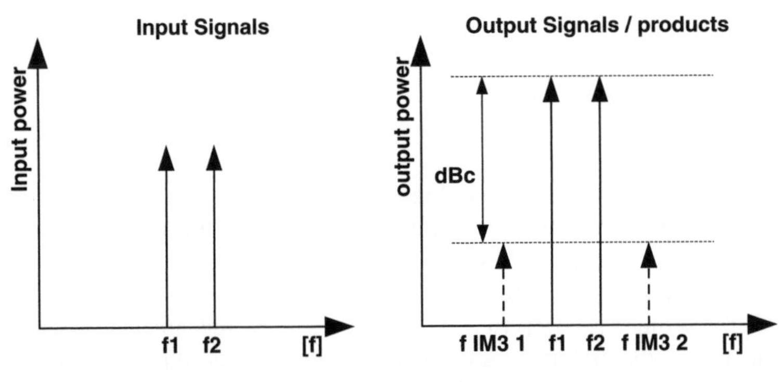

Figure 4.45 The two-tone test of IM3

4.12.6 Harmonic Distortion, Inter-modulation

The harmonic distortion (Figure 4.46) specifies the distortion products created at integers of the fundamental frequency; dBc means dB in relation to the carrier.

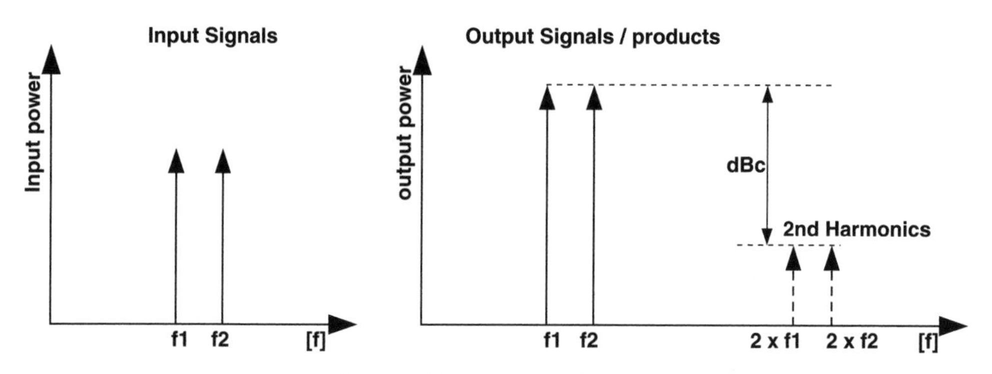

Figure 4.46 Harmonic distortion

4.12.7 Spurious Emissions

Spurious emissions (as shown in Figure 4.47) are emissions, which are generated by unwanted transmitter effects such as harmonics emission or inter-modulation products.

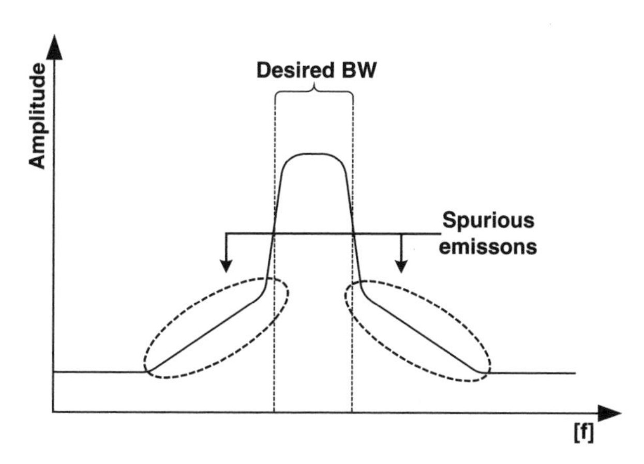

Figure 4.47 Spurious emissions from a transmitter

4.12.8 Noise Figure

The noise figure is the noise factor described in dB, and is the most important figure to note on the uplink of an amplifier system. The NF will affect the DAS sensitivity on the uplink.

The noise factor (F) is defined as the input signal-to-noise ratio divided by the output signal-to-noise ratio. In other words the noise factor is the amount of noise introduced by the amplifier itself, on top of the input noise.

$$\text{noise factor } (F) = \frac{\text{SNR}_{(\text{input})}}{\text{SNR}_{(\text{output})}}$$

$$\text{noise figure(NF)} = 10\log(F)$$

The effect of noise and noise calculations will be described in more details in Chapter 7.

4.12.9 MTBF

Failures are a concern when installing distributed active elements in a building. All components, active or passive, will eventually fail; the trick for the manufacturer is to insure that the expected failure is after the expected operational time of the system (as shown in Figure 4.48). Typically there will be an expected lifetime of a mobile system of 10–15 years. It makes no sense to design systems that can last for 130 years that are too expensive.

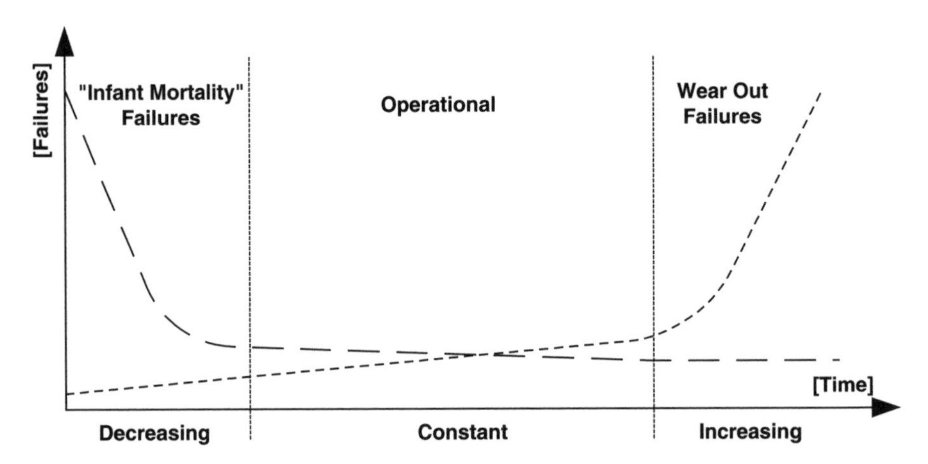

Figure 4.48 The MTBF curve – 'the bathtub curve' – of distribution of failures

'Infant Mortality'

The manufacture will perform a 'burn-in' test of the equipment in order to insure that the 'infant mortality' is cleared from the shipped equipment.

Operational Period

For an active element, it is assumed that during the useful operating life period the parts have constant failure rates, and equipment failure rates follow an exponential law of distribution. MTBF of the equipment can be calculated as:

$$\text{MTBF} = 1/(\text{sum of all the part failure rates})$$

Probability of Failures

The probability that the equipment will be operational for some time T without failure is given by:

$$\text{reliability} = \exp^{-T/\text{MTBF}}$$

Thus, for a product with an MTBF of 450 000 h, and an operating time of interest of 7 years (61320 h):

$$\text{reliability} = \exp^{-61\,320/450\,000} = 0.873$$

There is an 87.3% probability that the system will operate for the 7 years without a failure, or that 87.3% of the units in the field will still be working after 7 years.

This is a useful guideline to estimate number of spares needed for your installed base of equipment.

4.13 Electromagnetic Radiation

EMR is a concern for mobile users all over the world. From time to time there are heated debates in the media, but indoor radio planners need to stand above these often emotional discussions and try to be neutral and objective. Above all, you need to accept and respect that this is in fact a concern for the users, however unlikely you yourself believe the danger to. You need to comply with any given EMR standards and guidelines that apply in the region you work in and make sure that the indoor DAS systems you design and implement fulfill the approved regulations in the country where the systems are implemented. Often you will have to accept that a neutral party conducts post-implementation on-site measurements in the building, and certifies that the design is within the applicable EMR specification.

Different regions around the world use different standards and regulations, and these regulations change over time. Find out exactly what standard applies within your country.

4.13.1 ICNIRP EMR Guidelines

Currently many countries use the guideline laid down by the ICNIRP (International Commission on Non-ionizing Radiation Protection), who are recognized by WHO. This guideline specifies the maximum allowed EMR exposure for the general public 24 h/7 days/ 52 weeks of the year (higher levels are allowed for professional users).

A measurement specification (EN50382) specifies how these measurements should be conducted and that a mobile transmitter must radiate lower than:

- 900 MHz, maximum 4.5 W/m^2.
- 1800 MHz, maximum 9 W/m^2.
- 2100 MHz, maximum 10 W/m^2.

The measurement should be averaged over 6 min.

Guidelines are different for different countries, and will be adjusted over time. Check exactly what applies in your case; some countries apply a standard that is much stricter than the ICNIRP levels.

You cannot relate these values to a specific receiver level in dBm, due to the fact that the ICNIRP levels are power density, and therefore will be the sum of all powers on air within the measured spectrum. Measuring in dBm using a test mobile receiver will only indicate the specific level for that one carrier you are measuring, not the sum of power from all carriers.

Example

This is an example of a measurement at a real-life installation of a DAS, using DCS 1800 MHz 18 dBm radiated from the omni DAS antenna, four TRX with full load:

$$50\,\text{cm distance from the antenna} : 0.630\,\text{W/m}^2 \text{ (average over 6 min)}$$

$$200\,\text{cm distance from the antenna} : 0.0067\,\text{W/m}^2 \text{ (average over 6 min)}$$

The measurement clearly shows that, in practice, you are well below the maximum allowed ICNIRP levels.

DAS Systems Are Normally Well Below These Levels

Typical DAS systems, passive or active, are well below these levels. However, I have seen examples of high-gain outdoor sector antennas, connected directly indoors to a high-power base station, and users able get so close that they can touch the antenna! This is clearly not a correct design; this 'hot' antenna might blast away, providing a lot of coverage in the nearby indoor area, and most likely leaking out interference outside the building. In this extreme case you could exceed the limits. However, common sense must be applied both with regards to good indoor radio planning and for minimizing EMR from the indoor DAS.

When you are being questioned by the users in the building whether it will be safe to work every day underneath the installed indoor DAS antenna, you should always ask yourself the question whether you would want to sit underneath that antenna 24/7 and design accordingly.

Is it Safe?

Well that is the question; the current EMR guidelines are defined with a large margin to any level that might cause any known effect on humans. Often you will be asked if you can prove that these EMR levels are safe. This is an understandable question to ask, but science cannot prove a noneffect, only an effect, and until now no effect has ever been documented when adhering to these guidelines for the design levels.

To use an analogy, we cannot prove it is safe to drink a glass of milk each day for 50 years! Still we drink milk every day. We might be able to prove that milk was hazardous, if we could conclude an effect due to a level of toxin found in the milk, but with no toxin found, there is nothing we can do but to accept that it is most likely safe to drink milk.

4.13.2 Mobiles are the Strongest Source of EMR

Most users inside buildings are concerned about the radiation from the DAS antenna, but it is a fact that the main source of human EMR when using mobile phones is not the indoor DAS antennas, but the mobile handset. This is due to the proximity of the handset to the mobile user. Even if a high-power outdoor base station may generate 700 W and the mobile only 2 W, the determining parameter is the distance to the antenna. Having the mobile close to the user's head will expose the user to more power than would an outdoor base station even as close as 50 m! Indoors the margin is even clearer as shown by the measurement just documented in the example above.

It is a fact that the main source for EMR exposure is the mobile, and the trick is to keep the mobile at the lowest possible transmit power. This is the only way to minimize the exposure of users inside the building. The power control function in the mobile network will adjust the transmit power from the mobile automatically, in order to insure that the received signal level on the uplink at the base station is within the preset level window for minimum and maximum receive levels.

This power control insures that the more attenuation there is on the radio link (buildings, walls etc.) and distance between the mobile and the base station, the higher transmit power the mobile will use to compensate for this loss. Therefore the only way to minimize the EMR exposure is to make sure to keep the attenuation on the link as low as possible. The more indoor antennas you have in the building, the lower the link attenuation is to the users and the lower transmit power indoor users will be exposed to.

4.13.3 Indoor DAS Will Provide Lower EMR Levels

The effect is that, even in buildings very close to an outdoor base station, and where the mobile coverage seems perfect, the mobile will typically operate at or close to the full transmit power. The high downlink power from the base station might provide high signal levels received by the mobile, but the power control depends on the uplink level at the base station, and the mobile transmits at far lower levels than the high power transmitter at the base station.

Indoor antenna systems with low attenuation will help. By deploying an indoor DAS you can create lower path loss for the mobile to the base station, and the mobile will operate on lower transmit power.

Less Radiation with Active DAS

Installing a passive DAS inside a building will to some extent bring down the mobile transmit power, but it is a fact that the mobile still needs to overcome the attenuation of the passive DAS, thus operating on a relatively high power even on an indoor system close to the DAS antenna. The mobile transmit power obviously depends on the attenuation of the passive DAS, but with 20 dB of attenuation on the passive DAS, the mobile has to transmit 20 dB higher power compared with the active DAS from Section 4.4.2.

By deploying an active DAS inside buildings, the attenuation between the mobiles and the indoor antennas is low. The result is that the mobile will run at or close to minimum transmit power. This is due to the active DAS being a 'zero loss' system; it is an active system where all the losses in cables are compensated by small amplifiers close to the antennas. This only applies to pure active DAS; using hybrid DAS the mobile will still have to compensate for the passive losses prior to the HRU (as shown in Figure 4.18).

Note that power control is triggered by the received level, and the noise figure of the active DAS will not cause the transmit power from the mobile to be adjusted up.

Example, Lower Mobile Transmit Power with Active DAS

The low mobile transmit power, using an active DAS system, is evident in this graph (as shown in Figure 4.49). The graph shows the typical mobile transmit power inside an office environment vs the distance to the indoor DAS antenna.

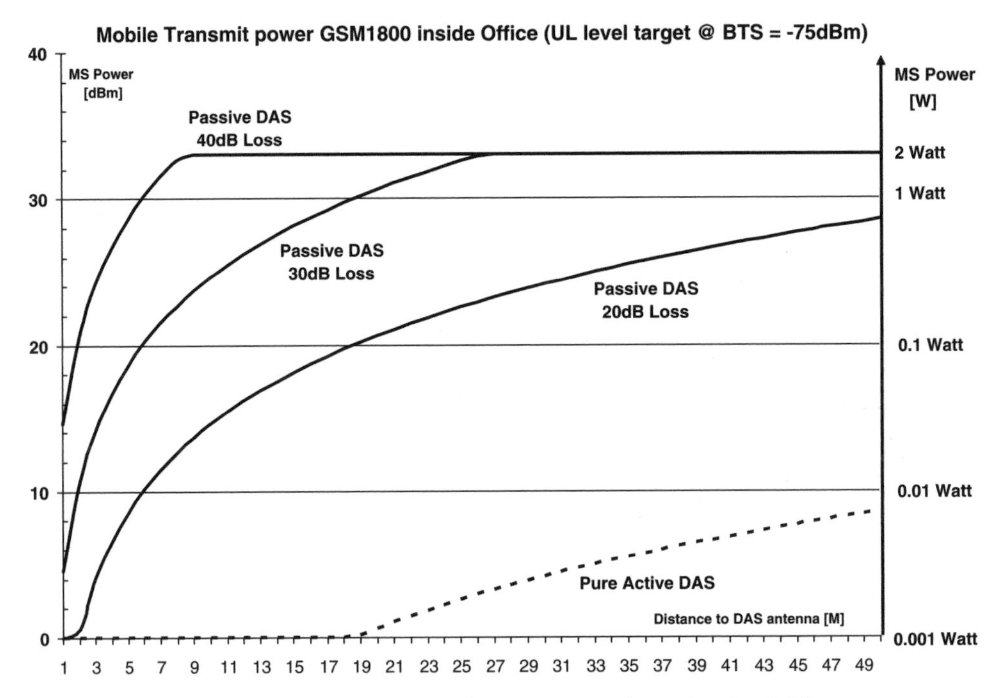

Figure 4.49 Mobile transmit power on passive and active DAS

This is a DCS1800 example, where the minimum received level on the base station is set to −75 dBm, so the base station will adjust the transmit power of the mobile in order to reach this uplink level.

On the graph (as shown in Figure 4.49), the pure active system is compared with typical passive systems with 20, 30 and 40 dB of attenuation from the base station to the DAS antenna.

As shown, the transmit power from the mobile covered by a passive DAS will only use the lowest power level when it is located very close to the antennas, whereas the same mobile in the same type of environment on an active DAS will stay on the lowest possible transmit power even up to a distance of 19 m from the indoor antenna, and stay at a low level compared with the passive DAS.

The mobile connected to the passive DAS will ramp up transmit power even close to the DAS antenna, and this is due to the passive attenuation. This is evident on the graph, even for the 'low loss' 20 dB attenuation of passive DAS.

4.14 Conclusion

It is evident that the mobile connected to the pure active system consistently maintains an output power below 0.01 W, and the mobile connected to a passive system can easily reach 1 or even 2 W (as shown in Figure 4.49). Using a traditional passive distributed antenna system will to some extent help with radiation from mobiles, especially mobiles being serviced by antennas with relatively low loss, close to the base station room. However, the fact is that, due to the losses in the passive system, the mobile has to compensate for the losses in the passive cable dB for dB, resulting in higher transmit power from the mobile and thus higher EMR exposure of the users.

Even if you often need to install an uplink attenuator between the active DAS and the base station to minimize the noise load of the base station, it is clear that the active DAS will keep radiation from the mobiles to the lowest possible power.

Both passive and active DAS will help bring down the transmit power from the mobiles, if the alternative is to rely on coverage via the outdoor macro net. All mobiles have to apply a certain radiation limit (SAR value), so even when operating on the highest power level, no mobile is dangerous.

5

Designing Indoor DAS Solutions

Before starting to design the first indoor solution, it is highly recommended to develop a well-structured and documented workflow for the task ahead. This procedure should include every aspect of the process from start to finish. This will insure a uniform workflow from design to design, and make sure that all these solutions (investments) follow the same process, helping to prioritize the projects and the investment.

5.1 The Indoor Planning Procedure

Operators need a well-structured procedure to evaluate the business case and implementation process of indoor solutions. Often the need for an indoor coverage solution in a particular building is initiated by the sales and marketing department of the operator responsible for that particular area or customer. The procedure must include all aspects of the process, in order to make it visible for all working within the operator: how, why and when an indoor DAS is to be implemented; who is responsible for what part of the process; what documents are needed; and the general workflow of the process.

There are many valid methods to use when organizing the workflow; one typical structure could be like the one shown in Figure 5.1

5.1.1 Indoor Planning Process Flow

Briefly, these are the main parameters: in- and output of the different parts of the process (as shown in Figure 5.1).

Input from the Sales Team/Key Account Manager

The process starts with the requirements from the potential customer, preferably at an early stage, before the customer starts to use mobiles on a wider scale. The sales team then provides an application to get approval to implement a dedicated solution. There must be a clearly defined revenue goal.

Indoor Radio Planning: A Practical Guide for GSM, DCS, UMTS and HSPA Morten Tolstrup

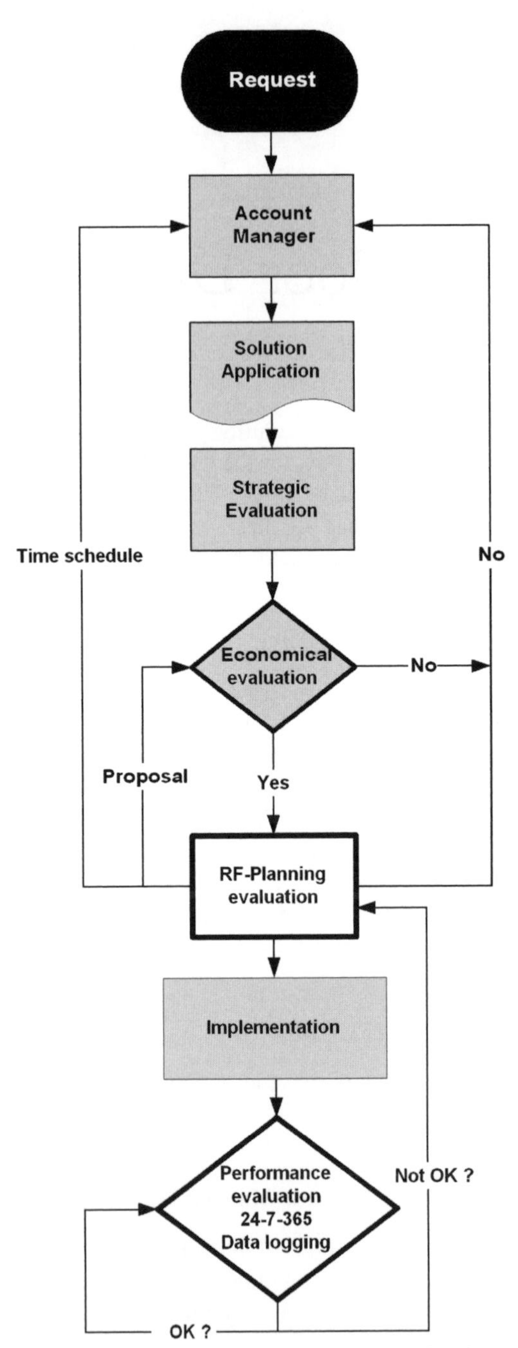

Figure 5.1 One way of structuring the indoor planning process

The application should consist of:

Business input

- Number of users.
- Types of users.
- Types of service requirements, data speeds, etc., needed.
- Duration of the contract.
- Expected airtime per user.
- Expected on-site traffic in the building.

Documentation of the building

- Floor plans.
- Markings on floor plan showing the coverage needed in the different grades and areas:
  - 100% coverage
  - 90% coverage
  - areas where it is 'nice to have' coverage.
- Markings on the floor plan showing different environment types:
  - dense areas with heavy walls
  - open office areas
  - storage rooms.
- Pictures of the building (often photographs can be found on the company's web page).
- Details and drawings of planned reconstructions or extensions of the buildings.
- Details for the contact person, who is responsible for approving the installation.

Floor Plans are the Base of the Design

Often it can be hard to obtain usable floor plans, but in most countries it is mandatory to have some sort of site plan in A3 format located at the central fire alarm, so that the fire department can find their way around the buildings.

These plans often provide excellent basis for the planning and can be used for:

- Design documentation.
- Installation documentation.

5.1.2 The RF Planning Part of the Process

The RF planner uses the input from sales to make a draft design. Often this draft design can be done by an experienced radio planner without doing a site visit. This of course depends on the quality and detail of the input, the size of the solution and the experience of the RF planner.

The RF planner provides the following output:

- Floor plans with suggested placement of antennas and equipment.
- Diagram of the DAS.

- Equipment list.
- Estimated implementation costs.
- Estimated project time.

These outputs from the RF planner are used as the final input to the 'Coverage Committee', which is responsible for a 'go' or 'no-go' implementation of the solution. The Coverage Committee, which is also responsible for the roll out budget for indoor solutions, has technical, sales and marketing representation.

5.1.3 The Site Survey

Prior to the site survey, the RF planner has done a draft design, using a link budget tool, RF propagation simulation and experience. To do the final design, the RF planner uses the draft design as a basis for a site survey, and adjusts the draft design according to the results of the site survey.

The purpose of the site visit is to:

- Get the solution approved by the building owner.
- Collect information regarding equipment rooms, installation challenges cable ducts, etc.
- Take necessary photographs for the installation team, and for the RF planner.
- Take photographs of rooftop 'line-of-sight' possibilities to other sites, if microwave transmission is to be used.

Participating in the site visit should be:

The RF planner

- The RF planner is the project manager.
- After the survey, the RF planner will provide the final design to be approved by the building owner, and used by the installer.
- The RF planner might need to take measurements of the existing coverage provided by the macro layer.
- The RF planner also might need to do RF survey measurements inside the building, to verify the draft design.

The acquisition manager

- The acquisition manager is responsible for the constriction permits, legal contract, etc.

The installer

- The installer is responsible for the implementation.
- He will provide 'as-built' documentation after implementation.

The local janitor and local installer of IT and utilities

- He knows all the details and cable ducts.
- He and the operator's installation team will be in direct contact regarding the installation, once the RF planner has provided the final design.

The building owner

- The building owner will approve the design, antenna placements, etc., as the team walks the site, together with the janitor and IT department.

Process Control

It is important that all the involved parties, all members of the Coverage Committee and the implementation team, are all working with predefined timeframes, and well-defined input/output documents and procedures in order to control the process.

5.1.4 Time Frame for Implementing Indoor DAS

For a normal indoor planning process, the typical timeframes would be:

Sales/key account meeting with the end-user, providing inputs and documentation
  1–2 weeks.
Draft design from RF
  1–2 days
Site visit
  2–6 hours
Final design and documentation
  1–4 days after site visit
Implantation start-up
  1–2 weeks after site visit

Transmission

Based on experience, it is often the implementation time of transmission to the BS that is the show stopper. Therefore it is very important that the transmission department is advised as soon as the solution is approved by the Coverage Committee, to avoid the typical situation where an implemented solution is still awaiting transmission 5–8 weeks after implementation.

5.1.5 Post Implementation

The RF designer or installer is responsible for doing a walk test of the system once it is operational. The coverage is documented on floor plans, using post-processing software. The RF designer should also contact the end-user, to check that the coverage is as expected, and monitor the performance of the system using the statistical tools available, evaluating the live network data and checking neighbor lists, quality, capacity, etc. These results (the walk test and the performance tools) are used to fine-tune the RF parameters of the cell, enhancing the system performance.

An important check point is that the traffic produced is within the expected range, based on the input from sales, and the RF planner provides feedback on the total performance to the account manager. After 2 weeks, the responsibility for the performance of the cell is handed over to 'operations'; after this it is their responsibility to monitor the cell. The RF designer receives and approves the invoices from the installers, so he knows (and can learn) if the estimated price was within scope, and be better placed to predict project costs for future systems.

5.2 The RF Design Process

5.2.1 The Role of the RF Planner

After we have had a quick look at the general process of the total indoor implementation process, we will have a closer look on the design tasks for the RF planner; after all, this is the purpose of this book.

Draft Design and Site Survey

Based on the design inputs provided by sales, the RF planner will do a draft radio link budget and prepare a draft design prior to the site survey. Thus the radio planner can check all the planned antenna locations, and adapt the design accordingly. Based on the experience from the site survey, the radio planner will be able to adapt the draft design to the reality and restrictions in the building, in order to make the final design. During the RF survey, it is important to check the type of walls, take notes on the floor plans of the different types, etc.

Take Photographs

The RF planner should bring a digital camera, take lots of photos and mark the position of each photo on the floor plan; this will help in making the final design. It is advisable to take a photo of each antenna location, in order to document the exact antenna location.

It is a good idea to bring a laser pointer, and point at the exact planned antenna location when you take each photo. The red dot from the laser will be very clear on the photograph. In the design documentation, each photograph is then named according to the antenna number, A1, A2, etc. This helps the installer to install all the antennas in the correct location and avoid expensive mistakes.

5.2.2 RF Measurements

RF measurements are a crucial part of designing and verifying indoor coverage solutions. It is important to know the 'RF-baseline' (the existing coverage) both inside and outside the building in order to establish the correct design level and parameters to use when designing the indoor DAS.

Log and Save the Data

It is highly recommended to always use a measurement system that allows you to log the measurement data on a PC for post analysis. You should preferably use a system that can navigate and place the measurements on a matching floor plan, and indicate the measurement result by color or text.

These floor plans, with plots of signal level, quality and HO zones are also crucial documents to prove that the system implemented fulfills the agreed design criteria, and are very useful as a reference for trouble-shooting at a later stage. These measurement results should be saved in a structured data base system. This can be very valuable experience in future designs.

5.2.3 The Initial RF Measurements

The measurement routes for the initial measurements needed in order to design the radio system are shown in Figure 5.2. The first measurement that is needed is the outdoor measurement 1 in Figure 5.2, in order to determine the outdoor level and the servicing cells. It is very important to determine the outdoor signal level in order to design the HO zone between the indoor solution to the outdoor network. This important HO zone will cater for the handover of the users entering and exiting the building.

Measurement 1 in Figure 5.2 is also used to estimate the penetration loss into the building, when compared with measurements 2 and 3 in Figure 5.2, and can be useful to calculate the isolation of the building. In this way you can estimate the desired target level for the indoor system.

Measurements 2 and 3 in Figure 5.2 serve the purpose of obtaining an RF baseline of the existing coverage levels present in the building. It is very important to establish this prior to the design and implementation of the indoor solution, in order to select the correct design levels for the DAS design, according to any interference from outdoor base stations.

Measure the Isolation

Using the measurement method described above, you can calculate the isolation of the building and plan accordingly. Typically you will need to perform measurement 2 at least on

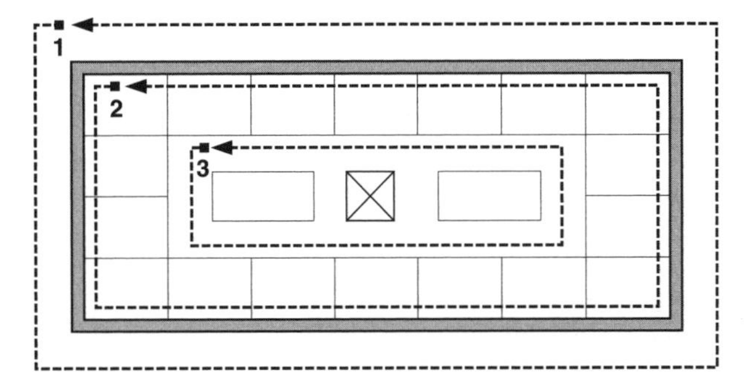

Figure 5.2 Initial RF survey measurement routes

the ground floor, middle floor and topmost floor. In a high-rise building it is advisable to repeat measurement 2 on every fifth floor.

When designing a GSM system, you should design for dominance of the indoor cell. When designing GSM, the indoor system should preferably exceed the signal level of any outdoor macro cells present in the building by 6–10 dB. This will ensure that the indoor cell is dominant and prevent the mobiles from handing over to the outside macro network. This can sometimes be a challenge to fulfill, especially with nearby macro sites adjacent to the building; there are tricks on GSM that 'lock' the traffic to the indoor cell, even if the indoor cell has a lower signal level (see Section 5.5.1), but it is preferable to solve the dominance problem with careful radio planning upfront.

For UMTS designs these measurements are also very important. Based on the measurements you should try to design the indoor system to be 10–15 dB more powerful, in order to avoid extensive soft handover zones. It is important to minimize the soft handover zones in the building to avoid cannibalizing capacity for more than one cell. In particular, hot-spot areas in the building with high traffic density should not have any soft handover zones.

5.2.4 Measurements of Existing Coverage Level

Measurements of existing coverage, penetration losses and verification measurements are a crucial part of the RF design. These measurements provide the RF designer with valuable information to be used for the design and optimization of the indoor solution. In addition to the initial measurements just described in Section 5.2.3, several other measurements are equally important. There are several types of measurements that needs to be done.

Channel Scans

You should always consider performing channel scans on the same floors as measurement 2, preferably close to the windows (or even open the window if possible), in each direction of the building. A channel scan can be performed by most measurement tools using a test mobile.

Typically the user defines the start and end channels, and the mobile will scan and measure all the channels in that specified range. For UMTS you use a code scanner that logs all the decoded cells, scrambling codes and CPICH levels. The purpose of this channel scan is to measure potential pilot polluters, find unexpected neighbors and establish the baseline for the noise level on the radio channel.

Example Channel Scan GSM

In the example shown in Figure 5.3 the test mobile has been programmed to scan from CH01 to CH33, all the GSM channels of the specific GSM operator CH1–CH32 planning the indoor system, including the first adjacent channel of the next operator in the spectrum CH33. It is evident that the high signal levels of CH29 (−43 dBm), CH06 and CH26 are powerful. They are probably nearby macro sites across the street from the building. You cannot isolate against these powerful cells, and so you need to take them into account when defining and optimizing the neighbor list and HO zones.

The scan is also useful to select appropriate frequencies for the indoor system. In this case CH20, CH02 and CH31 look like good candidates to use inside the building. However, be

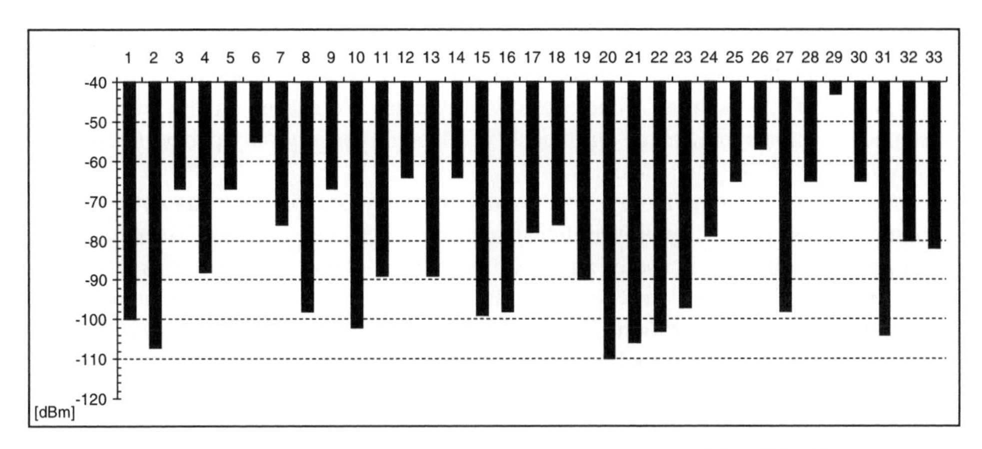

Figure 5.3 Example of channel scan measurement on GSM CH1-33

careful; the reason for these channels being low in level could simply be lack of traffic on these channels at that particular moment in time. You must always verify the channel you propose to use inside the building in the macro frequency planning tool.

5.2.5 RF Survey Measurement

In some cases you should also verify the RF model that is used to simulate the coverage in the particular building. This is especially important for the first 10–20 projects. By then you will have gained enough experience and measurement results to adjust and trust the propagation model you have selected (see Section 8.1.5) and your 'RF instinct' for that particular environment.

Performing the RF Survey Measurement

The typical measurement set-up can be seen in Figure 5.4. In this example two proposed antenna locations are verified in the draft design, A and B. The design concept with these

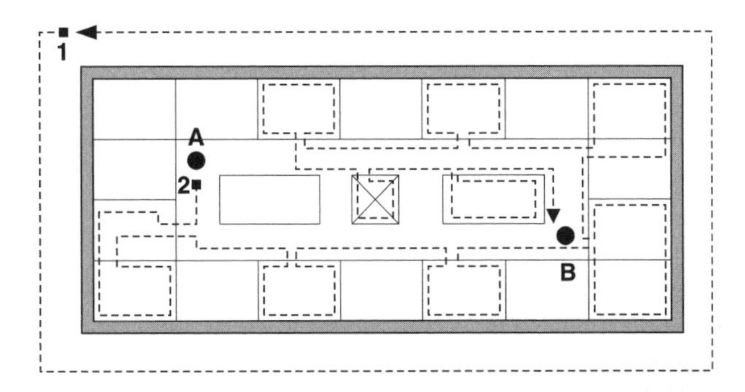

Figure 5.4 Typical RF survey route, measuring in the same area as the proposed antenna location

two antennas is to use the 'corridor effect' to a maximum; using the corridors to distribute the signal around the more solid core of the building and reach the users in the offices along the perimeter of the building. Since the two antennas are placed in the same type of environment, with a symmetrical placement, you would typically only perform verification measurement on one of the antennas.

In the example shown in Figure 5.4 we want to verify the location antenna A, and in order to do so we must place a reference signal source there, a test antenna. The test antenna should preferably the same type of antenna, transmitting at the same power as you plan to use in the final design. It is also very important to place the antenna as close as possible to the intended installation position. Preferably the RF survey transmit antenna should be placed in the ceiling in the actual position or at least as close as possible.

The location of the test antenna is very important because the performance of antennas will be highly influenced by their immediate environment. The performance can be different if it is placed on a tripod 1.5 m off the floor, compared with the real position in the ceiling, upside down. This is because the local structures and environment around the antenna will affect the directivity performance of the antenna and you want the RF survey to be as close as possible to the actual conditions of the final installation.

Preferably you should perform measurements on the adjacent floors under and above the floor with the survey antenna installed. This is to establish the floor RF separation, which is useful when estimating whether 'interleaving' of the antennas is possible (see Section 5.3.9). Normally you will not take measurements in the area very close to the antenna or the nearest rooms because:

1. You will know that the coverage in the area only 5–10 m from the antenna is good. If not, it will be evident in the measurement results further from the antenna.
2. In many cases, measuring close to the antenna will saturate the receiver of some test mobiles, and this will distort the real measurement results, e.g. one mobile might not be able to measure signals higher than −40 dBm (GSM example), and therefore measurement values higher than this value will only be logged as −40 dBm, and will skew the calibration of the model.

5.2.6 Planning the Measurements

The planned RF measurement route must represent all areas planned to be serviced by the antenna location. It must be planned to measure with a good overlap to areas where other antennas are intended to provide service. One example can be seen in Figure 5.4, where the purpose of the measurement is to verify the planned location of antenna A.

It is very important to conduct measurements close to the windows on each side of the building and in particular in corner offices. These areas are prone to 'ping-pong' handovers (GSM) and soft handover zones (UMTS) and this might lead to degraded quality due to interference and pilot pollution. Therefore you must carefully select a design level based on the measurements with special focus on these areas to make sure you can overpower the macro signals, and at the same time avoid leakage of signal from the building.

It is also important to conduct a measurement outside the building to estimate the leakage from the building. However, for obvious reasons, this can normally only be done during measurement of the ground floor antenna verification. It will not be possible to measure every antenna location on each floor; typically only one or two antenna locations are selected in each type of different environment to verify the model used.

These RF survey measurements are a crucial part of the experience you want to gain as indoor radio planners. After you have conducted a few of these measurements and analyzed the results, you will soon gain trust in your model and experience of when to trust the RF model. Over time you will be able to 'see' the RF environment, and will be able to decide when it might be necessary to perform RF survey verification measurements when designing future projects and when to trust your 'RF vision'.

Save your measurement results in a database, to gain a knowledge database of penetration losses for future projects, and for fine-tuning your simulation models. This will enable you to do statistical analysis of the results in a structured manner, which is useful for reference and for sharing experience.

Use Calibrated Measurement Tools

When conducting these RF reference measurements it is very important to use a calibrated transmitter and receiver-calibrated test mobile. A test mobile is similar to a standard mobile, but it has been calibrated and enabled with special measurement software. This software enables the user to perform detailed measurements on the network. These measurements can be conducted in both idle and dedicated modes. Typically the user will be able to read the basic RF measurement information in the display of the mobile. This will typically be channel number, cell-ID, RF-quality, information about neighboring cells, signal strength and data rates. The simplest test mobiles let you read this information in their display; more sophisticated models let you save the measurements for post processing.

Log and Save the Measurements

It is important to use a test mobile that is able to save the RF measurements on a connected PC. In addition to the propriety data format that enables the user to save the results in a file for post processing within the software package that comes with the test mobile, it is important that the measurement system software also enables you to export the RF measurement results in a text format so you can import the results in standard software like MS Excel. This enables you to do various post processing analyses of the measurements (this is how the measurements in Figures 4.26 and 4.28 were documented).

Log the Measurements According to the Measured Route

It is highly recommended to use a measurement software package that enables you to import the floor plan of the building where the measurements are being performed. This enables the user to mark reference points on the floor plan, and the software then distributes the measurement samples between the markers along the route. This might even be the same software platform you use for your RF planning tool, merging it all into the same tool.

The navigation works typically by marking the next point you are heading to on the floor plan, and clicking when you arrive. Then the tool distributes the measured samples between the points. Typically this will be the corner points all along the route. Be sure to keep a constant pace from way mark to way mark, in order to secure an even distribution of the measurement samples along the route.

Some of the simpler measurement tools only allow you input 'waypoint' markers in the measurement file for the position reference. These tools will be alright for a measurement in a tunnel, where you can use typical reference points such as 100 m, 200 m, 300 m, since you know the direction is from *X* to *Y* (there being no *Z* in a tunnel) and, by keeping a constant speed, you can log the results accurately to the position. However, for a measurement in a complex building it is close to impossible to use only waypoints for reference, unless you break the measurement up into many files, each covering a specific room in the building.

Do Not Bias the Measurement Results

Make sure that the measurement receiver for these measurements is in a neutral set-up. Do *not* carry the measurement receiver upside-down in your pocket when conducting the survey measurements. You need to be sure that the antenna is unobstructed in all directions, in order not to skew the measurements. During the post analysis of the measurement, you can always add body loss and other design margins, but the measurement must be done as neutrally as possible.

When performing RF-survey measurements be sure that you use a clean RF channel in the spectrum for the survey transmitter, so no interference from other base stations will distort the measurements.

5.2.7 Post Implementation Measurements

After implementation you will also need to perform a measurement in order to document the 'as-built' system. This will also help you to find any antennas in the system not performing as expected. You will need to measure all floors on all levels of the building, preferably measuring 'edge-to-edge', and do samples of all the different areas of the building.

These measurements can also help you calibrate your design tools and models, for future designs. Preferably the radio planner will do these measurements at least for the first 10–20 buildings, in order to gain experience of how the building, walls and interiors affect the signal and propagation of the signal. A copy of these measurement results should be kept as a part of the on-site documentation; this is useful for trouble-shooting on the system in the future.

The Simplest Measurements, Post Installation

The simplest measurement you can perform on an indoor coverage system is to stand below each and every antenna in the building, and average out about 20 s of samples, just using the display of the phone. Note this result on the floor plan as a reference for checking the radiated power from the antennas, using the free space loss formula to estimate if all antennas are performing as expected.

You will also need to perform measurements of more locations distributed throughout the building, especially in the areas where you expect to find the lowest signal. Typical this would be the corners of the building, the staircases and the elevators. Do not forget to measure the executive offices as it is nice to be absolutely sure everything is alright in that area – there is no point in claiming that the indoor system performs to the required 98% coverage area, if the last 2% is in the CEO's office! See Section 5.3.10 for more details.

5.2.8 Free Space Loss

Free space loss is a physical constant. This simple RF formula is valid up to about 50 m distance from the antenna when in line-of-sight inside a building. The free space loss does not take into account any additional clutter loss or reflections, hence the name. The free space loss formula is:

$$\text{free space loss (dB)} = 32.44 + 20(\log F) + 20(\log D)$$

where F = frequency (MHz) and D = distance (km). The free space loss for some standard frequencies will be as shown in Table 5.1.

Table 5.1 Examples on free space losses (rounded to the nearest dB)

Free space loss	1 m	2 m	4 m	8 m	16 m
950 MHz	32 dB	38 dB	44 dB	50 dB	56 dB
1850 MHz	38 dB	44 dB	50 dB	56 dB	62 dB
2150 MHz	39 dB	45 dB	51 dB	57 dB	63 dB

As we can see in Table 5.1, the loss on 950 MHz is 32 dB at 1 m, and each time we double the distance or frequency we add 6 dB more in free space loss (as shown in Figure 5.5). We can also see that the path loss on 1850 MHz is 6 dB more than on 950 MHz. The difference

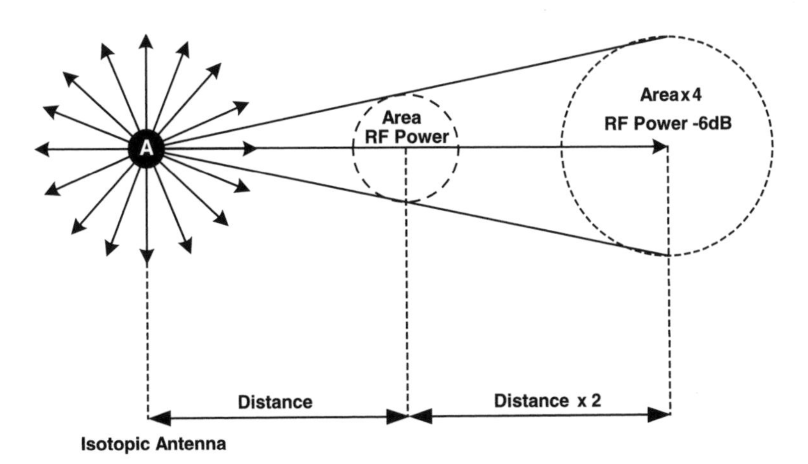

Figure 5.5 It is a physical constant that each time you double the distance, the free space loss is increased by 6dB

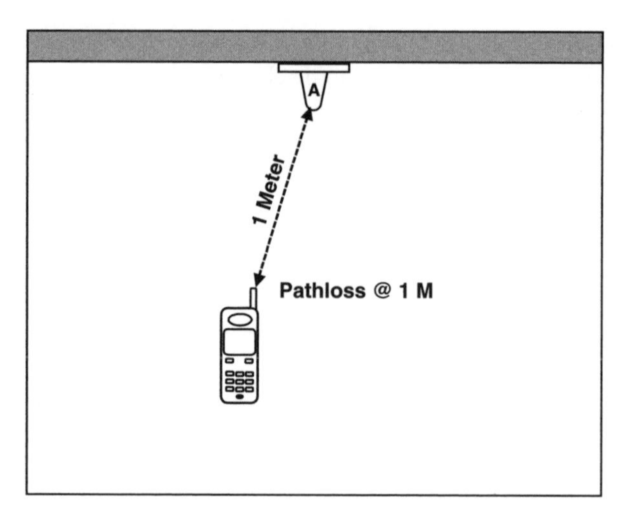

Figure 5.6 The '1 m test' is a fast way to estimate the antenna power

in frequency between DCS1800 (1850 MHz) and UMTS (2150 MHz) is only about 1 dB due to the relatively small difference in frequency.

5.2.9 The One Meter Test

After implementation of the solution, one can be in doubt whether an antenna is performing as expected or not. Often you want to connect a power meter or spectrum analyzer to the antenna system to check if the antenna is fed the correct power level. However this is often challenging due to the installation, with the antenna connector hidden above the ceiling and difficult to access.

Another issue is that this will still not check if the antenna itself is working or not. Often you will not consider the antenna to be a likely point of failure, Experience, however, documents that this is a common issue in indoor antennas systems. A very practical way to test if indoor antennas are performing correctly is to use the '1 m test', based on the free space loss.

Performing the One Meter Test

To test if the antenna is radiating the expected power is simple; all you need to do is to apply the free space loss. After all, this is a physical reality, and a good guideline for a fast, efficient test.

Example 1800 MHz
The test mobile measures −35 dBm at 1 m distance

The radiated power from the antenna = −35 dBm + 38 dB = +3 dBm

Note that some mobiles 'saturate' at higher signal levels. You will often need to 'average' some samples. The radiation pattern of the antenna also impacts the level, so be sure to measure in the expected 'main beam'.

The One Meter Test is Not 100% Accurate

This method is not to be considered 100% correct, but it is very useful for a quick verification, and one can easily estimate if the power is off by more than 6 dB. For an accurate measurement, however, it is recommended to connect a power meter or spectrum analyzer directly to the feed cable.

5.3 Designing the Optimum Indoor Solution

The optimum indoor solution exists only in theory. All implemented indoor solutions will to some extent be a compromise. It is the main task of the indoor RF planner and the team implementing the indoor solutions to make a suitable compromise between meeting the design goals, securing the system for the future and designing and implementing the system to maximize the business case. You must also design the DAS so the antenna system utilizes the features, possibilities and limitations in the particular building to a maximum, in order to make the system applicable for practical implementation.

5.3.1 Adapt the Design to Reality

Even with theoretical knowledge on how to design the perfect indoor solution, the main task for the RF planner will often be to know when and where to compromise in order to implement the system in practice and still maintain an economical and high performing solution.

The preferred approach when designing and implementing an indoor coverage solution would be based on site survey with measurements, experience and the link budget. However, reality often dictates that the architectural and installation limitations, and cost concerns, play a major role in the final design (as shown in Figure 5.7).

5.3.2 Learn from the Mistakes of Others

An old mountaineering saying goes like this: good judgment comes from experience but experience is often a result of bad judgment! This is also true for indoor RF planning. There is no need to repeat the mistakes of others, so let us learn from some of the most common errors. Here are a few tricks to share, most of them learned the hard way.

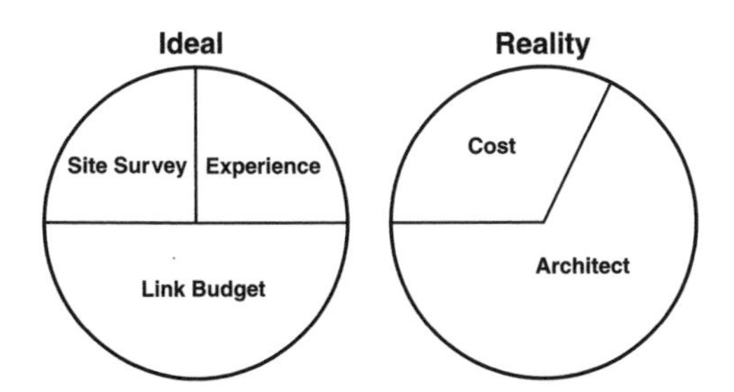

Figure 5.7 The RF planner needs to adapt to the reality of the building

Keep the Simple Things Correct

Indoor antenna systems are often complex and the radio services operating on these systems are getting more and more advanced, with higher data rates and multimedia applications. However, do not forget to focus on the simple and trivial things. These simple things tend to be underestimated, but can have a major impact on the performance of the implemented system.

Remember that designing indoor distributed antenna systems it is still 'only' RF planning, and good RF design and implementation is often a matter of understanding why you need to pay close attention to RF components that might seem trivial. You must also appreciate the importance of the craftsmanship needed to install these RF components and antenna systems so they perform as planned over time. Education, craftsmanship and common sense of the installers make the difference between success and failure. Even the best designed DAS will not perform better than the weakest point in the DAS, and a small problem with a connector, for example, will have major impact on the performance.

Use Only Visible Antenna Placement

Often the RF planner is asked if it is possible to hide the antennas above the suspended ceiling. This should be avoided at all costs and you should refrain from giving any guarantee of the performance of the implemented system if forced to accept hidden antennas.

The main reason is the consequent unpredictability. You don't know what is hidden above the ceiling tiles: fire sprinklers, ventilation ducts, heating systems, electrical wiring. Worst of all, you have no post installation control of objects close to the antenna after the installation of the indoor system has been completed. Furthermore you have no control of the RF properties of the ceiling tiles themselves. What happens if the ceiling is repainted, or changed for tiles of a different material? Aluminium ceilings are in fashion in some parts of the world, and would be disastrous for the performance.

Obtain a responsible architect or designer of the building to accept and appreciate that the antennas must be 'visible' for the mobiles. After all, mobile coverage is a crucial part of the buildings utility infrastructure. If you are forced to accept installation of antennas above the ceiling or behind walls, be sure to note it in the design document and contract that the performance of the system is 'best effort' and not guaranteed due to installation restraints.

Use Only Quality Certified Components

Again, the importance of the performance of 'simple' RF components is often underestimated. It is normal to focus on the performance of the more complex components in the indoor antenna system, the base station, repeater, the active distribution system etc. However, the quality of the simpler passive components – antennas, splitters, tapers, connectors and jumpers – may have a major degrading impact on the RF service performance if these components do not perform to specification.

Use Only High-quality Antennas

One example of how important the performance of passive components can be is the 'standard' 2 dBi omni antenna. This antenna is used in 99% of installations, and therefore will play a major role on the performance of all the indoor systems in the network.

This 'simple' antenna can be obtained from various suppliers but the different antennas are markedly similar looking – if you compare the data sheets they will most likely also be similar in terms of mechanical design. Even during an RF test the antennas perform the same, but how about 5 years from now?

Is there a reason for one of the 'identical' antennas to be more costly? Could some antennas be copies, where one manufacture has not taken into account passive inter-modulation effects, due to incompatible metals inside the antenna? Could there be degrading galvanic corrosion over time? This will cause bad connections, and passive inter-modulation (see Chapter 5.7.3), with serious degrading performance of the indoor system.

Small Things Matter

The same goes for other 'simple' components like an RF connector. Could there be a reason why one type of connector comes at a cost of 1 euro more a piece? Could the more expensive connector be a better option?

Please do not misunderstand me – there is not always a 1:1 relation between price and quality. The statement is only general; be careful, and choose a manufacturer that you trust. Look, feel, measure, test and judge if you trust the quality of the component to hand. Disassemble the antenna, have a look at the internal materials: does it appear to be good craftsmanship? It is not always easy to estimate, but in many cases you will get a feeling for the quality.

Use Only Educated Installers

The craftsmanship needed to install RF systems is often underestimated. Use only educated installers who are certified to work with RF components and cables. There is a huge difference in doing a quality installation of a mains power connector, and then doing a high-quality installation of a coax connector operating at 2.1 GHz. Did you know that manufacturers of coax connectors recommend a specific torque for tightening the connector to insure the performance?

Make sure that the selected installer only uses qualified and certified workers, who know what is needed to do high-quality RF installation. Also make sure that all the installation teams have the correct tools to hand, and not just one set of tools for the first team and 'knife and spanners' for the rest! This might seem obvious, but often you will have four to six parallel teams installing the same building to be able to do it overnight. It is very important that each of these teams is equipped with the correct tools, the correct connectors, etc.

Use of quality components, tools and installers will pay of in the long run. Yes, it might seem costly, but it is worth the investment. Believe me, I know how costly it is to have all connectors replaced in an operating metro system, after the metro trains are operational!

Design Documentation

This is another example of a very important issue when designing and implementing an indoor solution. For the RF planner it might seem straightforward to decide where to install what, and obvious what the correct orientation is for that directional antenna at the end of the hall. However, again, produce detailed design documentation and installation guidelines. This will also help with future upgrades, trouble-shooting and extension of the indoor system. It can be a hard task to reconstruct the documentation of an implemented DAS system four or seven years after implementation, even for the RF planner who initially designed the system. If the same RF planner is still around, that is!

The design documentation should be easily accessible on-site in hardcopy, preferably also in an electronic format that is easy to update with future changes.

5.3.3 Common Mistakes When Designing Indoor Solutions

Often new indoor RF planners will make the same initial mistakes. Let us try to avoid these.

Dimensioning Coverage on Downlink Only

It is a fact that GSM indoor systems are mostly downlink-limited, so most focus is on the downlink. However, it is important not to forget the uplink part of the design, especially on UMTS/HSUPA, depending on the service profile of the traffic on the cell. Remember that the Link Budget consists of an analysis of both the downlink and the uplink.

We will focus more on the Link Budget calculation later in this book, in Section 8.1.

Underestimating Passive Inter-modulation

Do not underestimate potential passive inter-modulation (PIM) problems. Especially with high-capacity passive DAS with many radio channels in service, the concentration of power in the cables, connectors, splitters, tappers and antennas close to the base stations can be high. If you use low-cost, low-quality components, the impact of inter-modulation can be high. We will take a closer look at this in Sections 5.7.3 and 5.7.4.

Not Accounting for Coax Loss

Many good indoor DAS designs have been ruined by the reality of the installation, often due to underestimating the shear length of the coax cable, and thus underestimating the passive losses in the system. The passive losses in the DAS have a big impact on the uplink and downlink performance of the system. It is very important to use realistic distances on the cable part of the passive system, in order to be able to calculate the link budget correctly. It is recommended to have a safety margin on the coax losses, by adding 10% to the length of the individual cables.

Underestimating the Costs of the Installed System

The business case is a crucial part of any indoor design. A major part of the cost of an indoor coverage system is installation costs, especially when installing a passive system using heavy rigid cables, which are time-consuming and labor-intensive to install.

In order to estimate the DAS cost when designing passive systems it is very important that you know the exact installation challenge, cable routes etc. upfront during the initial evaluation of the project. If you do not, the cost of the project is impossible to estimate. This means that the RF planner needs to 'walk the building' before he can do the cost estimate, and it is as much installation planning as it is radio planning.

There are many examples of passive systems costing in excess of several times the initial estimated costs once implemented. This is mainly due to lack of knowledge about the installation challenges of rigid passive coax cables. Other unforeseen installation costs, such as lack of installation trays, reconstruction of fire separation barriers in the installation ducts and access to the building only at night, will increase the labor costs significantly.

Not Accounting for the Total Cost of the Project

Mistakes in the business case are often caused by not including all the costs of the total indoor DAS. An indoor solution is much more than 'just' the cost of the indoor system itself:

- DAS costs.
- Installation costs.
- Maintenance costs.
- Site-support costs (power supply, transmission, etc.).
- Upgrade costs for future services and capacity.
- Planning costs.
- Backhaul cost.

5.3.4 Planning the Antenna Locations

The Inter-antenna Distance, Theory

Once you have calculated the link budget you can establish what the limiting link is, the uplink (from mobile to base station) or the downlink (from base station to mobile). For multilayer systems, i.e. systems where several radio services are sharing the same antenna installation, this could be GSM and UMTS on the same DAS. There might be some difference in the service range from the two systems, using the same DAS antenna. An example of service ranges from different mobile services could be: GSM DL = 28 m, GSM UL = 78 m, UMTS DL = 23 m, UMT UL = 21 m. In this example the determining factor of the system will be the UMTS UL, and the antennas should be placed accordingly.

For the theoretical example, imagine that the environment inside a building is like a uniform pulp of RF attenuation, with uniform attenuation in all directions and no 'corridor effect' or similar real-life behavior. One could think that the inter-antenna distance of omni antennas should be 21 m as calculated to fulfill the UMTS UL limit above.

However, if an area is to be covered '100%' (there is no such thing as 100% coverage, but this is a theoretical example), one could believe that the inter-antenna distance must be $2 \times 21 = 42$ m, but this is not the case.

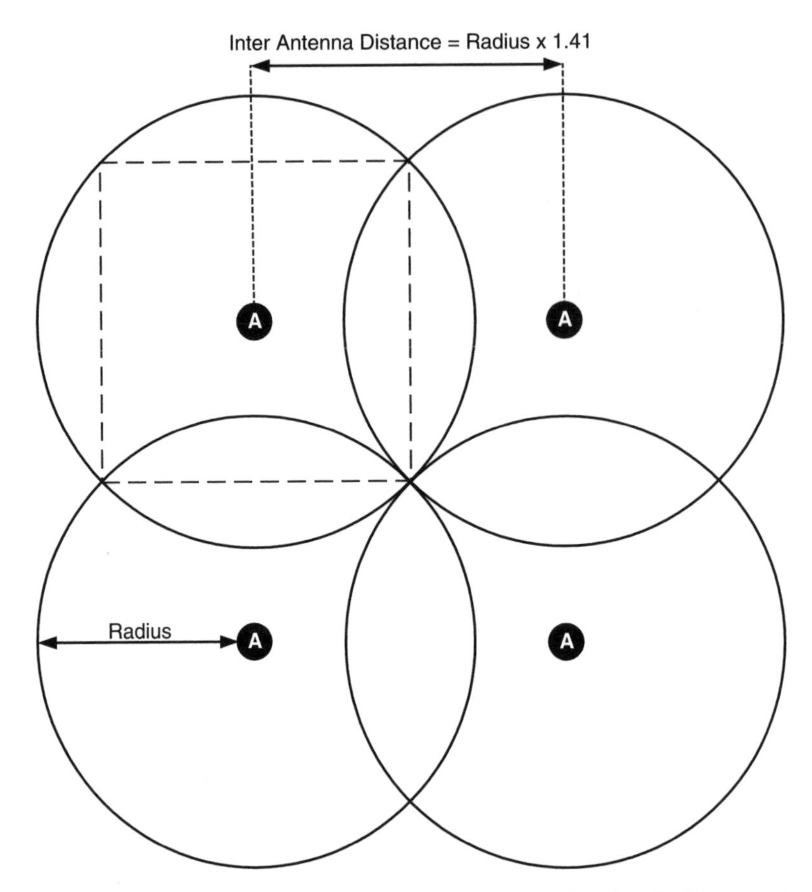

Figure 5.8 In order to provide 'full coverage', antennas need to be placed with a certain coverage overlap

In order to have coherent coverage throughout the area, the inter-antenna distance has to be shorter than twice the service radius; after all, the footprint of the ideal omni antenna is circular, not square (as shown in Figure 5.8).

In theory the correct inter-antenna distance for omni antennas is radius $\times\sqrt{2}$. This clearly shows us that there is a need for coverage overlap between the antennas in order to provide good indoor coverage.

Adapt the Antenna Placements to the Reality

In the real world inside a building, you will never have uniform loss in all directions from the antenna. In real buildings the antennas will typically serve many types of 'clutters' and areas, e.g. open office space, dense office areas, dense areas (stair cases), heavy dense areas (elevator shafts). Therefore you must adapt the antenna locations to these specific environments when applying the link budget-calculated service ranges from the antennas to the reality of the building.

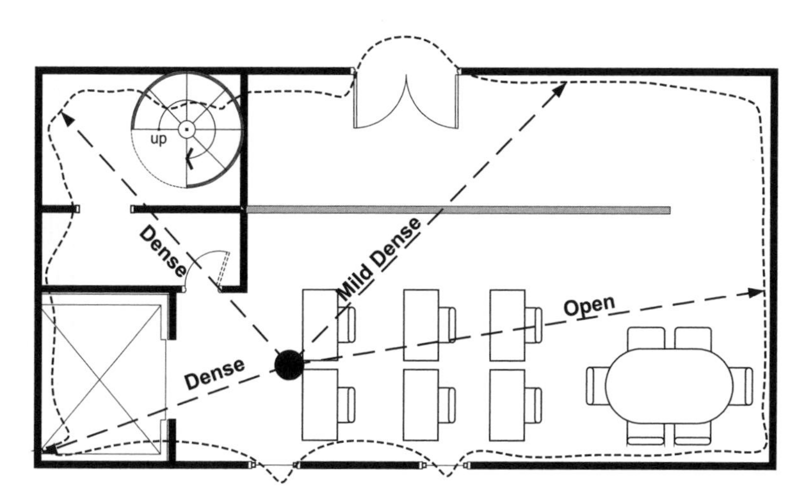

Figure 5.9 In reality the coverage from a typical indoor antenna will be uneven in different directions, due to the service of different environments

In practice, this means that you will have a shorter range from the antenna towards the denser areas, whereas you will have longer range in the more open directions (as shown in Figure 5.9). This is where the radio planner must use his experience and knowledge of the building gained during the site survey to adapt the design to reality.

The experienced RF planner will use the calculations from the link budget. The calculated service varies from the individual antennas throughout the different clutters and areas of the building, and the antennas should be placed so that the maximum footprint is obtained for each antenna. The antenna locations have to be adapted to the reality of the building, taking into account the installation limitations of the individual antenna placements, cable routes and antenna overlaps.

5.3.5 The 'Corridor Effect'

One effect of placing antennas inside a building that is very important to know and utilize is the 'corridor effect'. This is the distribution of coverage when placing an antenna in one of the most typical locations, the corridor (as shown in Figure 5.10).

This antenna location will give you several advantages:

- Typically there will be easy installation access to cable conduits in the corridors of the building, which will save implementation costs.
- Corridors are often 'static' when buildings are being refurbished, so the antennas are left in place with no impact of service degradation caused by refurbishment of the internal structure of the building.
- The users of the building are less concerned about radiation when they do not have antennas installed in the ceiling above their office desk.
- You can use the corridor to distribute the signal from the antenna; there is typically line-of-sight throughout the corridor, and this is what is known as the 'corridor effect'.

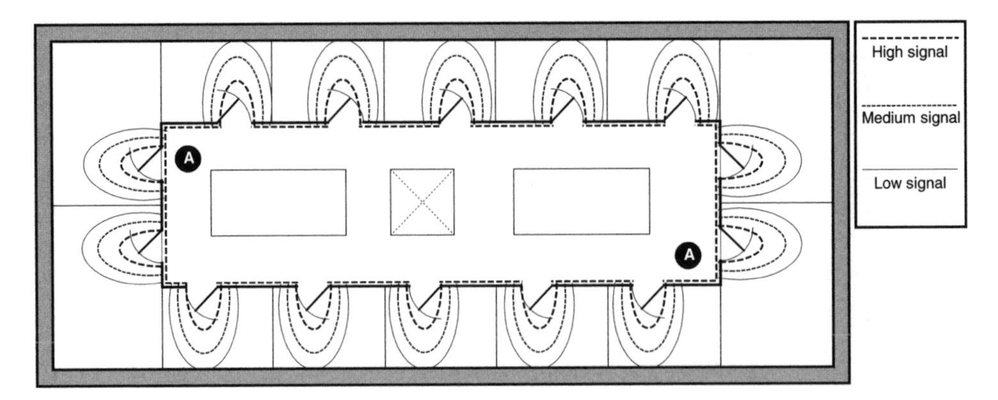

Figure 5.10 The 'corridor effect': the corridor in the building will distribute the RF signal

The downside is that it can be hard to dominate the area along the windows at the perimeter of the building. Be sure to measure the level from any existing signals along the windows, to be able to select your RF design target level. In many modern buildings metallic-coated 'tinted' windows are installed, this will help in solving this potential problem.

The corridor effect is a strong tool, but you need to be careful when you base the design on using this method. By choosing the optimum location for the antennas, the building will help distribute the RF signal and you can utilize the corridor effect to a maximum. However, if you do not choose the antenna locations wisely, the building can limit the performance of the system. You must pay special attention when designing buildings that are planned as multicell systems, and try to minimize the areas of the handover zones.

5.3.6 Fire Cells Inside the Building

It is important to remember that most buildings are divided into several 'fire zones'. These zones within the building are separated both vertically and horizontally by the use of heavy walls and metal doors. The reason for this is to contain any potential fire inside the building, thus minimizing the damage to property and people. The heavy materials used to construct these firewalls or cells will typically also attenuate the RF signal significantly when it needs to pass between these 'fire zones'. This has a major impact on the antenna layout in the building. In reality there must always be minimum one antenna inside each of these cells, or else the firewalls will attenuate the RF signal by 30–40 dB or maybe even more. The example in Figure 5.11 shows a typical building layout including the heavy firewalls and proposed antenna locations. Note that two antennas are placed near the elevator to provide coverage inside the lift. Lift shafts are normally heavily fire- and RF-isolated from the rest of the building, although an exception to this is the 'glass' lifts used in some open atria. Two antennas are placed on the edge between two fire zones and will provide service on both sides of the firewall, thus saving one antenna placement.

5.3.7 Indoor Antenna Performance

For more detailed information about the deep theory on antennas, please refer to Reference [6].

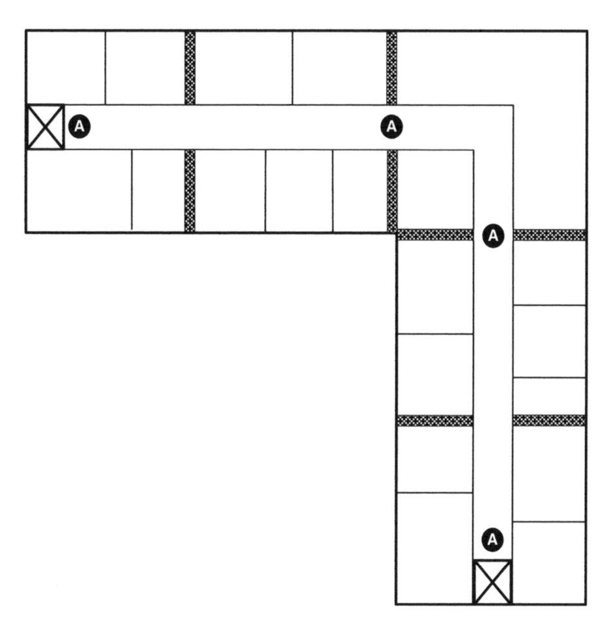

Figure 5.11 The inside of the building will be divided into 'fire cells', separated by heavy walls to contain any potential fire. These walls will attenuate the RF signal, and in most cases you will need an antenna within each of these 'fire zones'

To perform proper indoor radio planning, you must understand the basic function of the indoor antenna. The behavior and quality of the indoor antenna have major impacts on the performance of the indoor DAS system.

Antennas Have No Gain

I know this statement is a bit controversial, but nevertheless it is actually true. For the indoor radio planner it is very important to realize that antennas do not have any gain! How could the antenna have any gain – it is just a piece of metal? The gain of any system is defined by the difference between the power you feed to the system and the output power; the difference is the gain. Antennas have no gain; the antenna consists of no active elements. No power to the antenna is supplied besides the feed RF signal.

However in all datasheets you will find data relating to the 'antenna gain'. In reality this is not the gain of the antenna, but the directivity of the antenna, relative to the 'isotopic' antenna.

The Isotopic Omni Antenna

The 'perfect' theoretical omnidirectional antenna is often used as the reference for this 'gain' when evaluating the performance of real antennas. This isotopic omnidirectional antenna radiates its power equally in every direction, *X*, *Y*, *Z*, as a perfect RF sphere with 0 dBi 'gain' in all directions (as shown in Figure 5.5), hence the name 'omnidirectional' – but this antenna exists only in theory.

Where Does the 'Gain' Come In?

In reality the term 'antenna gain' refers to the directivity of the antenna, in the main beam direction of the radiation of the antenna. In practice the term 'gain' refers to the fact that the power in the main beam direction will be higher compared with an isotopic omni antenna. The power will be lower in other directions. This relative difference in performance compared with the isotopic omni antenna is the gain data that the manufacturer refers to as 'antenna gain', hence the name dBi (the 'i' stands for isotopic). It is used as a measure for stating the gain of an antenna. However the 'cost' of this 'antenna gain' is less directivity in other directions; the antenna is actually less sensitive in other directions compared with the isotopic omni antenna. Sometimes the antenna gain is stated in dBd, where the reference is a dipole antenna. The relative difference between dBi and dBd is 2.1 dB.

This principle is shown in Figure 5.12, where it is clear that the perfect theoretical omni antenna distributes the signal equally in all directions, and forms a perfect sphere. The +3 dBi gain antenna is directive, more sensitive in the main direction, at the cost of less sensitivity in other directions.

The Installation of the Antenna Plays a Role

The radiation and directivity of any antenna installed inside a building will be affected by the actual installation and local environment close to the antenna. Objects inside the building, walls and other objects will attenuate, reflect and diffract the radio waves radiated from the antenna.

In a large open hall the antenna radiation pattern from the omni antenna might look like that in Figure 5.13. In this example three different omni directional antennas are plotted: the isotopic omni, a medium-gain omni and an omnidirectional antenna with +7 dBi gain. The 'low-gain' antenna is actually more sensitive below the antenna, and the high-gain antenna is

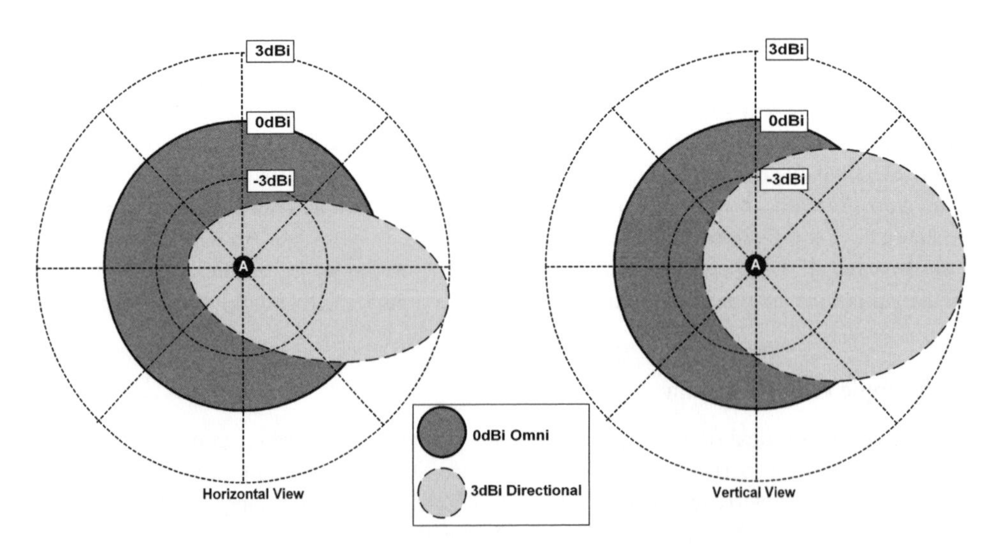

Figure 5.12 Example of vertical and horizontal directivity of an antenna

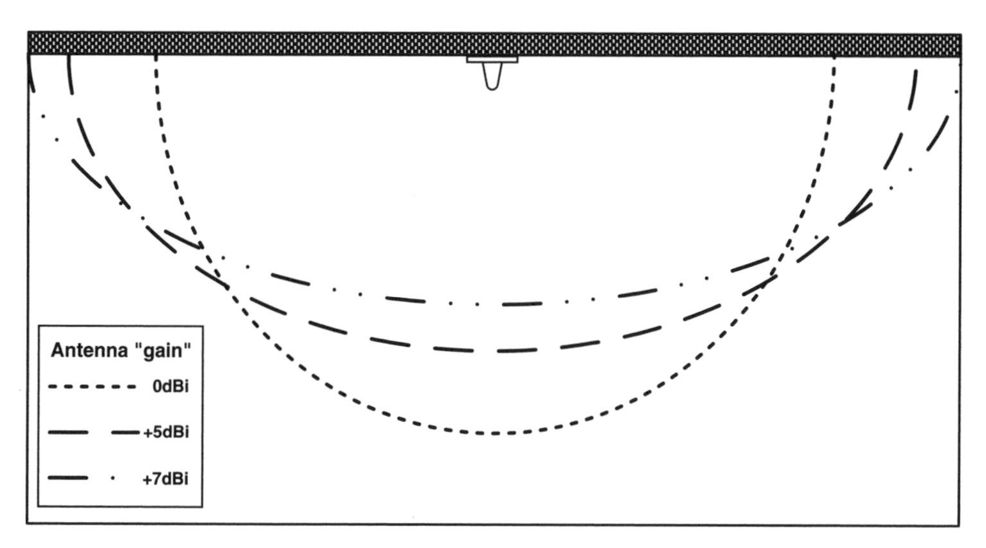

Figure 5.13 Omni antennas with different gains (directivity) in a large open space

more sensitive to the sides and will give less coverage in the large area underneath the antenna. However, the radiation pattern of the antennas will be influenced by the size and shape of the room in which the antenna is installed due to reflections from the walls. These reflections will to some degree decrease the relative difference between the antennas. However, in large open areas with high ceilings low-gain omni antennas should normally be considered, giving more direct power to the mobile users. This is important to remember when designing large convention halls, sport arenas, production facilities, etc.

Another typical example of influence on the antenna by the local environment can be seen in Figure 5.14, which shows three types of omni antennas, with three different gains in the most typical placement: a low ceiling corridor in a building. The directivity of the antenna is highly influenced by the physical constraints and shape of the corridor. The corridor masks out most of the antenna gain or directivity due to the reflections caused by the walls. Therefore, once the antennas are installed inside a building, it is often hard to measure the differences in gain of various types of low/medium/high-gain antennas that are stated in the data sheets.

Owing to installation restraints, it often makes a lot of sense to use a 'directional' antenna at the end of a long corridor. This is to save on installation costs by beaming the coverage down the corridor.

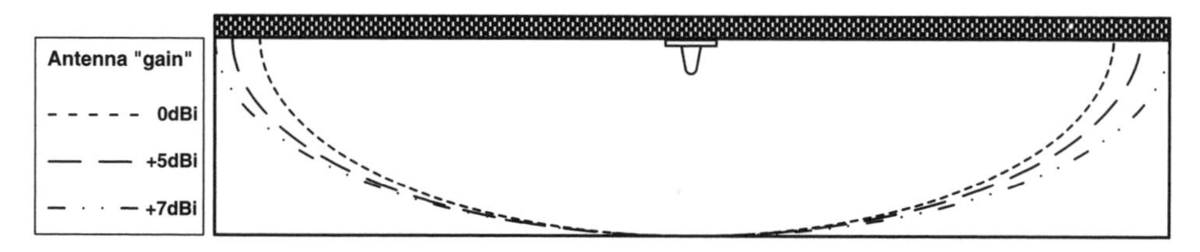

Figure 5.14 The building will 'shape' the directivity of the antennas, masking out most of the 'gain'

There is Only a Minor Effect of the Gain of Indoor Antennas in Reality

The previous examples show that, in a typical indoor installation (as shown in Figures 5.13 and 5.14), the shape and size of the room where the antenna is installed will have a big impact on the directivity. The installation masks out most of the 'gain' of different types of antennas; often there is not a major difference in level when comparing a '0 dBi' omni antenna with a '5 dBi' omni antenna, but installation might be less challenging for the small '0 dBi' type, due to the smaller size and less visual impact.

5.3.8 The 'Corner Office Problem'

When designing a typical office building, it can be tempting to install a centrally placed antenna in a typical small office floor (as shown in Figure 5.15). This will often save on installation and project costs, and the solution might even fulfill the RF design levels, but be careful as this approach has an inherent issue and potential problem: lack of dominance in the corner offices.

This lack of dominance in the total area inside the building will degrade the quality, cause handover loads on the network and even increase the risk of dropped calls. The centrally placed antennas will often compromise the isolation and quality in the corner areas of the building. In many cases the corner offices are occupied by the most important users in the building, those with the highest use of mobile services and highest data speeds. These are the high-revenue users, and often these will be the users who will decide to shift to another operator if the indoor service is not accommodating their needs in terms of performance and quality. The indoor radio designer should strive to design for perfect service inside the building, especially in these important corner offices.

Solving the Corner Office Problem

The solution to the problem is easy and inexpensive, especially when the extra cost is compared against the gain in performance of the indoor system, and the solution has several advantages. In order to provide a uniform, dominant indoor signal in the total area, the signal is split into more antennas, as shown in the example in Figure 5.15 to four antennas. The signal from each of these antennas will be attenuated by the 1:4 splitter by 6 dB and cable attenuation, but the gain by distributing the antennas is actually greater; they are now closer to the users with less free space loss, and the distribution is providing a more uniform coverage level and much better radio link performance. This strategy will provide:

- A more uniform signal.
- An improved RF link.
- Better data services on GSM, UMTS and HSPA.
- A more future-proof solution.
- Dominance, even in the corner offices of the building.
- Limited soft handover load on UMTS to the macro network.
- Lower radiated power from the downlink power close to the antenna, and so less EMR.
- Lower uplink transmit power from the mobile, lowering the EMR exposure of the mobile users.
- Extended battery life of the mobiles.

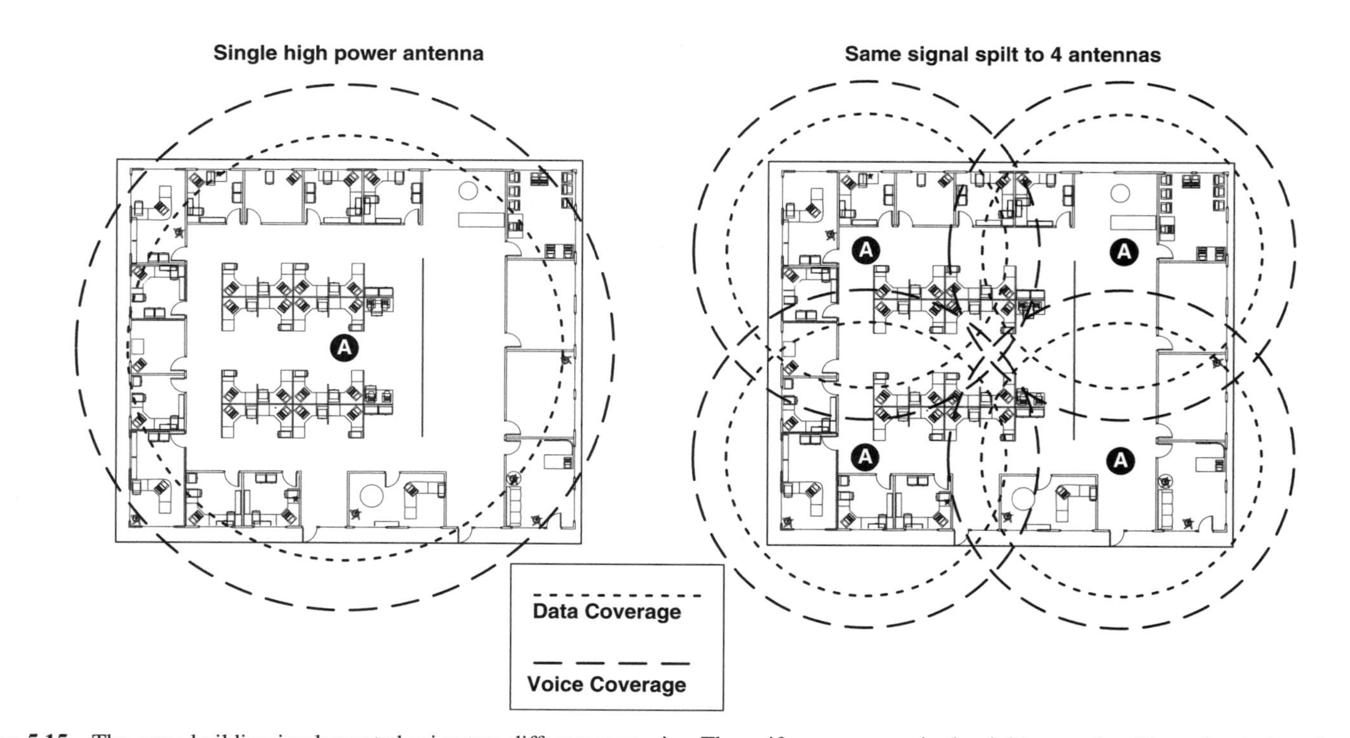

Figure 5.15 The same building implemented using two different strategies. The uniform coverage in the right example with perfect indoor dominance is to be preferred

5.3.9 Interleaving Antennas In-between Floors

If the floor attenuation in the building allows it (this must be confirmed by an RF survey measurement), one viable alternative to the solution shown in Figure 5.15 could be considered. The principle is to rely on coverage penetrating the floor separations, and thus only to use half the antennas per floor compared with the solution shown in Figure 5.15. In this example from Figure 5.16 it will take two antennas on every floor, and by using offset of the antenna placement on adjacent floors, you can interleave the coverage from the antennas in-between the floors. One antenna on one floor will leak signal to the adjacent floors above and below, and vice versa. The result of applying this method is that the antennas, in addition to the primary service area, will also be servicing the adjacent floors, as shown in Figure 5.17.

When considering this interleaving design strategy, it is very important to measure the exact attenuation of the floor separation. You need to take this measurement in order to estimate if this solution is viable at all. Only based on the actual measured attenuation through the floors are you able to evaluate this approach. However, even after analyzing these measurement results, it is highly recommended to add extra margin to the design, to be on the safe side.

Normally I would recommend covering everything from the antennas installed on the same floor as the intended primary service area, and then only using the coverage from the offset antennas installed on adjacent floors as ‘extra’ added coverage and dominance. Be sure that the antennas from adjacent floors that might service users on other floors are on the same logical cell to avoid handover load and decreased UMTS/HSPA performance.

Do not be tempted to crank up the radiated power from the antennas to boost the effect of fill-in coverage from adjacent floors; this might result in unattended leakage of signal from the building.

Example of a Measurement of the Floor-to-floor attenuation

Utilization of the coverage that might leak in-between floors is highly dependent on the attenuation of the individual floor separations; it is therefore highly recommended to measure the leakage from floor to floor, in order to evaluate this design approach. You must pay special attention to the layout of any handover zones between internal cells in the building. This is to avoid large soft handover zones between adjacent cells on adjacent floors and degraded HSPA service.

Example

One example of measurement results of isolation between floors can be seen in Figure 5.18. This is a GSM1800 indoor system in a typical modern office building with floor separation of 30 cm concrete slaps and suspended ceilings 50 cm under the slap with 3 m floor height. In this RF survey measurement the surveyed omni antenna is located in the leftmost part of the measurement route in Figure 5.18. Three measurements were taken, all using the same route on three different levels; one on the serving floor, one on the floor below and one two floors below.

The measurement samples were averaged and it was calculated that the attenuation of the floor separation was about 20 dB. This is valuable information when evaluating if you can use the effect when designing the solution by interleaving the antenna locations in-between floors, thus saving costs and increasing performance.

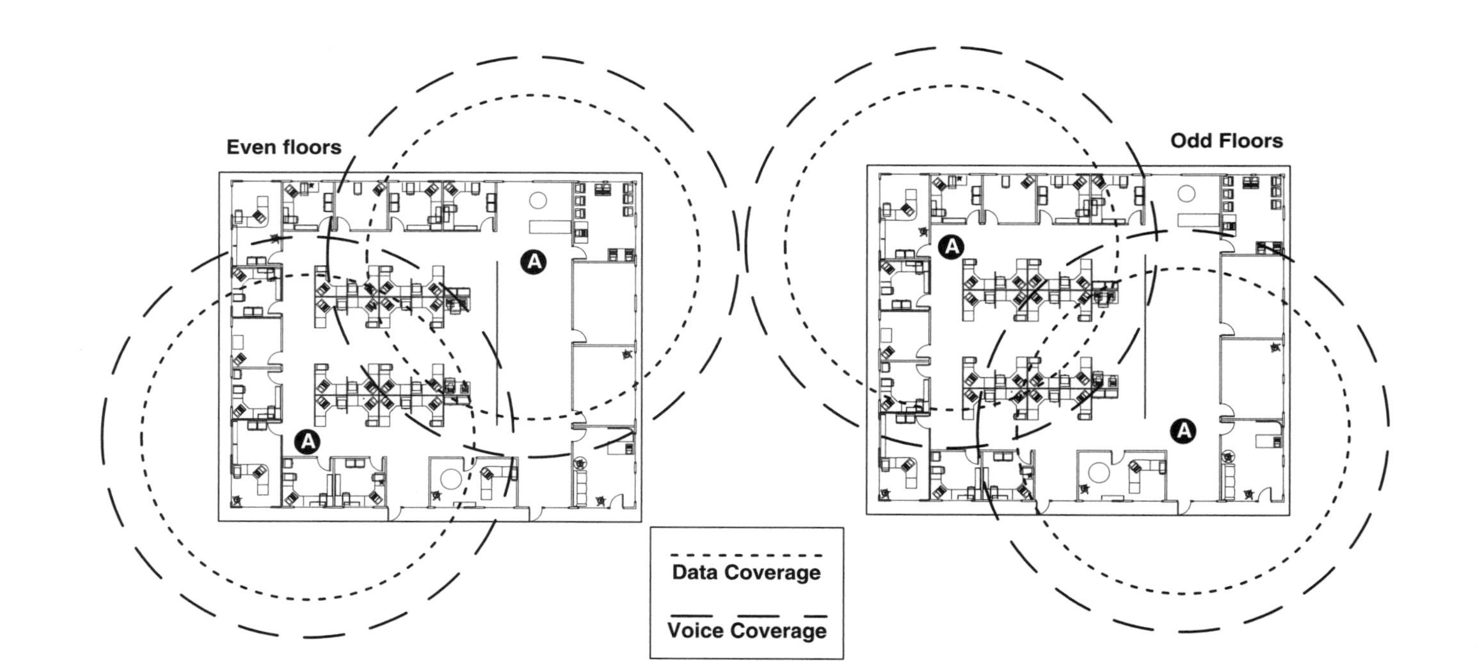

Figure 5.16 Often it is possible to interleave the layout of the antennas, in order to utilize the leakage between adjacent floors and to fill in the 'dead spots'

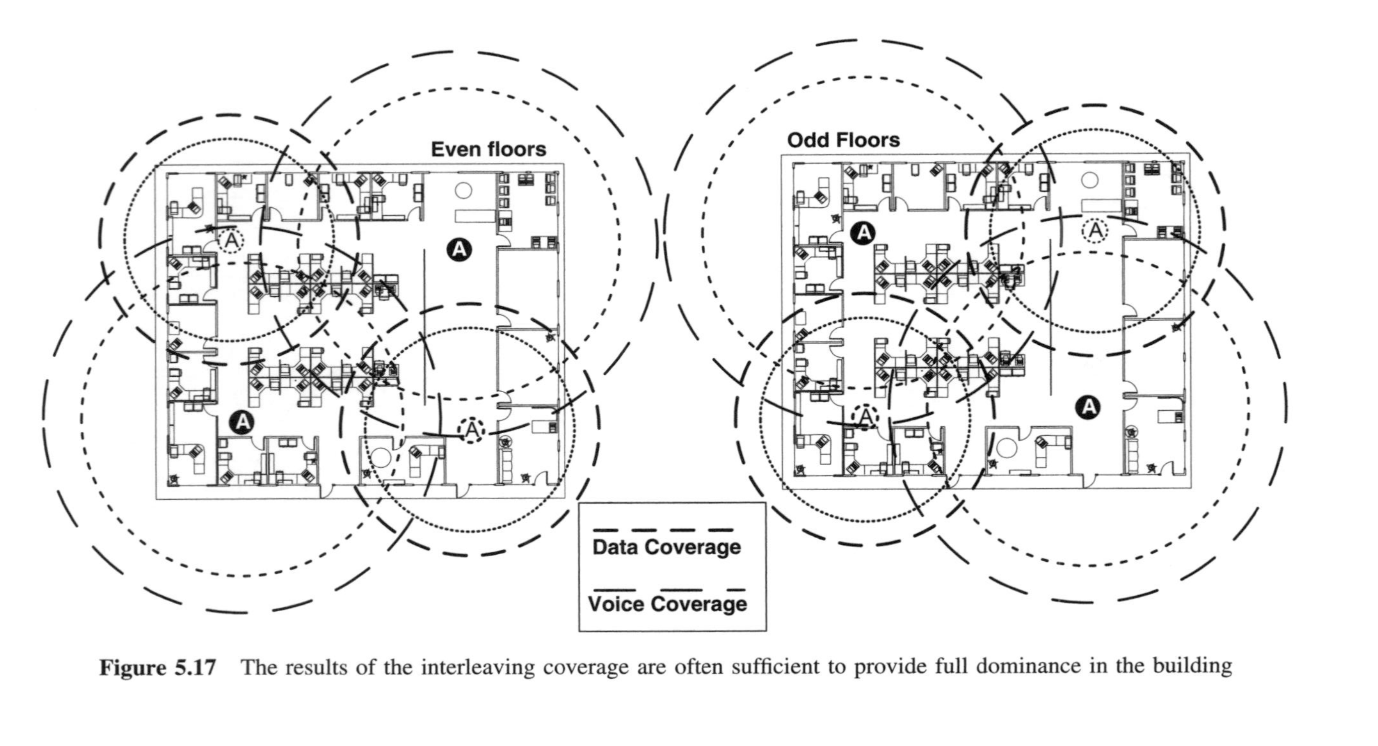

Figure 5.17 The results of the interleaving coverage are often sufficient to provide full dominance in the building

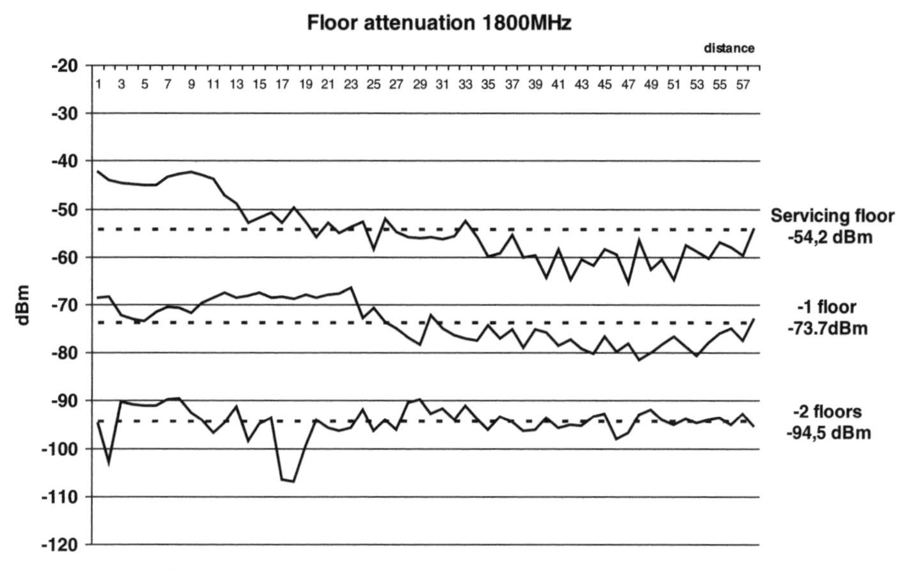

Figure 5.18 Measurement of floor attenuation on 1800 MHz

It is important to measure on all the frequency bands that are supposed to be supported by the indoor system. The penetration loss can be frequency-dependant, so make sure that you measure both GSM900 and UMTS if both services are to be used in the building.

Measure on all the Frequencies that are to be Used

If you only consider the free space loss (Section 5.2.8), it is a fact that, the lower the radio frequency, the less the link loss will be. Normally, when you analyze the loss through walls in a building, this effect would also be true, that the higher the frequency the higher the penetration loss. However, be careful with concrete walls and especially floor separations. The internal structure of the wall consists of a metal grid that has the purpose of reinforcing the structure of the wall or floor separation. Depending on the mask size of this internal grid structure, it will attenuate the RF signal in a nonlinear relation to the RF frequency. The reason for this is that a metal grid with a mask size of less than a quarter of the wavelength of the radio frequency will heavily attenuate the radio signal. Therefore it is very likely that a higher frequency UMTS/GSM1800 will penetrate floor separations and some types of wall better than GSM900. In practice, always measure all bands when performing this measurement.

5.3.10 Planning for Full Indoor Coverage

The behavior of radio waves and unknown factors inside buildings will to some degree result in unpredictable effects on the radio signals. Therefore radio planning is not 100% predictable. As we have seen in Section 3.2.2, the radio signal will have a fading pattern. This exact fading behavior depends on the environment and the speed of the mobile. The consequence is that, even with a very tight layout of antennas inside a building, it is impossible to guarantee '100%' coverage. You could get close to 100%, but it would be very costly.

The standard term for full mobile coverage indoors is often referred to as '98% coverage' defined at a given RF level. With this definition, it is possible to do a RF survey measurement post implementation to verify and confirm the design level. Based on these measurements, you are able to produce statistical evaluation of the measurements that hopefully backs up the 98% requirement.

It is very important to define whether the 98% is defined as area or time, or a combination. It is equally important to differentiate between the requirements in the different areas of the building. It might be alright to have 90% coverage in the basement of the building and then have 'full' 98% coverage in the offices floors.

Coverage of 98% Might Be Perfect, or it Might Not Be

The term 98% coverage is after all just statistics, and it is important to realize that even an indoor DAS design that fulfills the 98% level can be a bad design. Depending on where the last 2% not covered is located (Figure 5.20), it might leave the mobile users inside the building quite unhappy (as shown in Figure 5.19) and with a perception of far less than 98% coverage. It might be that most of the traffic is located in the last 2% of the building with low service (as shown in Figure 5.19).

The trick is to be very careful about where the last 2% is located. You can have perfect coverage throughout the building, but if the last 2% not covered is located in the executive area where all the senior management of the building is located, high-profile users with high airtime and high expectations of high-speed services (as shown in Figure 5.19), there will be a problem even if your statistics show 99.2% area coverage.

Therefore, always be sure to note on the floor plan where the heaviest users are located, in order to make sure they have perfect coverage and capacity. That one extra antenna installed from day one on the executive floor just might be money well spent and save you a lot of worries in the long run!

Also, in indoor radio planning, where statistics and reality can be different issues, always make sure you meet the demands of the users. Pay attention to 'service areas' like the storage room and the IT server room.

Achieving 100% Coverage

It is not possible to design 100% coverage, but if the users of the building can use the mobile service everywhere they expect, then the user perception can be 100% coverage. Therefore, a thorough and detailed input of the user expectations is very important. This input should preferably be a floor plan with clear markings of where the coverage is important, and where it is nice to have. This simple input can help in making a perfect '100%-covered' design, at a reasonable cost.

5.3.11 The Cost of Indoor Design Levels

Indoor coverage levels come at a cost – the higher the design level the higher the cost. It is crucial for the radio planner to select realistic indoor RF design levels when designing

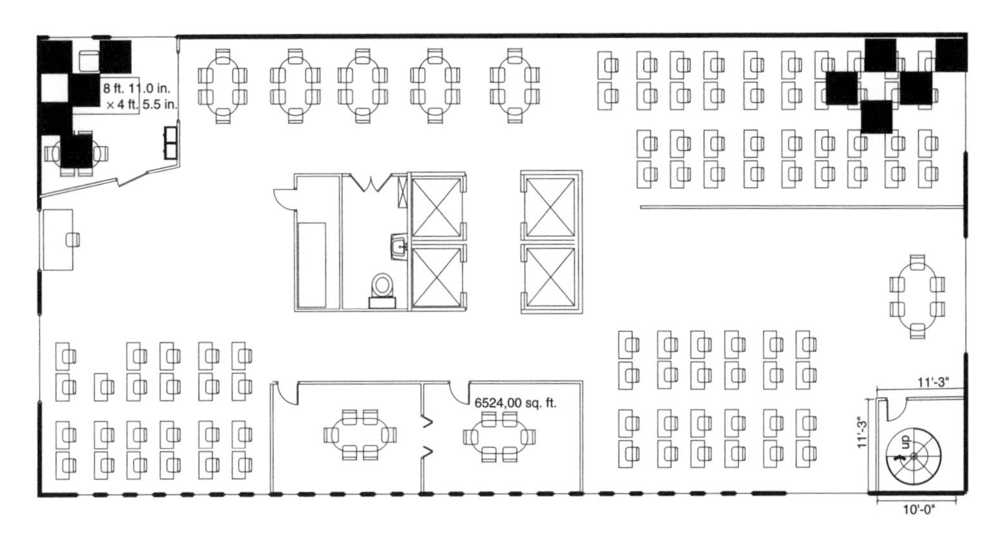

Figure 5.19 This building is covered 98%. This is verified by measurements, but the problem is that the part of the building with low signal is located in areas with heavy users. The users are not very happy with the performance

indoor DAS solutions, in order to make sure that the solution can service the needs of the mobile users. It is always wise to include some extra design margins to be on the safe side. It is also a good idea to look forward to the near future of the solution to make sure that the design can handle expected upgrades to 3G, higher data speeds or HSPA. This is a fine balance and the RF designer will need to make sure the business case is still positive. The

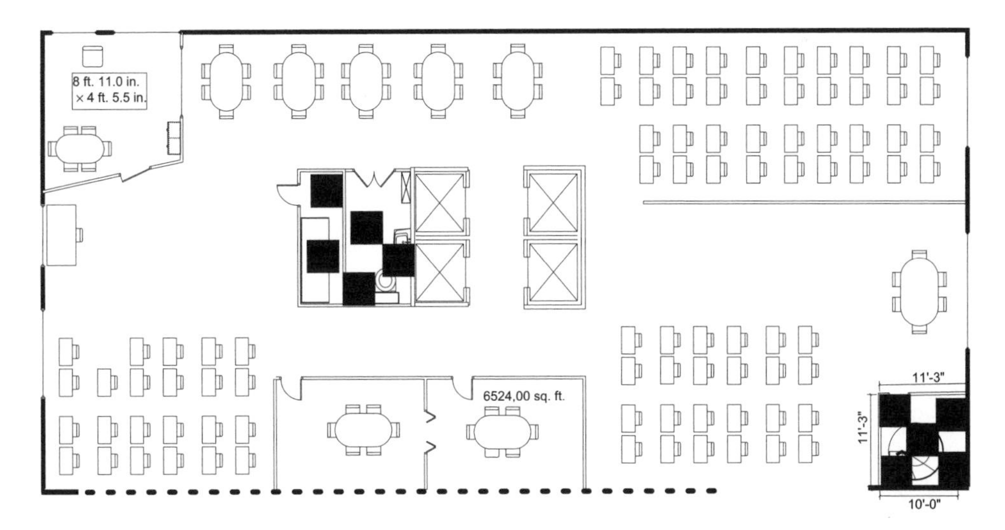

Figure 5.20 This building is also covered 98%. The RF designer has made sure that the areas with heavy users are covered and has placed the 'dead spots' in low traffic areas, thus providing perception of a quality solution

Table 5.2 Example of coverage level vs solution cost

10 000 m^2 dense office facility, 1800 MHz GSM 14 dBm EiRP			
DL level	Number of antennas	Coverage radius	Coverage area
−70 dBm	10	21 m	1385 m^2
−75 dBm	5	29 m	2642 m^2
−80 dBm	3	41 m	5281 m^2
−85 dBm	2	56 m	9852 m^2

cost of excessive design levels can be high. The example in Table 5.2 clearly shows that the radio planner has a big impact on the business case when selecting the design levels.

Do Not Under-design

Surely one should not under-design the solution, but it is recommended to use realistic design levels according to the specific building and the specific area in that building. One example could be that a 16-floor building designed for −70 dBm could be cost-optimized by using −85 dBm as the planning level in the basement and parking area. In these areas the interference from the outside network is much lower. You could have perfect quality using a 15 dB lower planning level in these areas, saving 10–20% of the system cost, just by selecting realistic design levels in these low-demand areas. Use 'the zone system' in Section 3.5.6 as a guideline, but always verify with measurements.

You can gain a lot in terms of performance boost and cost savings by pinpointing the high-use areas, the indoor hot-spots. Make sure you place the antennas in these areas where the users need higher-speed services and good quality, and plan the rest of the antenna placements with the hot-spot antennas as the base. By using that strategy, these areas might get really good service, −60 dBm or better when serviced by the nearby antenna. You can then without additional cost plan the rest of the DAS system so that you still keep an overall design level at −75 dBm in the rest of the area. The concept of 'hot spot' planning known from planning the macro network also applies indoors. See the example in the Section 5.4.1, where we provide really high data speeds on HSPA that can compete with WiFi.

5.4 Indoor Design Strategy

When all the measurement results have been analyzed and link budget calculations have been completed, you still need to place the antennas in the building. It is one thing to calculate the antenna service radius (see Figure 5.21) and the antenna overlap, but how do you actually use this information to implement the final design?

5.4.1 Hot-spot Planning Inside Buildings

The term 'hot-spot' is often used when planning the macro layer. A 'hot-spot' is a place of high traffic density of mobile users, needing special attention in terms of coverage, quality, capacity and data rates.

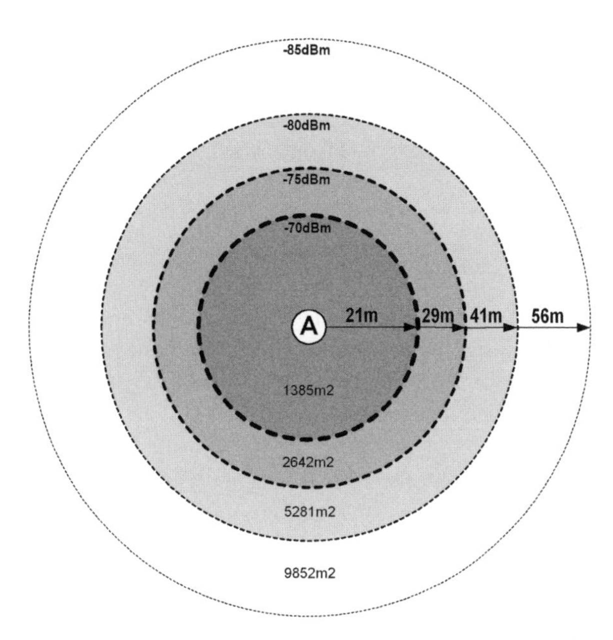

Figure 5.21 Indoor coverage radius and area vs design level from omni antenna

Hot-spots also exist inside buildings and, like in the macro network, these hot-spots can produce the major portion of the traffic in the cell. Indoor hot-spots will typically be areas where users sit down and work using their PC and mobile for an extended time. Most high-speed mobiles are actually data cards in PCs, so naturally the user will use a convenient sitting area where he can work in relative quietness. Examples of indoor hot-spots with high traffic density and high requirements for data speed performance could be business lounges in an airport, the food court in a shopping mall, the press area in a sports arena, the conference area at a hotel and executive and meeting areas in a corporate building.

It is highly recommended to place antennas in these areas, thus securing a good design margin for future data service. This strategy does not have to be costly – often it is just a matter of using the hot-spot areas as the base for the antenna placements, and aligning the placement of the remaining antennas accordingly.

UMTS/HSPA Can Easily Compete with Wi-Fi

Hot-spots in public buildings will often already be covered by Wi-Fi service, due to the high concentration of data users in these areas. In many cases the GSM/UMTS indoor DAS implemented by the mobile operator will be a direct competitor to Wi-Fi service. Therefore the radio designer must be sure that he places the DAS UMTS/HSPA antenna in these hot-spot areas. By doing so, the UMTS/HSPA data service and even the GSM EDGE can compete with the speed of Wi-Fi.

In theory, the data speed on Wi-Fi is quite high, in many cases stated as better than 54 Mbps. In reality the speed servicing the user will be limited by the ADSL backhaul from the Wi-Fi access point to the internet, not the radio speed on the air interface. Therefore in reality the speed is often less than 200–300 kps. Sometimes, however, it is up to more than 2 Mbps.

Mobile data services are also more user-friendly. There is no need for individual charging in the local Wi-Fi hot-spot and all charging and roaming on mobile data are settled over the normal billing system. The mobile GSM/UMTS/HSPA data service will also provide the user with total mobility, with handovers of data service providing a global coherent data service.

5.4.2 Special Design Considerations

Even though most of the design methods and considerations are the same no matter what type of building you are designing, special attention needs to be addressed to the type of building you are designing the DAS for. These are some of the points we need to address, in addition to all the standard RF considerations.

- Make sure you prepare for more capacity or sectors for future upgrades.
- Make sure you cover the executive floor 100%.
- Is there a need for elevator coverage?
- Are there special installation challenges (e.g. fire proofing)?
- Pay attention to the service rooms or areas (e.g. IT server rooms).
- Are there special EMR concerns (like in a hospital)?
- What type of services might be needed in the future – 3G, 3,5G?
- Are there any hot-spots in the building that need special attention?

5.4.3 The Design Flow

When designing large projects, like campus areas, airports or shopping malls, it is worthwhile structuring how you do the design. Done correctly, you will get the most performance of the system at the lowest cost and will be able to do a rapid, future-proof design.

Let us do an example of a shopping mall (as shown in Figure 5.22), assuming that the site survey has been done and the link budget has been analyzed and calculated. Now you have

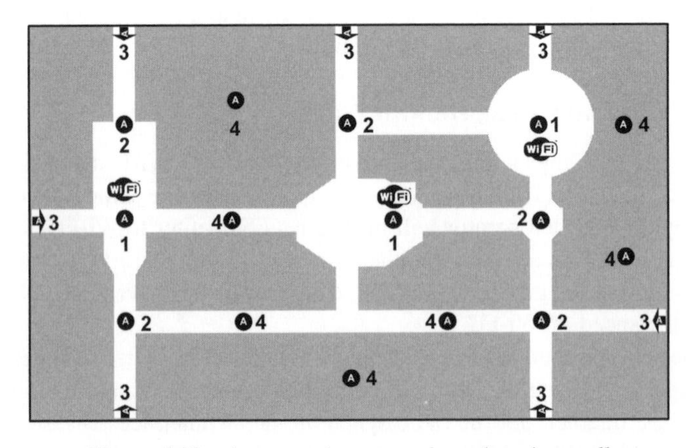

Figure 5.22 Antenna placements in a shopping mall

an idea of the coverage range of the antennas in this particular environment and can start placing the antennas on the floor plan. First consider the characteristics of the particular building. In this example, being a shopping mall there will be some basic characteristics:

- All the shops are likely to have 100% glass facades facing the internal streets; therefore there is close to no wall attenuation from the street into the shops.
- Be careful about open skylights with regards to leakage and isolation to the macro layer. There might be nearby macro base stations leaking inside the mall in this area.
- Often you are only allowed to place antennas in the internal 'streets' of the mall.
- Capacity load can be extreme at peak shopping time, e.g. during sales.
- Prepare the system for capacity upgrades by dividing it into more sectors in the future.
- Remember the parking area, often subground level, and do not forget overlap to the outdoor network for the HO zone where the users enter and exit the mall.
- Try to avoid placing antennas near the PA speaker system to avoid interference.
- Is it necessary to cover the areas with no public access, e.g. storage rooms or offices? Those areas can generate quite high revenue.

5.4.4 Placing the Indoor Antennas

The recommended breakdown of the actual antenna placements in the shopping mall is given in Figure 5.22.

1. *Place the hot-spot antennas and maximize data performance.* Locate the hot-spots for data users; they will be the most demanding areas with regards to good design margin and level. In this case of a shopping mall, the hot-spots for data and voice are typically the food court, internet cafes and sitting areas. If there is a Wi-Fi operator present in these areas, you can compete against Wi-Fi on data by placing hot-spot antennas in these areas. Therefore all the first placements of DAS antennas, marked '1' on the floor plan, are placed in these areas, forming the reference for the next step in the design phase.
2. *Place the 'cost-cutting' antennas.* After the hot-spot antennas are placed you must place all the antennas that will maximize the coverage per antenna, marked '2' on the floor plan. By using the 'corridor effect', you can maximize the coverage of each of these antennas. You place these antennas in all the intersections of the internal streets of the mall so that the coverage will be spread in the directions of the internal streets, which will give good value for money.
3. *Isolate the building.* Dominance and well-defined HO zones along the perimeter and the entrances of the shopping mall are secured by the placement of the antennas marked '3'. These antennas will often be directional antennas pointing towards the center of the building as the preferred solution. This will isolate the building from even very close outdoor sites.
4. *Fill in the gaps.* The last placement of antennas will be 'filling the gaps' between the antennas just placed. It is often necessary to place an antenna in big shops inside the mall, if you are allowed that is. Normally installation of antennas in shopping malls is restricted to the internal streets. Therefore it can be hard to cover deep inside the larger shops; try to cover them as much as possible by having more antennas in the internal streets of the shopping mall.

5.5 Handover Considerations Inside Buildings

It is important to make sure that the indoor DAS system implemented in the building is prepared for future traffic growth. The best way to prepare this is to have a sector plan for future sectorization of the system. Even if the system is implemented as one sector, you need to look ahead, especially for UMTS, in order to prepare for more sectors.

Well-defined HO zones are important for GSM and UMTS/HSDPA to avoid 'ping-pong' HO on GSM, extensive soft HO zones on UMTS and degraded HSPA performance. The focus should be on well-defined and controlled handover zones, preferably placed in areas with low traffic in the building. When the DAS is designed and implemented correctly, the dominance and isolation of the indoor system will insure well-defined handover zones to the outdoor macro network. However, in extreme cases, like a rooftop site on the neighboring building, you will sometimes have to deal with some signal leaking inside the building, even with high signal level from the indoor system.

As a general rule you must try to avoid having the handover zones in large open areas inside the building. Here it can be difficult to design and control the handover zone. Try to take advantage of the natural isolation provided by the building to separate the different sectors or cells. This can be done by using the floor separations as the handover border, or the fire separation zones inside the building (Section 5.3.6). Typically these fire zones will be divided by heavy walls with high RF isolation, which are perfect for giving a well-defined handover zone.

5.5.1 Indoor GSM Handover Planning

It is important to realize that the handover control parameters are a crucial part of the indoor design and, in order to optimize the implemented solution, these parameters must be tuned. The typical handover scenario in a GSM multicell building is shown in Figure 5.23. This is a multicell indoor system with a total of five cells. Well-defined handover zones between the internal cells in the building are a must, and in this case the primary handover between the internal cells are done via cell-5, which also serves the elevator, so no handovers will occur when using the elevator. Even with cell-5 as the normal internal handover cell, there are defined handovers to the internal adjacent cells, just in case cell-5 is full and cannot tender for the handover between adjacent cells.

Limit the Number of Macro Handover Candidates

Normally the internal cells in the building are limited to only having defined handovers to nearby adjacent indoor cells within the building. This to prevent those mobiles typical in the topmost part of the building starting to make handovers to distant macro sites via nearby macro sites. This is a common problem. The mobile might finally end up on a distant macro cell that has no neighbor relations back to the indoor cell. The consequence is often dropped calls when the users move from the perimeter at the windows back to the center of the building.

Keep Idle Mode Traffic on the Indoor Cells

In GSM idle mode the mobile can be controlled with special cell offset parameters. You must make sure that the mobile will camp on the indoor cell, even if nearby macro sites are leaking

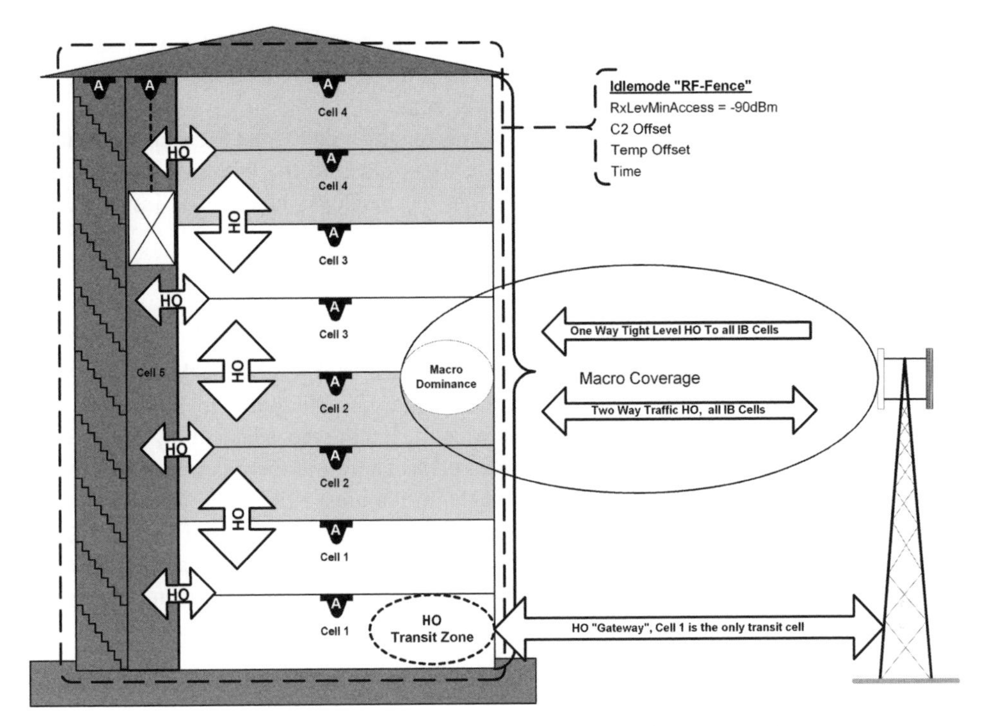

Figure 5.23 The typical GSM handover scenario in a building

high-level signal inside the building. In practice, you use a combination of the lowest allowed Rx access level on the mobile (RxLevAccessMin); in this example in Figure 5.23, it is set to −90 dBm. Then by applying a cell offset (C2), for example, of 40 dB, the mobile will add 40 dB to the evaluation of the cell once the cell is received more strongly than −90 dBm. The result is that the mobile evaluates the cell to be $-90 + 40 = -50$ dBm receive level.

This is very useful in keeping the mobiles on the indoor cell in idle mode; you can thus dominate in areas with really high macro level leaking in the building. In addition, you can use temporary offsets and time penalty periods; this also makes it a very useful tool to control the traffic if the indoor cell leaks out onto the nearby street.

Define 'Emergency' HO Candidates

It will always be a good idea to make sure you define a 'one-way' connection back to the indoor cell if mobiles should camp unintended on the nearby outdoor cell. In extreme traffic situations where the indoor cell might be congested, it can be a good idea to have traffic-controlled HO to the nearby macro site, in order to unload the indoor cell and avoid capacity blocking of calls.

Handover Zone to the Macro Network

It is very important to have a well-defined transit handover zone to and from the building. Normally this zone is around the main entrance and the entrance to the indoor parking area.

Only the cell(s) covering the entrance or exit of the building should have normal HO connections to the outdoor network and preferably to a very limited number of outdoor cells in order to ease the optimization of the handover zone.

For GSM applications it is recommended to make sure you do not pick up any outdoor traffic on the indoor system, by outdoor mobile users close to the building. The handover parameter should be tuned in order to insure that the handover takes place just inside the building.

5.5.2 Indoor UMTS Handover Planning

In GSM there is some margin of offsets that can be applied in order to tune the HO zones and areas. Adjacent UMTS cells will typically use the same frequency; the mobile has to be in soft HO as soon as it is able to decode more cells. If not, the adjacent cells will cause interference, degraded performance and dropped calls. Therefore there is not a large margin for tuning the UMTS handover and only good radio planning will do the job.

The typical handover scenario in a UMTS multicell building is shown in Figure 5.24. This is a multicell indoor system with a total of five cells. Well-defined handover zones between the internal cells in the building are a must to limit the soft handover load; in this case the primary handover between the internal cells is done via cell-5. Cell-5 also serves the elevator as the only cell, avoiding soft handovers in the elevator.

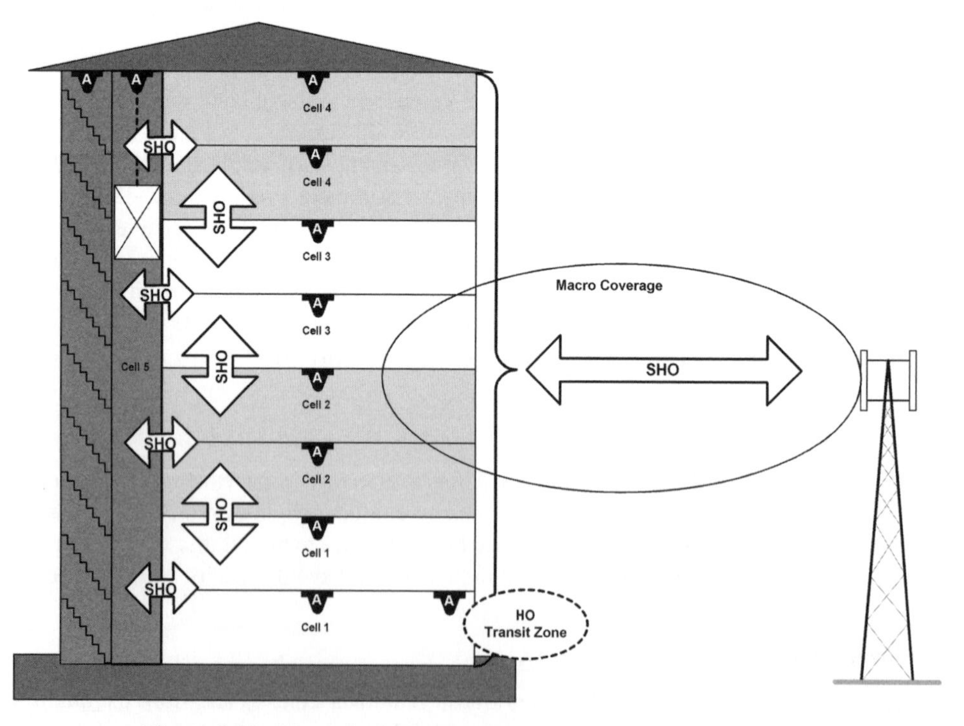

Figure 5.24 The typical UMTS soft handover scenario in a building

However, you must also be sure to define handovers to the internal adjacent cells, in order to perform soft handover in the limited areas where two cells might cover the same area. Keep those areas to a minimum.

Good UMTS RF Planning is the Solution

The best way to optimize the handover zones on UMTS is by good RF planning, correct antenna placement and careful attention to the handover zone planning from the initial design phase. The solution is once again to provide dominance of the indoor cells and isolation between the cells. This is the correct way to control the handover performance and to limit soft handover zones.

To some extent the CPICH (pilot channel) power can be used in order to offset the cell sizes, thereby moving the soft handover zone slightly. However, the effect is fairly limited, compared with the freedom of parameter settings for tuning the handovers on GSM.

The use of separate frequencies for the outdoor and indoor network can be a solution in some extreme cases, but these hard handovers are often a challenge for the network as well as many mobiles. Operators are typical assigned two or three RF channels for UMTS, so spectrum- and capacity-wise this is a very expensive solution and should only be considered as a last temporary resort, until the real problem can be solved or the macro sector removed.

Furthermore you might not be able to allocate a new RF channel for HSDPA, and HSDPA will take a big hit from co-channel interference from the macro cell, degrading the speed and performance.

The HO Zone to the Outdoor Network

Obviously there needs to be a handover zone to the outdoor network designed in the entrance areas of the building. A well-defined HO zone to the outdoor network is important, and on UMTS it can be wise to consider having the soft HO just outside the entrance of the building. The reason for this is that, if the soft HO occurs inside the building, it is likely the mobile will be running on high transmit power in order to reach the outdoor cell. This 'noise' generated by the high power mobile will raise the noise level on the UL of the indoor cell, thus cannibalizing the capacity of the indoor cell. This can be a big issue when designing high-capacity indoor solutions with many users at the entrance of the building or shopping mall.

5.5.3 Handover Zone Size

When designing handover zones it is important that you take the speed of the traffic into account. Furthermore, you need to make sure that you include a safety margin in order to make several handover attempts, should the first attempt fail. The handover zone must accommodate this safety margin, and you must make sure that you keep the coverage design level throughout the full size of the handover zone. The last thing you want to encounter is problems with handover retries once you are below design level. That is a sure recipe for disaster and dropped calls.

Example, HO Zone Size

If you use a HO time of 4 s as an example, you can evaluate the size of the needed HO zone. If a car moves at 30 km/h out of the underground parking area (8.33 m/s), you would need a HO zone of 33 m.

Example, Handover Zone Safety Margin

As pointed out earlier in this chapter, it is not enough to design the handover zone to the minimum size. You must provide the mobile enough time to perform cell decoding, measurement, evaluation and handover execution of both cells at the same time.

The signaling load of the cells can also be an issue. This is normally not a concern when designing indoor solutions: indoor cells will typically only have to perform a few handovers simultaneously. However, there are cases that might demand consideration of the signaling load of many simultaneous handovers; consider a tunnel scenario where a train is moving at relatively high speed, and inside the train there might be 40 users in traffic performing handover at the same time. In this case you must certainly include a safety margin in the handover zone size. In addition GSM offers some presynchronization functionalities or 'chained' cells. On GSM, UMTS and HSPA it is recommended to have the two cells in the critical handover zone serviced from the same base station. This will increase the handover success rate and load the network far less for these critical handover types.

5.6 Elevator Coverage

Mobile users inside a building expect a good quality coherent service level throughout the building, the elevators included. It is a major challenge to service elevators with RF coverage, and requires special consideration. Most elevator lift-cars are a virtual metal enclosure with very high RF attenuation, often exceeding 60 dB. The speed of the elevator adds to the challenge, especially in very high buildings with high-speed express elevators. However, the demand for mobile coverage inside elevators is growing, primarily motivated by the normal requirement for voice and data coverage everywhere. The mobile is also considered as an extra security line by the users, even though there is an emergency phone installed inside most lifts. Therefore mobile coverage inside elevators is a must.

There are several options on how to provide mobile coverage in the elevator; the optimum approach depends on the individual solution and the constraints of the actual elevator installation. The best mobile performance is achieved if the RF design is done so that there will only be one dominant cell covering the elevators throughout the building, thereby avoiding handovers inside the lift. However providing sufficient coverage inside the lift-car can be a challenge and sometimes close to impossible, due to the metallic enclosure of the lift-car.

In small to medium-sized buildings it might be preferable not to have a dedicated cell for the elevator only, but to use one of the existing cells in the building. In that case it is recommended to use the topmost cell in the building to service the elevator shaft as well as the indoor area.

5.6.1 Elevator Installation Challenges

To obtain approval and permission to place any equipment in the shaft that is not related to the operation of the elevator itself can be difficult. If you want to install antennas inside the elevator shaft or in the elevator car itself, you will often need special approval of the cable types and all other equipment used, and issues related to fire rating and mechanical stress need to be cleared and certified. Always check and make sure you follow the installation rules and guidelines in the specific elevator in the specific building.

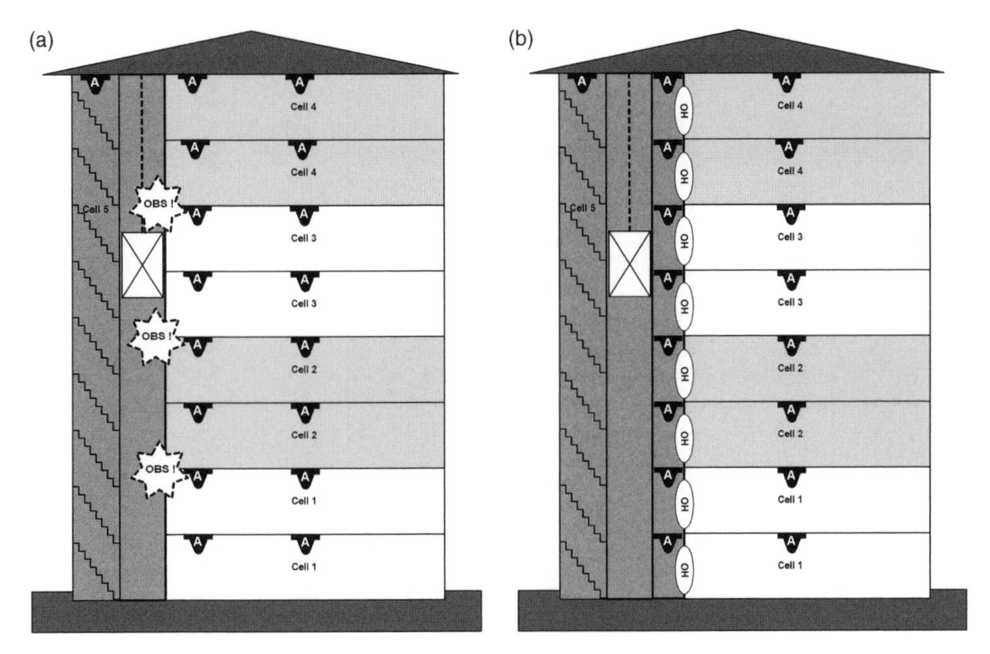

Figure 5.25 The typical way to provide elevator coverage

5.6.2 The Most Common Coverage Elevator Solution

The most used design when covering elevators is to place antennas close to the elevator shaft on each floor, preferably in the lift lobby with the antenna 1–2 m from the lift door [as shown in Figure 5.25(a)]. Often this design will work fine, especially in small buildings with only one sector/cell covering the building. For larger buildings with more sectors you must be careful using this approach, and be sure not to create a handover zone inside the lift.

The situation is shown in Figure 5.26(a), where the potential problem is evident. The handover zone must be designed just outside the lift shaft [as shown in Figure 5.25(b)], in order to avoid handovers inside the lift-car when moving at high speed.

5.6.3 Antenna Inside the Shaft

This design is shown in Figure 5.26(a), with the antenna mounted in the lift shaft, most commonly at the top, but it is also possible to have the antenna in the bottom and beam upwards. Sometimes a bidirectional antenna mounted in the middle of the shaft can be considered; it all depends on penetration losses into the lift-car and the height of the building. For larger buildings, a combination of these antenna placements can be considered.

Often the antenna inside the lift shaft has to be placed on a small bracket mounted on a door to the elevator shaft. Thus you access the antenna by just opening the door from outside the shaft. This provides easy access to the antenna without accessing the shaft itself; in that case it is easy to install cables and other equipment on the outside of the shaft. This will limit

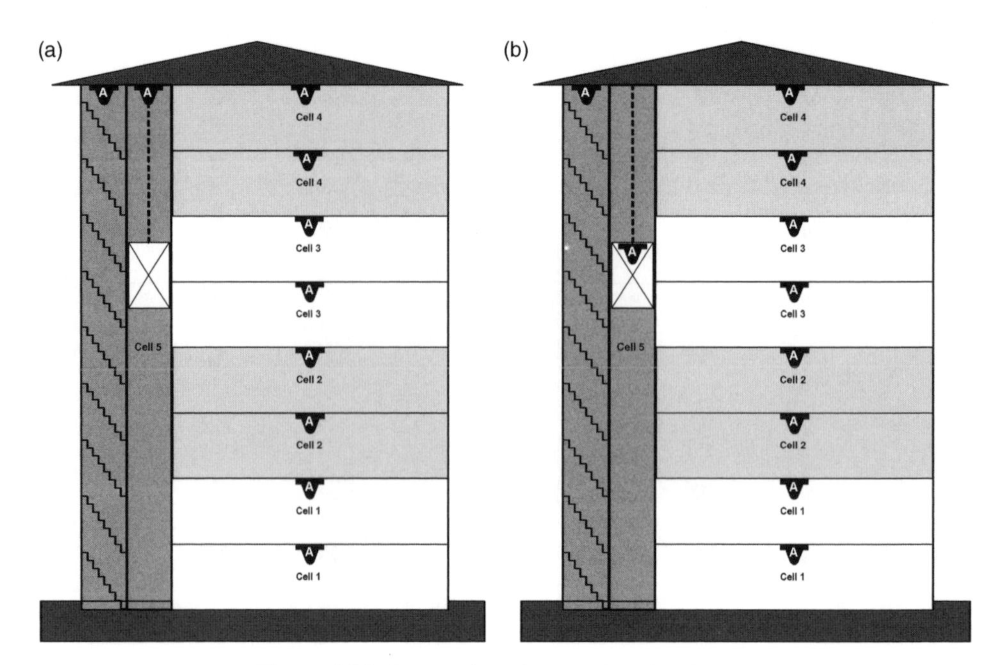

Figure 5.26 Two options for covering the elevator

the equipment inside the shaft to only the antenna. This approach might make it easier to achieve the approval for implementing coverage in the shaft.

Be careful in very high buildings, as sometimes elevators come in two levels! One elevator-car might actually be serving both even- and uneven-numbered floors at the same time; you will have two levels of users inside the elevator. In that case it will be very difficult to cover the lower-most lift-car from an antenna in the top of the lift-car. A new variant of this type of twin elevators is to have two cars moving independently in the same shaft. In this case you need to cover the shaft from both the top and bottom of the shaft.

Special consideration needs to be given to the other cells in the building, and dedicated attention to the layout of the handover zones is advisable. Often the antennas in elevator shafts will be designed to radiate high power levels in order to penetrate the elevator-car. However, be careful that this high level of signal inside the lift shaft does not leak out to the office area nearby the elevator lobby.

I have seen problems with this type of design, where all users within 20 m of the elevator lobby handed over to the elevator cell once the elevator doors opened, and were then dropped once the doors closed again, due to the abrupt loss of signal and lack of time to handover back to the office cell.

5.6.4 Repeater in the Lift-car

You could also consider installing a RF-repeater on the lift-car, using an external antenna on the top of the car and a small antenna inside the car. In combination with the shaft antenna installed in the top of the shaft, this can be a good solution, but you need to be careful with the gain settings on the repeater. Preferably you should configure the repeater settings so it just compensates for the penetration loss of the lift-car. If you use too high a gain you might saturate the mobile and the base station when the lift-car is close to the shaft antenna on the topmost floors. On UMTS this might have a serious impact on all the traffic in the cell, so caution is advised.

5.6.5 DAS Antenna in the Lift-car

Recently it has been possible to install an antenna inside the moving lift-car [as shown in Figure 5.26(b)], as a part of the active DAS (Section 4.4.2). This antenna will be connected to the active DAS system of the building like all the other antennas. This option, however, is only possible when using pure active DAS. The active DAS relies on thin cabling to the remote antenna unit that services the antenna. This cable will typically be CAT5 (LAN cables) or CATV (cable TV cable), similar to the cables that are already being used to service the lift-car, for control, communication and CCTV. By using similar cables to those already installed to service the elevator, it is easier to get permission to install the DAS antenna in the lift-car. The big advantage of this approach is full control of the RF environment inside the lift and control of the zones.

The DAS antennas must be set to a very low output signal level, and you must be sure to use a version that can handle mobiles not in power control close to the antenna radiating at high power. Remember that some mobiles inside the lift are not being serviced by the elevator antenna, but by the nearby macro site of the competing operator. Hence these users are likely to transmit at full power and saturate the antenna unit in the active DAS installed in the elevator. Therefore it is recommended to install an attenuator between the antenna unit and the DAS antenna inside the lift. This attenuator (20–30 dB) will add sufficient link loss and prevent this potential problem.

5.6.6 Passive Repeaters in Elevators

Sometimes you get the question, 'why not just mount two antennas back-to-back just connected via the cable, one antenna on top of the lift-car, the other inside the lift?' Let us have a look at the set-up of this concept in Figure 5.27. First of all, we know that the free space loss in this example of GSM 1800 MHz will be 38 dB at 1 m from an antenna. Applying the free space loss, we can calculate the system gain of the passive repeater system. From 1 m in front of the donor antenna to 1 m in front of the service antenna on the other side of the passive repeater, you will have a 'system gain' of $2 \times -38\,\text{dB} +$ antenna gain – feeder loss.

Example

You can try to do a link calculation of a 30 m-high building. According to the free space loss (Section 5.2.8), the free space loss would be about 67dB at 30 m distance, and then 38 dB from the antenna in the lift-car to the user.

$$\text{Total free space loss} = 67\,\text{dB} + 38\,\text{dB} = 105\,\text{dB}$$

If we use a donor antenna at the top of the shaft with 7 dBi gain beaming down the elevator shaft, a 7 dBi pick-up antenna, 2 dB cable loss and 2 dBi antenna gain for the antenna in the

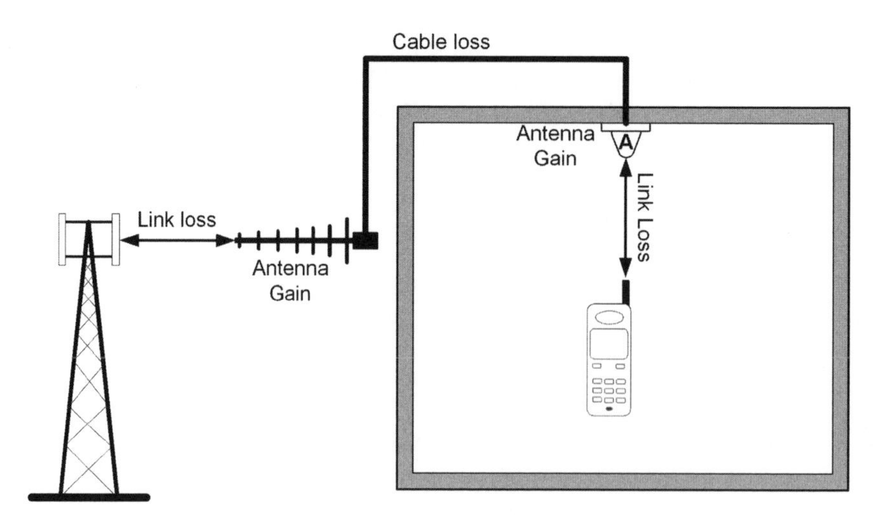

Figure 5.27 Two antennas 'back to back' might work as a passive repeater, but it takes a high donor signal

lift, then we can calculate how much power we need to feed this antenna in order to achieve the −85 dBm inside the lift-car when it is in the bottom of the shaft:

$$\text{Total link loss} = \text{total free space loss} + \text{cable loss} - \text{antenna gain}$$
$$\text{Total link loss} = 105\,\text{dB} + 2\,\text{dB} - 2\,\text{dB} - 7\,\text{dB} - 7\,\text{dB} = 91\,\text{dB}$$
$$\text{Required power at the transmit antenna} = 91\,\text{dB} - 85\,\text{dBm} = 6\,\text{dBm}$$

5.6.7 Real-life Example of a Passive Repeater in an Elevator

The concept of passive repeaters actually works fine for elevator coverage in many cases. In a real-life elevator shaft we placed a DAS antenna, with 7 dBi pointing down in the top of the lift shaft, a feed to the antenna of 10 dBm and using a similar 7 dBi pick-up antenna at the top of the lift-car and one 2 dBi antenna inside the lift-car. The signal from the passive repeater inside the lift was better than −80 dBm at distances of up to about 80 m.

We can try to verify this by doing a fast link budget for the downlink:

$$\text{free space loss (dB)} = 32.44 + 20(\log F) + 20(\log D)$$

We can calculate the free space loss in the shaft at 75 m:

$$\text{free space loss (dB)} = 32.44\,\text{dB} + 20(\log 1850) + 20(\log 0.080)$$
$$\text{free space loss (dB)} = 32.44\,\text{dB} + 65.3\,\text{dB} - 22\,\text{dB} = 75.74\,\text{dB}$$

We need to add the free space loss at the inside of the lift (38 dB) and the antenna gains [7 dBi + 7 dBi + 2 dBi + 2 dBi (mobile antenna gain)] and the cable loss (−2dB). The system gain between the connector at the DAS antenna and the mobile user inside the lift can then be calculated:

$$\text{system loss} = \text{free space loss shaft} + \text{free space loss lift} + \text{cable loss} - \text{antenna gain}$$

$$\text{system loss} = 75.74\,\text{dB} + 38\,\text{dB} + 2\,\text{dB} - 18\,\text{dBi} = 97.74\,\text{dB}$$

$$\text{calculated user level inside the lift} = \text{DAS antenna power} - \text{system loss}$$

$$\text{calculated user level inside lift} = 10\,\text{dBm} - 97.74\,\text{dBm} = -87.74\,\text{dBm}$$

You can see from the calculations that the measured level inside the lift is actually about 8 dB higher than the calculated level – why? Actually what happens here is caused by the 'tunnel effect' on the radio signal. Both the service antenna on top of the lift shaft and the donor antenna on top of the lift-car are 7 dBi gain antennas. According to the data sheet, the radiation pattern of the antenna is about 70° beam-width (opening angle), but the 'gain' of any antenna is related to the directivity of the antenna – the more 'narrow' the beam-width, the higher the gain.

The narrow elevator shaft has a big impact on the directivity; the radio waves are forced to stay inside the elevator shaft, focusing all the energy down inside the shaft. Therefore the 'gain' (directivity) of the antenna is no longer 7 dBi, but slightly higher. The mobile used inside the lift may also have been slightly closer to the service antenna, adding to the signal.

Using a passive repeater solution is definitely possible for small to medium-sized buildings for servicing the elevator, provided you can get a DAS antenna installed at the top of the lift shaft as the donor antenna to the system.

5.6.8 Control the Elevator HO Zone

The best performance is achieved if the HO zone can be placed outside the elevator, i.e. in the lobby in front of the elevator. This can only be achieved if we make sure that the same cell is covering the elevator from top to bottom, as in Figure 5.25(b). Often this cell will be one of the existing cells in the building (the topmost cell).

In order to perform a successful handover, there must be some overlap of coverage from the two cells. In a situation where you do not have antennas installed in the lift shaft and are covering the elevator only from antennas in the elevator lobby for the different floors, the RF signal has to overcome both the penetration loss into the lift-car and the penetration loss through the floor separation in order to provide cell overlap for the handover to succeed. Therefore, the speed of the elevator is an important parameter of concern when designing this type of elevator coverage for multicell buildings. You need the overlap of the two cells in order for the handover to succeed. Typically you need a minimum of 3 s to perform handover measurement, decoding, evaluation processing and signaling time in order to make the handover a success. In addition, you need to add an extra margin of time in case the first handover attempt fails, and time to perform a new handover attempt. This makes this approach in multicell buildings almost impossible in practice for use in high-rise buildings with fast-moving elevators. The only solution is to dedicate a cell to cover the elevator shaft and sometimes include the elevator lobby [Figure 5.25(b)].

5.6.9 *Elevator HO Zone Size*

When designing handover zones with relative fast-moving users, attention to the size of the HO zone is important. When designing indoor solutions, there are typical scenarios where the speed of the users plays a major role in the size of the HO zone; elevators used in the building and cars driving to and from the parking area inside the building are some examples. It is recommended to have minimum of 3–4 s to allow for decoding, measurement evaluation and signaling for the handover, especially for GSM. The handover might fail; therefore you must include a safety margin for at least one handover retry to cope with extreme cases where many mobile users will perform handovers between the same two cells at the same time.

Example, Elevator HO Zone Size

Using 4 s as an example, you can evaluate the size of the needed HO zone. If an elevator moved at 5 m/s, you would need a HO zone of 20 m. If the floor separation in the building was 3 m, you would have to provide an overlap of about seven floors! This clearly underlines the potential problem in multicell buildings in Figure 5.25(a).

5.7 Multioperator Systems

Often there is a need for the indoor DAS to support multioperator configurations, where more than one operator or band is to share the same DAS. This will typically be the case in large public buildings, airports, convention centers and tunnels, where there is a need for high capacity and in many cases more than one type of mobile system on air over the same DAS, GSM/DCS + UMTS.

From an economic perspective there is much to be saved by the operators if they share the DAS. Seen from the building owner's perspective, it is also preferable to have only one DAS, one installation, one set of antennas, one project to coordinate and one equipment room.

In order to connect more operators to the same DAS, you need to combine several base stations, repeaters and bands into the same system. Combining the operators and bands into the same DAS is not a trivial issue; you need to pay close attention to many potential problems that might degrade the service if you are not careful.

The principle of the active DAS, where DL and UL are separated at the interface to the main unit, as well as the low power levels required at the input (typically less than 10 dBm), will ease the requirements and the design of the combiner. If considering a passive DAS for a multioperator solution, then the combiner becomes an issue. Be careful when combining several operators or bands at high power using many carriers on air. This has the potential for some really big problems. However, if you select high-quality components and a careful design, preferably using a cavity filter system tuned to the individual bands, you succeed.

In many cases multioperator solutions will be installed in large buildings, airports, hotels and shopping malls. Therefore active DAS is often the preferred choice due to the improved performance on data services and ease of installation and supervision.

5.7.1 Multioperator DAS Solutions Compatibility

There are important RF parameters to take into account when designing multioperator DAS, but if you pay attention to the most critical parameters, then you will succeed.

Rx/Tx Isolation

The Rx/Tx signal from each individual base station must be separated according to the specification of the base station supplier. When using the combined Rx/Tx port on the base stations, this will normally not be a problem, due to the internal duplex filter in the base station, which will separate the uplink and downlink. However you must also make sure that the downlink signal from one base station will not reach any of the other base stations receivers, according to the isolation specification.

When using an active DAS where the main unit needs separate uplink and downlink signals, it is preferable to use the Rx diversity port for the uplink from the DAS to the base station, and use the combined Tx/Rx port for the downlink only. This will help in designing a combiner that can provide excellent isolation of the UL and DL signals. However, you must be aware that most base stations will trigger a 'main receiver fault' alarm, due to the lack of uplink signal on the main receiver port, the combined Rx/Tx. As long as you cancel this alarm trigger in the network surveillance system, then you can avoid the alarm problems. A typical value for Rx/Tx isolation requirements is 25–30 dB (without including the internal filter and combiner in the base station), but refer to the specifications and guidelines for the specific base station used.

Return Loss

This is the reverse power, the reflected power from the DAS–combiner system. In order not to trigger any VSWR alarm on the base station, you need to keep the reflected power below the trigger value. A typical value is >10 dB (the difference between forward and reverse power).

Inter-band Isolation

This is the isolation between GSM–DCS–UMTS. In order not to de-sensitize the receiver in the BS it is important to achieve maximum out-of-band rejection. The basic specification on this issue is specified by ETSI. The exact value depends on the base station manufacture, how much you will allow the base station to be desensitized and the band configuration. Look up the value in the specification of the base station. A typical value for inter-band isolation is better than 50 dB.

Passive Inter-modulation

PIM is one of the biggest potential problems in combiner systems. It is very important to select a combiner system with a low PIM figure (-155 dBc at 2×20 W). PIM is a major concern for combiner systems, especially when combining many high-power (typical 20 W/43 dBm) carriers in multioperator systems. The power density of several operators operating with many carriers can be high, as we shall calculate later in this chapter.

5.7.2 The Combiner System

There are many ways to combine operators and bands into the same DAS, from constructing your own combiner using discrete broadband components (as shown in Figure 5.28) to cavity filter combiners with filters tuned to the specific UL/DL band of each of the operators (as shown in Figure 5.29).

The broadband system in Figure 5.28 is ideal for use when combining low power signals into an active DAS. Note that the downlink attenuator, used for leveling the downlink signal

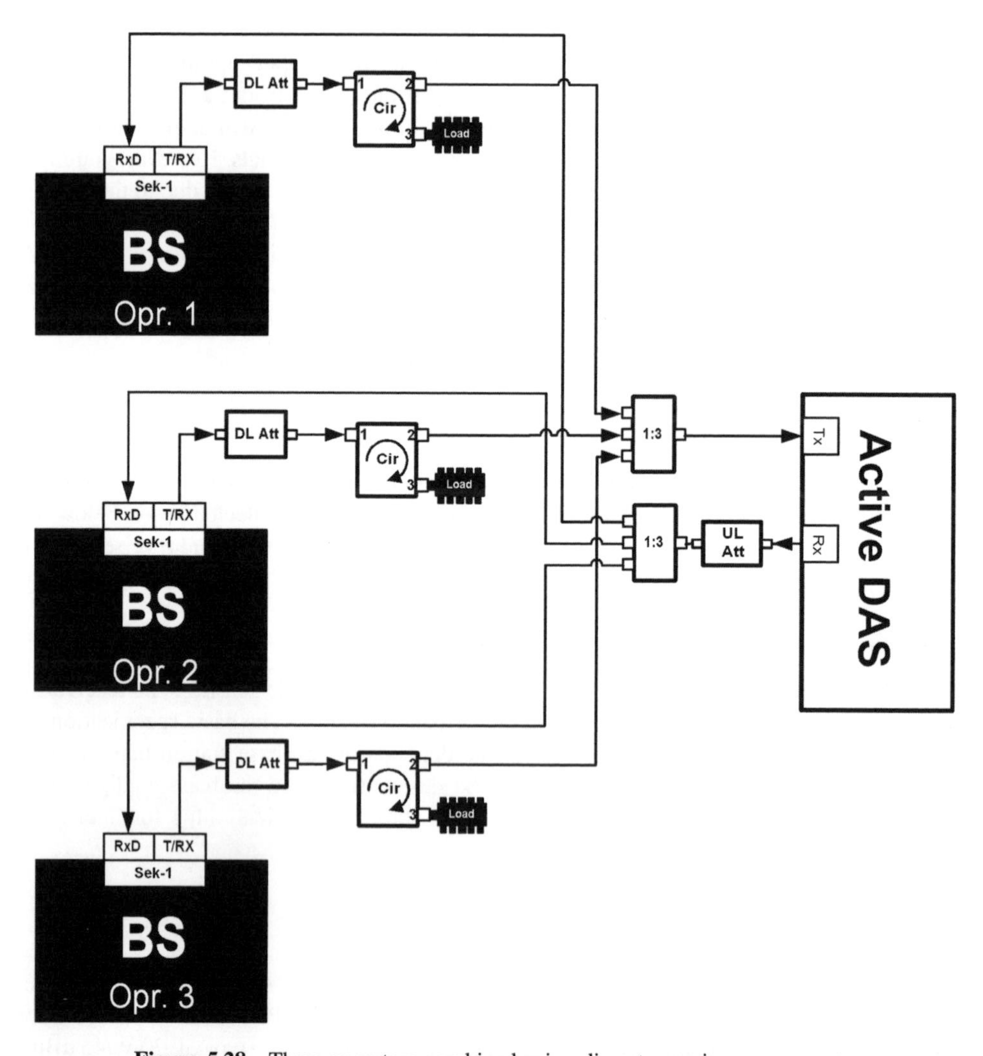

Figure 5.28 Three operators combined using discrete passive components

to the correct value of input to the active DAS, is installed prior to the circulator. Standard attenuators have rather poor PIM performance 120–140 dBc; therefore it is recommended to use a low-PIM cable attenuator. This will keep power levels in the circulator to a minimum, thus minimizing the potential PIM in the circulator. Note that this example is used for an active DAS, where the downlink power requirement at the input at the main unit is about 5 dBm. Therefore low-power base stations are used with about 30 dBm output power.

You can keep the uplink and downlink separated by using circulators; this will make sure that one base station's transmitters downlink signal is not feed to the other base station's combined Tx/Rx port, via the 1:3 splitter–combiner, thus securing isolation. Note the specification of the circulator; it might be an idea to use a double circulator with high reverse attenuation.

The custom tuned filter combiner in Figure 5.29 will typically perform much better. This combiner must have low PIM and high isolation, even at high power, which is typically used when combining high power systems into a passive DAS. The combiner is tuned individually to the specific frequency bands of each operator, and there is good inter-operator and inter-band isolation. Only high-quality components are recommended for use in any combiner system.

5.7.3 Inter-modulation Distortion

When two or more signals are mixed in a nonlinear component, passive or active, there will be generated other signals as a product of the two or more original input signals. Inter-modulation

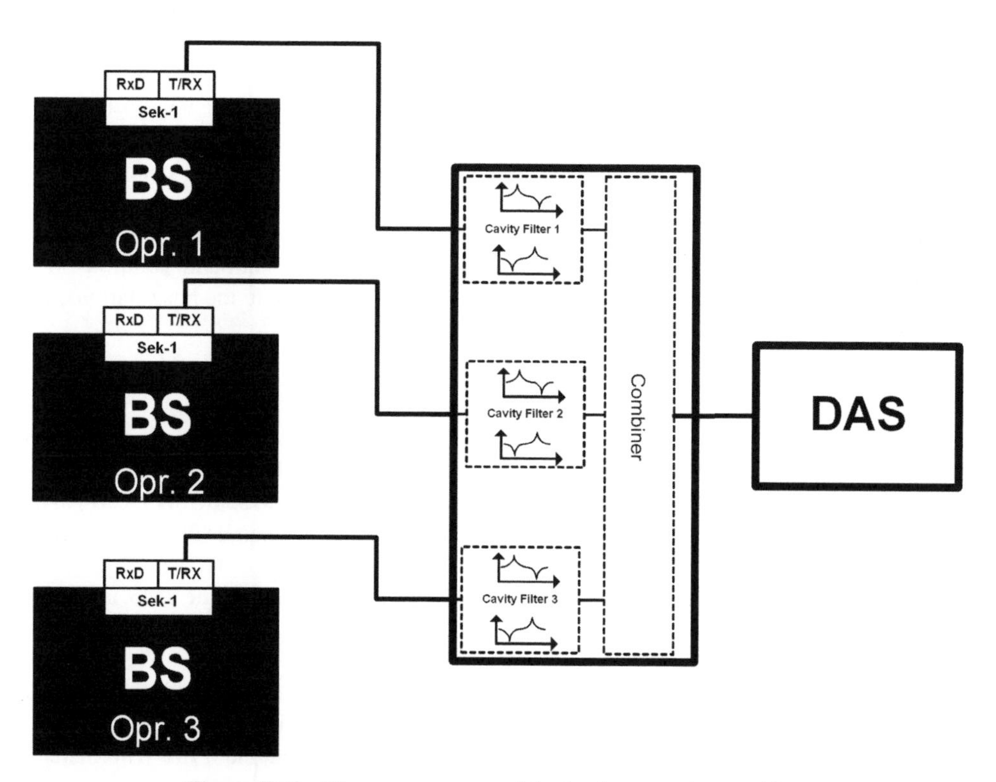

Figure 5.29 Three operators combined using a cavity combiner

distortion (IMD) will occur in amplifiers when operating in nonlinear mode. Therefore it is very important to always operate amplifiers according to their specification, and to keep below a certain limit, e.g. ETSI −36 dBm at GSM900. Make sure you stay within the linear operating window of the amplifier, and the amplifier will have linear performance with low distortion. If you overdrive the amplifier, you might get few dB more output power, but the side effect will be IMD problems that will generate inter-modulation interference.

What is Inter-modulation Interference?

There are three basic categories of inter-modulation distortion:

1. *Receiver produced IMD*: when two or more transmitter signals are mixed in the receivers RF amplifier.
2. *Transmitter produced IMD*: when one or more transmitted signals are mixed in a nonlinear component, in the transmitter.
3. *Passive IMD*: normally radio planners are only concerned about inter-modulation problems caused by active components like transmitters, amplifiers and receivers. However, you must realize that passive components like cables, splitters, antennas can also produce inter-modulation – passive inter-modulation. Typically the source of PIM is the junction between different types of materials, an example could be where the cable connects to the connector. The connection between any internal or external point of contact in a passive component can generate PIM.

The PIM performance is often the big difference between quality and low-cost passive components. Normally good quality passive components, antennas included, will be constructed using the same base material, with only a few or no internal connections or assembled parts and a high level of craftsmanship.

The PIM problem is often underestimated, especially when designing passive systems for large, high-capacity passive DAS, multioperator solutions and multiband solutions. The concentration of the high power in the passive DAS, especially close to the base stations, can be high.

Example

Let us look at a typical example of a multioperator solution in an airport, with this configuration: four GSM operators are each using eight carriers, transmitting 40 dBm (10 W) from their base stations. This is a total of 32 carriers. The total composite power is up to 55 dBm (320 W) in the first part of the passive DAS. Even if the first splitter in the DAS can handle 700 W, you might have a problem. How high is the generated inter-modulation power?

A typical value for a 'standard' splitter is −120 dBc, meaning that the IM3 is 120 dB below the c (carrier). With 40 dBm of power for one carrier to the splitter, the IM3 could be up to 40 − 120 = −80 dBm.

Depending on the frequencies used, this will generate a problem: it could hit the uplink of one of the radio services on the DAS. If you exceed the power rating of the components, or there is a bad connection somewhere in the passive DAS, then the PIM will increase dramatically.

5.7.4 How to Minimize PIM

Passive inter-modulation occurs when two or more signals are present in a passive device (cable, connector, isolator, switch, antenna, etc.) and this device exhibits a nonlinear response. The nonlinearity is typically caused by dissimilar metals, dirty interconnections or other anodic or corrosion effects. Bad connections are also a typical source, and often the effect does not appear at low power levels, but increases exponentially at higher power levels. Then the passive device starts acting like a frequency mixer with a local oscillator and an RF input, generating its own unwanted signals.

There are some rules of the thumb on how to design for low PIM:

- All passive components must fulfill a minimum specification of -155 dBc at 2×20 W.
- Cable absorbers should be used as terminations.
- The 7/16 type of connectors should be used.
- All connectors should be tightened according to the specifications, using the correct torque and tools.
- Low PIM cables should be used, with all connectors fully soldered.
- It is vital to maintain disciplined fitting of connectors to the cable, craftsmanship, use of correct tool and all metal cleaned before fitting the connector.
- Tools, cables and connectors that match and are from the same manufacture should be used.
- Proper installation means:
  –no loose cables, everything strapped with cable binders;
  –no mechanical stress on RF parts;
  –all RF interconnections cleaned as specified by the manufacturer.

5.7.5 IMD Products

The frequencies that are generated by inter-modulation can be found by mathematical calculations of the performance of nonlinear circuits (as shown in Table 5.3). The terminology used to define inter-modulation products classifies their order as second-order, third-order, fourth-order, etc.

The frequencies generated are calculated as the sum or differences between these inter-modulation products (as shown in Figure 5.30). In theory there are no limits to the number of them; however, typically just a few can result in serious consequences.

The main concern in indoor DAS solutions is the IM3 product,

$$2f_1 + f_2,\ 2f_1 - f_2$$

and you must strive to minimize that problem. The powers of products of higher orders, such as fifth to ninth orders and even higher, are usually so low that they do not create serious problems, but be aware of PIM and check.

The products of even second, fourth or sixth orders usually have limited impact due to the fact that the resulting PIM frequency is attenuated by the filters the base station.

Table 5.3 The inter-modulation components and results

Inter-modulation class	Result
Second order	$f_1 + f_2$
2 CH	$f_1 - f_2$
	$2f_1$
	$2f_2$
Third order	$2f_1 + f_2$
2 CH	$2f_1 - f_2$
	$f_1 + 2f_2$
	$2f_2 - f_1$
Third order	$f_1 + f_2 - f_3$
3 CH	$f_1 + f_3 - f_2$
	$f_2 + f_3 - f_1$
Fifth order	$3f_2 - 2f_1$
2 CH	$3f_1 - 2f_2$
Fifth order	$2f_1 + f_2 - 2f_3$
3 CH	$f_1 + 2f_2 - 2f_3$
	$2f_1 + f_3 - 2f_2$
	$f_1 + 2f_3 - 2f_2$
	$2f_2 + f_3 - 2f_1$
	$2C + f_2 - 2f_1$
Seventh order	$4f_1 - 3f_2$
2 CH	$4f_2 - 3f_1$
7th order	$3f_1 + f_2 - 3f_3$
3 CH	$f_1 + 3f_2 - 3f_3$
	$3f_2 + f_3 - 3f_1$
	$3f_3 + f_2 - 3f_1$

5.8 Co-existence Issues for GSM/UMTS

When deploying different radio systems and frequency bands using the same DAS for distribution, there are concerns that need special attention. You need to make sure that one system does not cause IM problems that will degrade the performance on its own or on other bands.

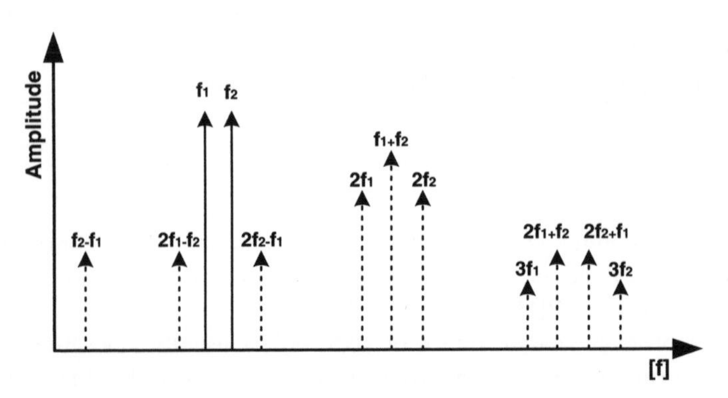

Figure 5.30 Example of second- and third-order IMD products

The problem rises exponentially with the power you feed to the DAS. The more power, the bigger the issue, mainly due to PIM (see Sections 4.3 and 5.7.3).

5.8.1 Spurious Emissions

A major source of co-existence problems is spurious emissions from the GSM transmitters (as shown in Figure 5.31). Therefore it is highly recommended *always* to use high-quality pass band filters on all transmitters, in a multioperator or multisystem DAS solution, when combining GSM and UMTS. Those filters will normally be a part of the base station, the combiner, but check the specifications of the base station to be sure.

Spurious emissions from GSM are restricted in the UL UMTS band (1920–1980 MHz) to a maximum −96 dBm measured in 100 kHz bandwidth (equivalent to −80 dBm in a 3.84 MHz channel). Any multioperator or multisystem DAS should be designed accordingly, in order to minimize the impact on the UMTS UL.

Big Impact on UMTS

As we know, the UMTS system is noise-limited; any noise increase on the UL of the UMTS system will severely impact the performance. Just a slight increase in the UL noise load will offset admission control; high noise increase will collapse the cell and cause the admission control to block for any traffic.

The number of GSM carriers in the system plays a significant role: when doubling the number of GSM carriers (with same level), the spurious emission is increased by 3 dB.

5.8.2 Combined DAS for GSM900 and UMTS

When combining GSM900 and UMTS on the same DAS, you must pay attention to the second harmonic from GSM900 that might fall into the uplink band of UMTS (TDD) (as shown in Figure 5.32).

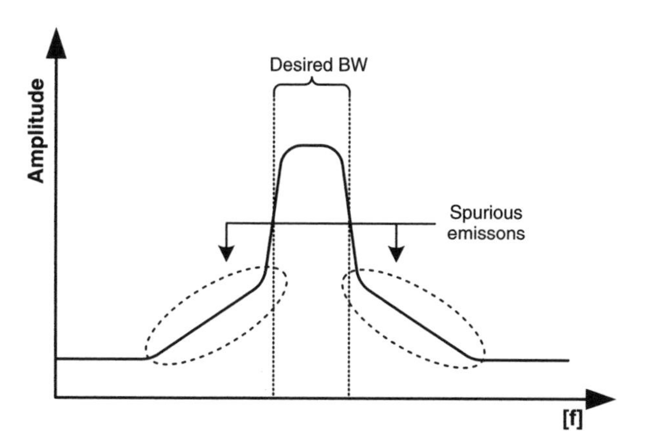

Figure 5.31 Spurious emissions from a transmitter

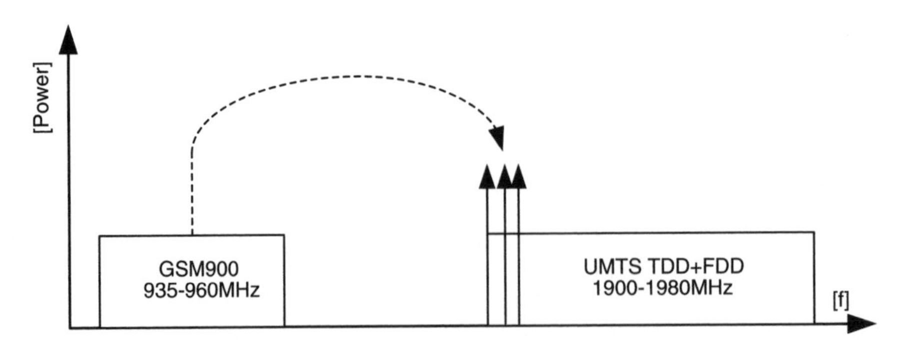

Figure 5.32 Example of second-order IMD products from GSM900 hitting UMTS UL

5.8.3 Combined DAS for GSM1800 and UMTS

When combining GSM1800 and UMTS into the same DAS, you need to pay attention to the third-order inter-modulation products. The third-order IMD from GSM1800 can fall into the UL band of UMTS, if you do not pay attention to the frequency allocation on GSM1800 (as shown in Figure 5.33). When doing the frequency planning on GSM1800, attention to this problem is advised in order to minimize it. This can be a challenge in multioperator solutions, where frequency coordination between competing operators must be carried out.

Example

Let us look at one example with 1800 MHz third-order IM. Using CH520 (1806.8 MHz) and CH825 (1867.8 MHz) will produce this IMD3 (refer to Section 5.7.3):

$$2f_1 + f_2 = 5481.4\,\text{MHz}$$
$$2f_1 - f_2 = 1745.8\,\text{MHz (will hit GSM1800 CH690 UL)}$$
$$f_1 + 2f_2 = 5542.4\,\text{MHz}$$
$$2f_2 - f_1 = 1928.8\,\text{MHz (will hit UMTS UL)}$$

It is highly recommended to plan the frequencies on GSM1800 so they do not fall in the UL band of any UMTS channels that are in use. The fact is that UMTS has good

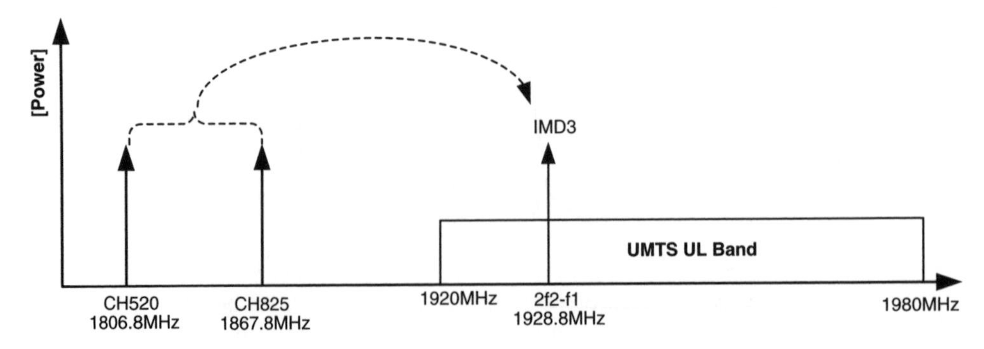

Figure 5.33 Example of third-order IMD products from GSM1800 hitting the UMTS uplink

narrow-band interference immunity, but a high narrowband signal on the UL of the UMTS base station can cause receiver blocking.

Knowing that the first DCS channel used is CH520 = 1806.8 MHz and that the problem is the third IMD ($2f_2 - f_1$), we can calculate what the highest frequency can be, in order to make sure that the third IMD will 'hit' just below the UMTS FDD frequency band 1919.8 MHz, and not cause any problems (if UMTS TDD is not used). The highest allowable frequency is:

$$1919.8\,\text{MHz} = 2^{*}f_2 - 1806.88\,\text{MHz} = (1919.88\,\text{MHz} + 1806.88\,\text{MHz})/2$$
$$= 1863.38\,\text{MHz} = \text{CH802}\,(1863.2\,\text{MHz})$$

With CH520 as one of the frequencies in operation, we must make sure that we do not select frequencies above CH802 as the other channel.

5.9 Co-existence Issues for UMTS/UMTS

In an indoor DAS system it is very likely that operators will have adjacent channels active on the same DAS. Because of the limited selectivity of the filters used in the mobile and base station, power from one channel will leak into the adjacent channels.

5.9.1 Adjacent Channel Interference Power Ratio

This is the major issue when co-locating UMTS operators in the same building. It is specified that the adjacent channel suppression must be better than 33 dB (as shown in Figure 5.34), but this is in many cases not enough, especially indoors where you typically have users close to the antennas. A user being serviced by another operator (and therefore not in power control of the cell) is likely to cause adjacent channel interference power ratio (ACIR) when this mobile is close to the antenna, having low path loss. Conversely, the adjacent operator might impact your user when he is close to the other operator's antenna.

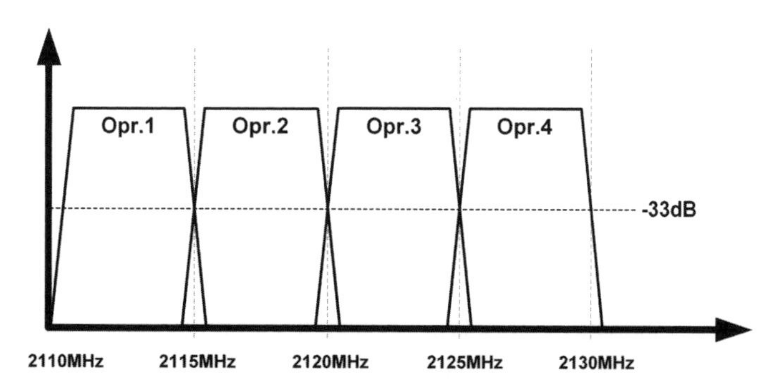

Figure 5.34 Channel allocation on UMTS

Noise Increase on Node B

The adjacent channel interference will cause noise increase on the UL of node B, degrading the capacity with noise load.

Example

We can calculate how much noise increase a mobile will generate and the noise load on our cell, when this mobile is being serviced by an adjacent operator and when the mobile is close to our antenna. The mobile on the adjacent channel is transmitting 21 dBm. The mobile (UE) is 10 meters from our antenna (free space loss = 59 dB), and then we have 10 dB loss from the node B to the antenna and 2 dBi antenna gain; the path loss is $59 + 10 - 2 = 67$ dB:

$$P_{\text{RxAdj}} = P_{\text{TxUE}} - \text{ACIR} - \text{PL}$$

Where P_{RxAdj} is the adjacent channel power level; P_{TxUE} is the mobile transmit power on adjacent channel; ACIR is the adjacent channel interference rejection; PL is the path loss from the adjacent mobile to our node B.

$$P_{\text{RxAdj}} = 21\,\text{dBm} - 33\,\text{dB} - 67\,\text{dB} = -79\,\text{dBm}$$

Now we can calculate the new noise level (assuming a noise floor on node B of -105 dBm). We need to sum up the noise, by adding the power contribution from each different source (the P_{RxAdj} + noise floor on node B of -105 dBm):

$$\text{power total} = \text{power 1} + \text{power 2} \ldots \text{power}\,(n)$$

We now need to convert the dBm to mW:

$$P\ (\text{mW}) = 10^{\text{dBm}/10}$$
$$-105\,\text{dBm} = 31.6 \times 10^{-12}\,(\text{mW})$$
$$-79\,\text{dBm} = 12.58 \times 10^{-9}\,(\text{mW})$$

Then we can add the two powers:

$$\text{combined noise power} = 31.6 \times 10^{-12} + 12.58 \times 10^{-9} = 12.62 \times 10^{-9}\,(\text{mW})$$

dBm

We then convert the noise power back to dBm:

$$\text{noise floor (dBm)} = 10\log(12.62 \times 10^{-9}) = -79\,(\text{dBm})$$

The -105 dBm from the BS plays only a minor role; it is after all 26 dB lower than the power from the adjacent mobile. This adjacent channel noise causes the uplink of the cell to collapse, disabling the cell from carrying any more traffic, and dropping all ongoing calls!

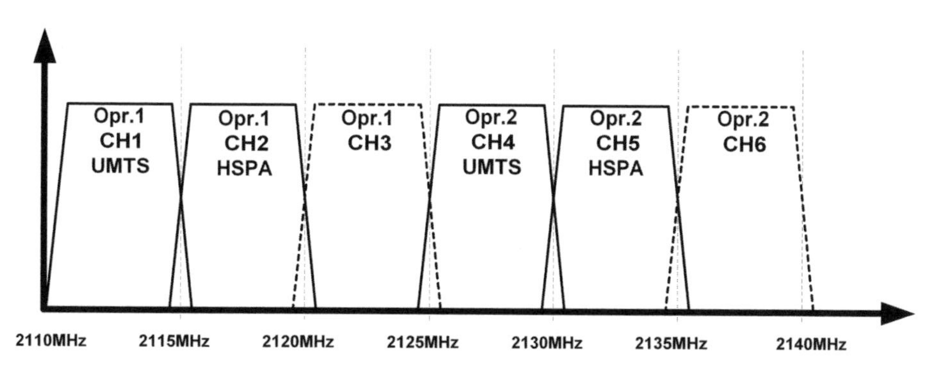

Figure 5.35 Typical channel usage of two operators with UMTS/HSPA deploys, and a third channel for future use

5.9.2 The ACIR Problem with Indoor DAS

ACIR is a major concern inside buildings, when the operator uses adjacent frequencies and separate DAS systems. The problem is demonstrated in Figure 5.35 and 5.36. Two separate DAS systems are deployed within the same building; one for operator 1 (OPR1) and one for operator 2 (OPR2).

When we look at the signal power as a function of the location relative to the antennas (as shown in Figure 5.36), it is clear that, when you are close to an antenna from the other operator and have low signal from the serving cell, there is a real potential for ACIR.

The Potential ACIR Problem in the Future

There are many of these potential ACIR problems in real buildings around the world, but the problem is not evident until the operators deploy more channels. This is due to the fact that most UMTS operators have two or three UMTS channels allocated and if they deploy the first carrier only, there is still 10–15 MHz distance to the next channel from the other operator, minimizing the ACIR problem. Most operators deploy UMTS (R99) traffic on the first carrier, HSPA on the second carrier and leave the third channel for 'future upgrades'.

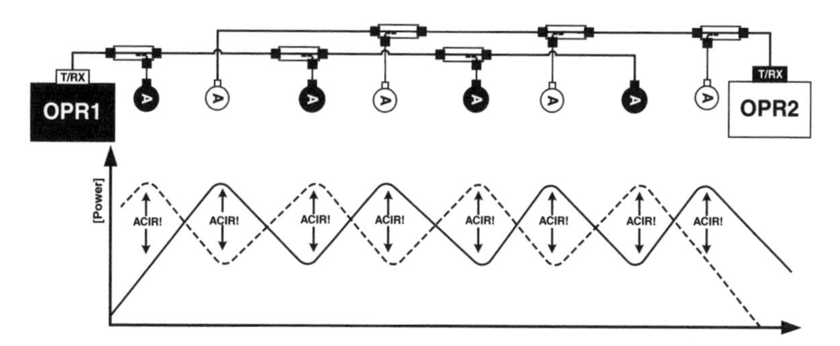

Figure 5.36 Adjacent interference problems in a building with two operators on separate DAS systems

This usage of the 'future' channel could be a potential problem in many buildings (as shown Figure 5.35).

5.9.3 *Solving the ACIR Problem Inside Buildings*

There are means to solve the ACIR issue; the best way is for the operators to agree between themselves to use the same DAS and antenna locations, and coordinate the roll-out in these buildings.

It is always a good idea to place the indoor antenna so that the user cannot get too close to it; however this is not always possible inside a building. After all, if the ceiling height is 3 m, and the user is 180 cm tall, the mobile will come as close as 120 cm to the antenna (with a path loss of about 40 dB).

Multioperator Systems will Help

When using the same DAS, the scenario can actually be controlled by the fact that both the wanted and the unwanted signals track the same relative level. If both operators use the same antenna locations (as shown in Figure 5.37), then the two signal levels will track and have the same relative difference in level throughout the building, thus avoiding any potential ACIR problems, on both UL and DL.

Offsetting the WCDMA Frequency

It is possible for the mobile operator to offset the WCDMA frequency in steps of 200 kHz, thus minimizing the problem. If CH1 is used for UMTS, CH2 for HSPA and CH3 is unused, then the two channels can be offset 1 MHz, improving the adjacent channel isolation in the building (as shown in Figure 5.38).

5.10 Multioperator Requirements

The complexity of designing a multioperator solution is often underestimated. To design, implement and operate a multioperator solution is much more than 'just' the RF design

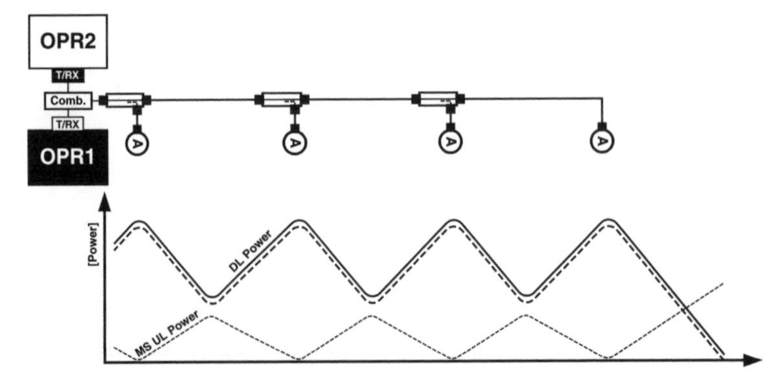

Figure 5.37 No adjacent interference problem in a building with two operators on the same DAS systems

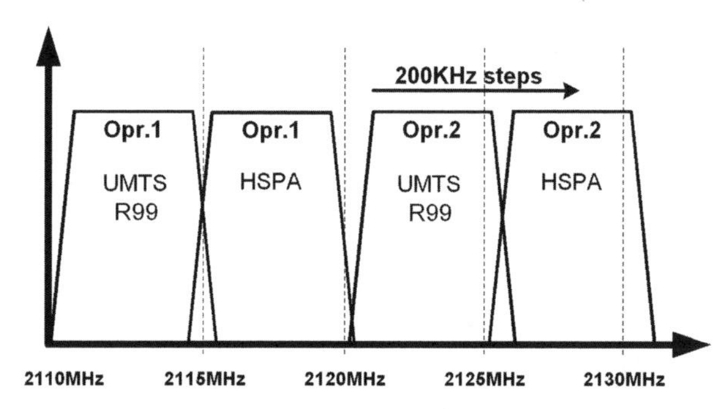

Figure 5.38 Operators might offset their WCDMA frequency to solver the ACIR problem

issues. It is highly recommended that all the mobile operators in each country or region work towards a mutual accepted document that defines all the parameters, interfaces and issues with regards to multioperator DAS solutions.

Often RF planners focus on the RF design only and go ahead with the project. However costly mistakes and performance degrading problems can be avoided if you plan more than just one step ahead and pay attention to the whole process. It is recommended to prepare, develop and agree on a multioperator agreement before the first multioperator project is decided. This agreement takes time to develop and obtain agreement on; therefore it is highly recommended that you complete the multioperator agreement before the first multioperator DAS solution design is actually needed. If you wait until the first project, it will probably be delayed.

There are many issues, technical, political and legal, that must be clarified in the process of developing the multioperator agreement. In most cases it is worthwhile hiring a neutral consultant to help the operators with this agreement. Compromises sometimes need to be made, and it can be much easier for an independent consultant to make ends meet and come to a mutual agreement and compromise.

5.10.1 Multioperator Agreement

The multioperator agreement (MOA) document must be agreed and must cover all aspects of the process and lifetime of the DAS system. There must be clear definitions of all aspects of the project, including technical, logistical, installation and legal aspects.

The MOA should ideally be a universal document that can be used for all projects in the future and constantly be updated and adapted. Clear responsibilities must be defined for all aspects of the project. It needs to be precise and accurate, preferably containing document templates to handle and control the process. The best way to handle any multioperator project is to get the MOA in place before anything else is done. Even if you currently do not have any multioperator projects in mind, start now to define the MOA, so it will be ready when the first multioperator project is initiated.

In many aspects we consider other operators as competitors, but it is important to agree on a common MOA to handle these projects.

5.10.2 Parties Involved in the Indoor Project

We, RF planners, like to think that the whole indoor project revolves around us. In fact, many other important persons and departments are involved, internally at each operator but also quite a few external parties. Here are a few examples of parties involved in the typical project:

- The building owner, and also the architect of the building.
- The local IT consultant or personnel responsible for the site.
- Radio planners from all operators.
- Co-location managers from all operators.
- Site hunters.
- Transmission planners.
- Parameter optimizers.
- Equipment procurement departments.
- Commissioning teams.
- Contractors and installers.
- Optimizing team.
- Operations and maintenance team.

5.10.3 The Most Important Aspects to Cover in the MOA

Indoor DAS projects are complex and many parameters and aspects need to be taken into account. Obviously there is much focus on all the radio parameters, but many other parameters and issues need to be taken into account when designing, implementing and operating a multioperator system. Here are some examples of the most important aspects.

The Radio Design Specification

This is the basis of the design. It is very important to absolutely crystal clear on the RF specifications:

- RF design levels – UL and DL level and quality or BER levels.
- Data service levels – DL and UL speeds and EDGE or HSPA.
- Noise levels – UL noise power and DL noise level.
- Delay – maximum allowable delay of the system, end to end.
- Differentiated design goals – special design levels in specific areas.
- Link budgets – complete link budget for all services, all bands must be documented.
- General design guidelines – other, non-RF design parameters.
- Performance merits of the used DAS components – PIM specifications, power rating and type approvals.
- Handover zones – HO zone size and traffic speed, and GSM or UMTS zones.
- Capacity – number of channels needed for voice and data requirements.
- Maximum downlink transmit power at the DAS antenna – to avoid near–far problems, especially on UMTS and HSPA.

Co-existence Parameters

It is important to insure compatibility between all the operators and radio services, to make sure that the radio services can co-exist on the same DAS without any service degradation of any of the radio services.

- RF interface specification – type of connectors, power levels and isolation between bands and between operators.
- Inter-modulation optimization – frequency restrictions, spurious emissions, and ETSI and other requirements.
- Public radiation safety guidelines (EMR) – WHO/ICNIRP and local limits.
- Co-existence issues with other equipment – EMC and RFI compliance.
- Leakage from the building.

Future of the Indoor DAS

We need to look into the expected future of the building and DAS project. We must be sure that we define how to react on any changes in DAS:

- Future upgrades – more sectors, capacity, how and when new services are added to the DAS, 3G, 3.5G, WiMAX, etc.
- Connecting new operators to the DAS in the future – legal issues and headroom in the link budget.
- Disconnecting from the DAS – how this is handled if an operator wants to disconnect.
- Discontinuing the DAS – how this is handled when the DAS has to be removed.

Logistics

Clear definition of who is responsible for what in the project is very important. Gentlemen's agreements might be nice, but clearly defined contracts and agreements in writing are a must. These documents well help resolve any misunderstandings about responsibility in the future:

- Responsibility matrix – all parties included.
- Selection of DAS system – vendor and supplier, and request for quotation (RFQ) process.
- Selection and certification of installers – education, tool requirements and certification.
- Legal agreement and contracts – between operators, between operators and building owner, between operators and installer or DAS supplier, on who owns the DAS and insurance issues.
- Documentation – specify function, in- and output of all documents, documentation control and tracing updates, and design and as-built documentation.
- Installation guidelines – building code and fire retardants.
- Implementation plan.
- Acceptance testing and measurements – pass or levels, test method specification and operations.
- Be sure to define how and who takes care of maintenance of the DAS in the future – site access, alarm monitoring, spare parts, operation and maintenance service level agreement.

6

Traffic Dimensioning

One of the most important design parameters for indoor DAS solutions is to define how many speech channels you must provide for the users of the building in order to cater for their need for voice traffic channels. Voice traffic is time-critical. Many data services can be degraded in speed, offset slightly or delayed in time without any degradation of the user experience. It does not matter if your mail or SMS is delayed by 30 s, but the voice connection must be maintained 1:1 in time; one voice call needs one permanent traffic channel.

How do you calculate the number of channels needed to cater for the capacity requirements of the users in a specific building? And when you know the number of users and the type of users, how do you then use this information to calculate the number of channels? In the early days of telephony over fixed lines, a Danish mathematician faced similar challenges when dimensioning how many telephone lines were needed to interconnect fixed telephone switches.

6.1 Erlang, the Traffic Measurement

Agner Krarup Erlang (Figure 6.1) was born in Denmark in 1878 (close to where I live). He was a pioneer in the theory of telecommunications traffic and he proposed a formula to calculate the percentages of users served by a telephone exchange that would have to wait when attempting to place a call. He could calculate how many lines were needed to service a specific number of users, with a specific availability of lines for the users.

A. K. Erlang published the result of his study in 1909: *The Theory of Probabilities and Telephone Conversations*. Erlang is now recognized worldwide for this work, and his formula is accepted as the standard reference for calculating telecommunication traffic load. The unit we use for telephony traffic load is 'Erlang' or E. Erlang worked for the Copenhagen Telephone Company (KTAS) for many years, until his death in 1929.

6.1.1 What is One Erlang?

An *Erlang* is a unit of telecommunications traffic measurement. One Erlang is the continuous use of one voice channel. In call minutes, one Erlang is 60 min/h, 1440 call

Indoor Radio Planning: A Practical Guide for GSM, DCS, UMTS and HSPA Morten Tolstrup

Figure 6.1 The Danish statistician Agner Krarup Erlang

min/24 h. In practice, when doing mobile capacity calculations, an Erlang is used to describe the total traffic volume of 1 h, for a specific cell.

Erlang Example

If a group of 20 users makes 60 calls in 1 h, and each call had an average duration of 3 min, then we can calculate the traffic in Erlangs:

$$\text{total minutes of traffic in 1 h} = \text{duration} \times \text{number of calls}$$
$$\text{total minutes of traffic in 1 h} = 3 \times 60$$
$$\text{total minutes of traffic in 1 h} = 180\,\text{min}$$

The Erlangs are defined as traffic (minutes) per hour:

$$\text{Erlangs} = 180/60 = 3\,\text{E}$$

Knowing the number of users (20), we can calculate the load per user:

$$\text{user load} = \text{total load}/\text{number of users}$$
$$\text{user load} = 3/20 = 0.115\,\text{E} = 150\,\text{mE per user}$$

Then, if we have the same type of users inside a building with 350 mobile users, we can calculate what capacity we need:

$$\text{total load} = \text{number of users} \times \text{load per user}$$
$$\text{total load} = 350 \times 0.150 = 52.5\,\text{E}$$

Yet what do these 52.5 Erlangs mean in terms of number of voice channels needed? Do we then need 53 channels in order to service the users? Not exactly. It depends on the user behavior: how the calls are distributed and how many calls are allowed to be rejected in the cell. In theory – if you do not want any rejected calls due to lack of capacity, you should provide 350 channels for the 350 users, and then you would have 100% capacity for all users at any given time. However, it is not likely that all users will call at the same time, and users will accept a certain level of call rejection due to lack of capacity. To estimate the number of channels needed, we need to look at maximum call blocking we allow in the busiest hour during the day.

6.1.2 Call Blocking, Grade of Service

We need to calculate how many lines, or in our case how many traffic channels, we need to provide in order to carry the needed traffic. In this example we calculated that 350 users, each of whom load the system with 150 mE, would produce a total load of 52.5 E. What if all the users call at the same time? How are the calls distributed? How many call rejects due to lack of channels can we accept? This was exactly the basis of Erlang's work; he calculated how many channels you would need to carry a specific traffic load, with a specific service rate or blocking rate, for a given distribution of the traffic arrival.

The blocking rate (grade of service or GOS) is defined as the percentage of calls that are rejected due to lack of channels. If the users makes 100 calls, and one call is rejected due to lack of channels (capacity) the blocking rate is 1 in 100, or 1%. This is referred to as 1% GOS. Operators might differentiate the GOS target for different indoor solutions, with a strict GOS of 0.5% in an office building but allowing a GOS of 2% in shopping malls.

6.1.3 The Erlang B Table

Provided that the calls are Erlang-distributed, you can use the Erlang B formula to calculate the required number of channels at a given load rate, and a given grade of service. The Erlang distribution is a special case of gamma distribution used to model the total interval associated with multiple Poisson events. The Erlang B formula is complex; it is more practical to use an Erlang B table based on the Erlang B formula (see Table 6.1).

Using the Erlang Table

First of all we need to decide the quality of service in terms of how many calls we will allow the system to reject in the busiest hour. This is expressed as the grade of service. If we design for a GOS of 1%, a maximum of 1 call out of 100 will be rejected due to lack of traffic channels. Using the Erlang table is straightforward; you simply select the GOS you want to offer the users, and a typical number for mobile systems is 1%. Then you look up in the table how many voice traffic channels (N) you need to support this traffic.

Example

If we have calculated that the traffic is 5 E, and we will offer 1% GOS, we need 11 voice channels (5.1599 E) to support the traffic. We can also use the Erlang to work in reverse, and conclude that having only 10 traffic channels to support 5 E will give a GOS (blocking) of about 2%. How do we determine what load in Erlang an indoor system will give? First of all we need to know the traffic profile of the users in the building.

Table 6.1 Erlang B table, 1–50 channels, 0.01–5% grade of service

Number of channels	Blocking (GOS) 0.1%	0.5%	1%	2%	5%	Number of channels
1	0.001	0.005	0.010	0.020	0.052	**1**
2	0.045	0.105	0.152	0.223	0.381	**2**
3	0.193	0.349	0.455	0.602	0.899	**3**
4	0.439	0.701	0.869	1.092	1.524	**4**
5	0.762	1.132	1.360	1.657	2.218	**5**
6	1.145	1.621	1.909	2.275	2.960	**6**
7	1.578	2.157	2.500	2.935	3.737	**7**
8	2.051	2.729	3.127	3.627	4.543	**8**
9	2.557	3.332	3.782	4.344	5.370	**9**
10	3.092	3.960	4.461	5.084	6.215	**10**
11	3.651	4.610	5.159	5.841	7.076	**11**
12	4.231	5.278	5.876	6.614	7.950	**12**
13	4.830	5.963	6.607	7.401	8.834	**13**
14	5.446	6.663	7.351	8.200	9.729	**14**
15	6.077	7.375	8.108	9.009	10.633	**15**
16	6.721	8.099	8.875	9.828	11.544	**16**
17	7.378	8.834	9.651	10.656	12.461	**17**
18	8.045	9.578	10.437	11.491	13.385	**18**
19	8.723	10.331	11.230	12.333	14.315	**19**
20	9.411	11.092	12.031	13.182	15.249	**20**
21	10.108	11.860	12.838	14.036	16.189	**21**
22	10.812	12.635	13.651	14.896	17.132	**22**
23	11.524	13.416	14.470	15.761	18.080	**23**
24	12.243	14.204	15.295	16.631	19.031	**24**
25	12.969	14.997	16.125	17.505	19.985	**25**
26	13.701	15.795	16.959	18.383	20.943	**26**
27	14.439	16.598	17.797	19.265	21.904	**27**
28	15.182	17.406	18.640	20.150	22.867	**28**
29	15.930	18.218	19.487	21.039	23.833	**29**
30	16.684	19.034	20.337	21.932	24.802	**30**
31	17.442	19.854	21.191	22.827	25.773	**31**
32	18.205	20.678	22.048	23.725	26.746	**32**
33	18.972	21.505	22.909	24.626	27.721	**33**
34	19.743	22.336	23.772	25.529	28.698	**34**
35	20.517	23.169	24.638	26.435	29.677	**35**
36	21.296	24.006	25.507	27.343	30.657	**36**
37	22.078	24.846	26.378	28.254	31.640	**37**
38	22.864	25.689	27.252	29.166	32.624	**38**
39	23.652	26.534	28.129	30.081	33.609	**39**
40	24.444	27.382	29.007	30.997	34.596	**40**
41	25.239	28.232	29.888	31.916	35.584	**41**
42	26.037	29.085	30.771	32.836	36.574	**42**
43	26.837	29.940	31.656	33.758	37.565	**43**
44	27.641	30.797	32.543	34.682	38.557	**44**
45	28.447	31.656	33.432	35.607	39.550	**45**

Table 6.1 (*Continued*)

Number of channels	Blocking (GOS) 0.1%	0.5%	1%	2%	5%	Number of channels
46	29.255	32.517	34.322	36.534	40.545	**46**
47	30.066	33.381	35.215	37.462	41.540	**47**
48	30.879	34.246	36.109	38.392	42.537	**48**
49	31.694	35.113	37.004	39.323	43.534	**49**
50	32.512	35.982	37.901	40.255	44.533	**50**

6.1.4 User Types, User Traffic Profile

Each user in the network will have a specific load profile; the profile may vary and some days one user will load the mobile system more than other days. You need to use average load numbers, based on the type of user, in order to make some assumptions to help us design the capacity need (see Table 6.2). They can only be assumptions, but these assumptions will be adjusted over time when analyzing the post traffic on the implemented solutions. Thus you will build up a knowledge database to help you be more accurate in future designs. As a general guideline the typical numbers shown in Table 6.2 could be used.

Table 6.2 Typical user load in Erlang

User type	Traffic load per user
Extreme user	200 mE
Heavy user	100 mE
Normal office user	50 mE
Private user	20 mE

If you are designing an indoor system for a building with 250 users and the users comprise 40 heavy users, 190 normal office users and 20 private users, you will need a capacity of:

$$(40 \times 0.1) + (190 \times 0.05) + (20 \times 0.02) = 4 + 9.5 + 0.4 = 13.9\,\text{E}$$

However we must remember that this load is only present when all users are in the building at the same time. It may be more likely that on average 80% of the users are in the building, and thus the traffic load would be $13.9 \times 0.8 = 11.12\,\text{E}$.

If we design for a GOS of maximum 1%, we can look up in the Erlang table the necessary number of traffic channels. In this case we will need 19 channels in order to keep the GOS below 1% (11.23 E).

Mobile-to-mobile Calls in the Same Cell, Intracell Mobile Calls

In a totally wireless environment where users generate a lot of internal traffic in the same cell, you must remember that to support one call where a user is calling another user in the same cell will actually take up two traffic channels. Therefore the 'internal call factor' must be applied. This is an important parameter to take into account, when analyzing the capacity need for the specific building.

In practice, if 10% of the load is internal mobile-to-mobile traffic, the load per user can be multiplied by a factor of 1.10; a heavy user that normally loads the network with 100 mE must be calculated as 110 mE if the internal traffic is 10%.

6.1.5 Save on Cost, Use the Erlang Table

If it is an indoor GSM cell you are designing, then typically the number of traffic channels is seven per transceiver (radio channel). The exact number of traffic channels per transceiver depends on the configuration of the logical channels in the cell, but seven is often the average.

To calculate how much capacity you need, take the calculated total traffic in the cell, in this case 13.9 E, and multiply it by the 'intracell' factor, 1.10 in this case. To calculate the total capacity requirement:

$$13.9\,\text{E} \times 1.10 = 15.29\,\text{E}$$

Using the Erlang table, you can determine that you would need a total of 24 traffic channels to make sure you can keep the voice service below 1% GOS.

You know that each radio channel comes with seven traffic channels, so you would need to deploy four TRX with a total of 28 traffic channels – in fact an overcapacity of four traffic channels. The four TRX would be able to carry 18.64 E, or an overcapacity of 3.35 E, or 30 heavy users. Thus some extra margin is integrated for capacity growth in the solution. However, there is also another option.

Instead of deploying four TRX, you could save the cost of one TRX and only deploy three TRX with a total of 21 traffic channels. You can now look up in the Erlang table the GOS of a load of 15.29 E on 21 channels and see that you would have a GOS of 2–5%. Depending on the solution, you might decide to go with three radio channels and save the cost of one transceiver unit, and then analyze the call blocking performance once the system was on air, over a period of 4–6 weeks, to make sure that the capacity was acceptable. In fact, the Erlang table and traffic calculation can help you cost-optimize the design.

You must remember that the cost of an extra TRX is not only limited to the HW price of the TRX itself, there might be combiner and filter costs, BSC costs, A-bis interface and transcoder costs, as well as annual license fees to the network supplier.

6.1.6 When Not to Use Erlang

There are cases where we cannot use Erlang for voice calls; the most common example is when calculating the capacity for a metro train tunnel system. Even though the traffic inside the trains is Erlang-distributed and has normal telephone user behavior, you must realize that the trains moving the users are not Erlang-distributed! In practice, the arrival of traffic in cells in a metro train system is distributed by the train schedule, not by Erlang distribution, thus Erlang cannot be applied.

In practice, these trains are 'capacity bombs' that drift around the network, so be careful. In practice you might not even discover the real blocking of calls in a tunnel system; most network statistics will not be able to detect the call blocking in this case due to the fact that the blocking occurs only in a very short moment in time, maybe a few seconds when all the users inside a train hand over (or try to hand over) to the next cell when traveling in the train. To make sure there is no blocking, you might have to analyze the specific network counter indicating blocked calls for the specific cell, not the Erlang statistics, which might be averaged out over 15 min. Signaling capacity in this case is also a major concern when a 'train-load' of calls are to handover within the same 2–5 s.

6.1.7 GSM Radio Channels and Erlang

Now you know how to use the Erlang table, you can easily produce a small table that can help in the estimation of how many voice users you are able to support for a given GSM configuration of radio channels in the cell (see Table 6.3).

Table 6.3 Typical user load in Erlang and number of users vs TCH

Number of radio channels	Number of traffic channels (at seven TCH/TRX)	Number of office users at 50 mE/1% GOS	Number of private users at 20 mE/2% GOS
1	7	50	147
2	14	147	410
3	21	256	702
4	28	372	1007
5	35	492	1322
6	42	615	1642

Bear in mind that you also need to be able to support the required data services, and in some cases you will dedicate specific time slots (traffic channels) to data traffic.

It is interesting to note that the capacity of two radio channels is more than double the capacity of one! This effect is called trunking gain, and is explained in detail later in Section 6.1.9.

Half-rate on GSM

Note that GSM can utilize 'half-rate' where two voice users share the same TSL, with half the data rate. This will result in double the voice capacity for each radio channel. In cells with high peak loads this is normally implemented as 'dynamic half-rate', where the voice calls in normal circumstances gets one normal TSL per call. However, if the load on the cell rises to a preset trigger level, the system starts to convert existing calls to half-rate, and assigns new calls as half-rate, until the load fall below another trigger level.

The use of half-rate voice coding will to some extent degrade the perceived voice quality, but in normal circumstances the voice quality degradation is barely noticeable, and is to be preferred considering that the alternative is call rejection due to lack of voice channels.

6.1.8 UMTS Channels and Erlang

The voice capacity on a UMTS cell is related to many factors. How much noise increase will we allow in the cell? How many data users and at what speed do we need to support these users? How much processing power does the base station have (channel elements) to support the traffic? Typically one channel element can support one voice connection. It also depends on how much power is assigned to the CPICH and the other common channels.

Many factors interact, but assume that the only traffic on the cell is voice calls and the other parameters are set to:

- Sum of power assigned to CPICH and other common channels is 16% (−8.1 dB).
- Maximum noise rise is 3 dB (load 50%).
- Voice activity factor is 67%.

Then one UMTS carrier can support 39 voice channels of 12.2 kbps. This corresponds to 28.129 E at 1% GOS, enabling the cell to support 563 office users at 50 mE per user, or 33.609 E at 2% GOS, enabling the cell to support 1680 private users at 20 mE per user.

However we must remember that this is pure voice traffic, and data calls will take up capacity. Using the same conditions, the same cell will only be able to support one 384 kps data call and 27 voice channels, or three voice channels and three 384 kbps data calls. This is provided that there are channel elements to support the traffic installed in the base station.

6.1.9 Trunking Gain, Resource Sharing

Trunking gain is a term used in telecom traffic calculations to describe the gain from combining the load into the same resource (trunk). The effect of the trunking gain can also be seen in the Erlang table, and in Table 6.3. If you look up in the Erlang table the traffic capability of a GSM cell using one radio channel with seven traffic channels and 1% GOS, you can see that the capacity for the cell is 2.50 E. Then two separate cells with this configuration of seven traffic channels will each be able to carry $2 \times 2.50 = 5$ E.

However if we combine the resources, all of the 14 traffic channels into the same cell, we can see in the Erlang table that 14 channels can carry 7.35 E. This is 2.35 E more with the same resources, 14 channels. This is a gain of 47% with the same resources, but combined into the same cell; this effect is called 'trunking gain' (see Figure 6.2).

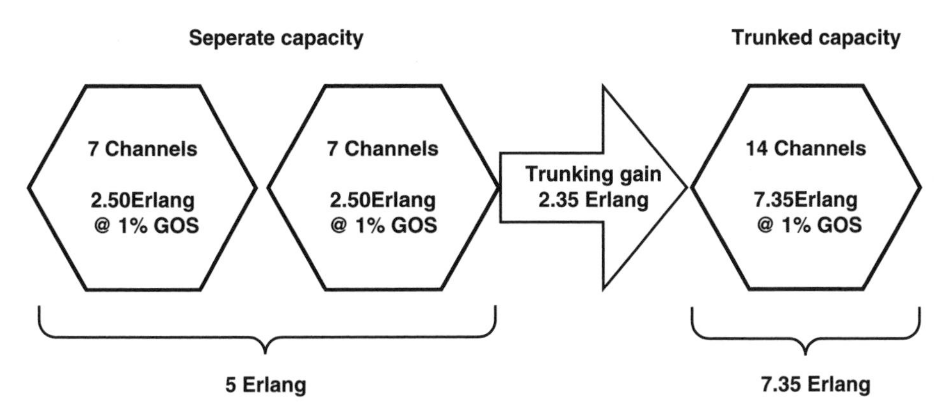

Figure 6.2 Trunking gain when combining the resources into the same cell

Traffic Profile of the Cell and Shared Resources

The traffic in different cells will often have different traffic profiles, the profile being the distribution of the load throughout the day. A shopping mall will have low traffic until opening hours and almost no traffic at night time. An airport might have a relatively high load constantly. This offset in load profile can be used to improve the business case of indoor solutions.

Example

If we take two traffic profiles, one from an outdoor cell, and the traffic from the cell covering inside an office building and combine the load from the two areas (see Figure 6.3), we can actually get 'free' capacity. This is the case for the solution shown in Section 4.6.3, where we are tapping into the capacity of the outdoor macro cell, distributing the same capacity

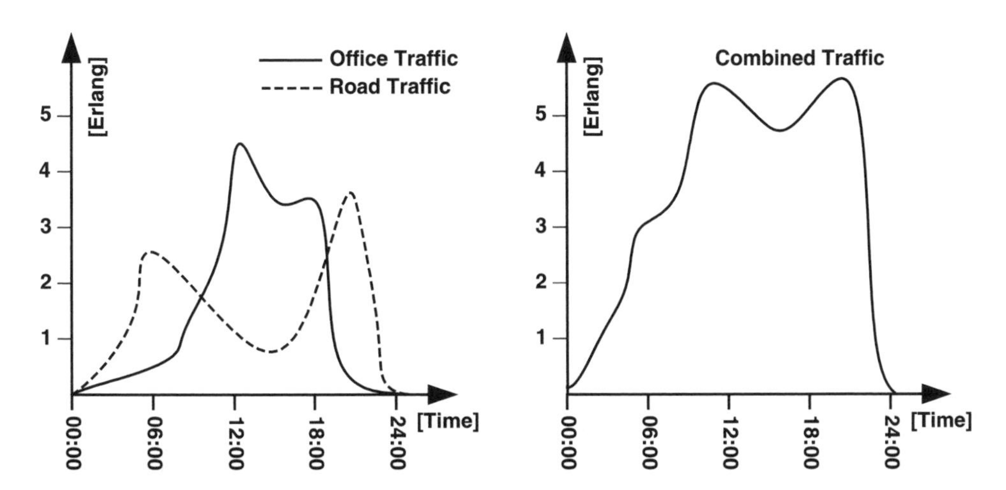

Figure 6.3 Trunking gain when combining different daily traffic profiles into the same cell

resource inside the building. By combining the two load profiles, a high trunking gain is obtained due to the offset in the traffic profile between the two cells.

Sharing Load on a Weekly Basis

The offset in traffic profile should not only be evaluated on a daily basis, but also over a week (see Figure 6.4) or even longer periods. It might be that the capacity of a big office building next to a stadium could be shared if the game in the arena occurred outside shopping hours. Thus you can save a lot of roll-out cost if you can share the load from different areas with noncorrelated traffic profiles.

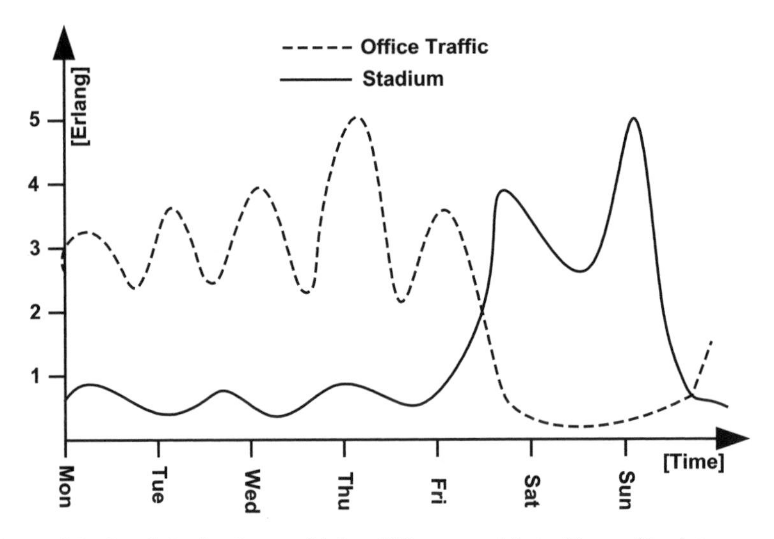

Figure 6.4 Load sharing by combining different weekly traffic profiles into one cell

6.1.10 Cell Configuration in Indoor Projects

Trunking gain has a direct effect on the business case of the indoor system. The more traffic you allow to be serviced by the same cell, the better the trunking gain is, and the lower the production cost per call is. Why then do you need to divide large buildings into more sectors?

UMTS

When designing UMTS indoor solutions it is evident that you cannot just deploy a new scrambling code (sector) in the same cell. This would be an overlap of the two cells in 100% of the area, and all the traffic would permanently be in soft handover, using resources from both cells, so actually nothing would be gained.

Dividing the building in to more sectors is the only way to add capacity in the building, but UMTS needs physical isolation between the sectors in order to minimize the soft handover zones and decrease cannibalization of the capacity of the cells.

GSM

On GSM you can add new radio channels to the cell; the limit is the number of channels the operator has in the spectrum and restrictions with regards to frequency coordination with other adjacent cells. Also, hardware restrictions on the base station will demand a channel separation of typically three channels in the same cell. However, in theory you could deploy 16–24 channels or even more in the same cell. In practice it is preferable to keep the number limited to a maximum of about 12 for an indoor project. In a high-rise building you should use a maximum of four to six carriers in each of the topmost sectors, to avoid 'leaking' too many carriers to the nearby macro area. However, there are other reasons – for the passive part of the DAS it is a matter of keeping the composite power at a relative level when combining several high power carriers to minimize inter-modulation problems. For active DAS it can be a matter of a too low a power per carrier, due to the limited resources of the composite amplifier, where all carriers must share the same power resource, and this will limit the downlink service range from the antennas.

6.1.11 Busy Hour and Return on Investment Calculations

Busy Hour

You must bear in mind that the capacity calculation for estimating the needed traffic channels must be based on the busiest hour in the week, whereas the business case calculation is based on the total production of traffic, or call minutes in the building. The number of traffic channels needed will be dependent on the GOS you want to offer in the busy hour.

Return on Investment

Return on investment (ROI) is mostly the motivation for implementing any indoor solution. Usually an operator will know what the revenue is per call minute, but how do you calculate the expected traffic production for the system? It is actually quite easy: 1 E is, as we know, defined as a constant call. Therefore, you can calculate that 1 E is $24 \times 60 = 1440\,\text{min/day}$. This will mean that a system with a traffic load of 11 E will produce $11 \times 1440 = 15840\,\text{call min/day}$. Knowing the revenue per call minute, you can then calculate the revenue for the system, and do the business case evaluation.

Example, Shopping Mall

Sometimes it is a little more complex to calculate the traffic and the needed traffic channels. If you have a case with a multioperator solution, you will need to divide the traffic according to the market share.

Inputs

The input is the statistics from the shopping mall: an average of 18 000 shoppers per day each in the mall for an average time of 2.5 h. There are six shopping days; opening hours are Monday to Friday 9.00 to 19.00 and Saturday 10.00 to 14.00. The busiest day of the week is Saturday, when 25 000 shoppers are in the mall. The mobile penetration in the country where the shopping mall is located is 85%. The market share for the operator we are designing for is 55%. The operator allows 2% GOS in the shopping area.

Calculations
First of all we need to understand that there actually are two calculations of the traffic: one for the business case and one for the traffic design. The business case should be based on the total call minutes, but the traffic design should be done to accommodate the busy hour.

Busy Hour Calculation
From the data we can see that the busiest day is Saturday, with 25 000 shoppers in the mall, and we assume that in the busy hour there are 20 000 shoppers present. We can then calculate the traffic load in the busy hour:

$$\text{BHL} = \text{users} \times \text{LPU} \times \text{MP} \times \text{MS}$$

Where BHL is the busy hour load; users is the number of users; LPU is the load per user (assumed to be 20 mE); MP is the mobile penetration; and MS is the market share for the operator.

$$\text{BHL} = 20\,000 \text{ shoppers} \times 0.02\,\text{E} \times 0.85 \text{ mobile penetration} \times 0.55 \text{ market share}$$

$$\text{BHL} = 18.7\,\text{Erlang busy hour traffic}$$

The GOS was 2%, so using the Erlang table we can see that we need 27 traffic channels in order to support this traffic.

The Business Case Calculation
In order to evaluate the business case, we need to sum up all the traffic for the week. On weekdays the mall is open for 10 h, and on Saturday for 4 h. This is an average of 1800 shoppers per hour on weekdays and 6250 on Saturdays. Using the average of 20 mE per shopper, we can calculate the traffic production per hour: we know that 1 E produces 60 call min/h; 20 mE will then produce $60 \times 0.02 = 1.2\,\text{call min/h}$ on average.

We need to incorporate the market share of 55% and the mobile penetration of 85%, so for the operator we have an average of $1800 \times 0.55 \times 0.85 = 841.5\,\text{users/h}$ on weekdays, and $6250 \times 0.55 \times 0.85 = 2921.8\,\text{users/h}$ on Saturdays. Using the 1.2 call min/h, we can calculate the production:

$$\text{daily production} = \text{users} \times \text{call min per user} \times \text{opening hours}$$

$$\text{total all weekdays} = 842.5 \times 1.2 \times 10 \times 5 = 50\,550 \text{ call min for all five days}$$

$$\text{Saturdays} = 2921.8 \times 1.2 \times 4 = 14024.6 \text{ call min per Saturday}$$

The total production per week is then $= 50\,550 + 14\,024.6 = 64\,574$ call min.

We could also have done the calculation in Erlang:

$$\text{Erlang} = \text{users} \times \text{E per user} \times \text{opening hours}$$

$$\text{All five weekdays} = 842.5 \times 0.02 \times 10 \times 5 = 842.5 \text{ E per week.}$$

$$\text{Saturdays} = 2921.8 \times 0.02 \times 4 = 233.7\,\text{E per Saturday.}$$

$$\text{One week is a traffic production of } 842.5 + 233.7 = 1076.2\,\text{E}$$

We know that one E is 60 call min, so we can easily convert back to call minutes:

$$\text{call min} = \text{E} \times 60$$
$$\text{call min} = 1076.2 \times 60 = 64\,574\,\text{call min}$$

6.1.12 Base Station Hotels

Load sharing can be used on a large scale, and be a type of 'base station hotel', where a central resource of base stations distributes coverage and capacity to many buildings in the same area, typically a campus environment. This has several advantages but the main points are:

- Very high trunking gain.
- Low cost rollout – only needs one base station and site support.
- Only needs one equipment room.
- Easy upgrades in capacity.

Base stations are designed by deploying optical transmission between active DAS solutions. One central point will have the base station and main unit located, and from there coverage and capacity are distributed to nearby buildings. One must be careful with the total delay introduced by the distribution system, and not exceed the maximum timing advance in GSM; the synchronization window in UMTS must be tuned accordingly (see Chapter 10)

6.1.13 Data Capacity

Unlike voice service, data traffic is normally not time-sensitive: short delays of SMSs, or emails or slightly degradation of data speed will normally not impact the user perception of the quality of service. Therefore, when designing capacity for data services there are more dynamics, in terms of 'blocking', etc.

We cannot user Erlang to estimate the capacity need for data, but we must realize that the amount of data traffic will take up capacity that could have been used for voice. Normally mobile operators will design the data capacity per user in kbps. Advanced traffic control parameter settings in the network allow us to adjust various margins to control the amount of data traffic, and the remaining capacity for voice service.

Design targets for indoor data service are normally expressed as maximum data speeds and the maximum number of users for a particular data speed. Guarantees for quality of service are relaxed.

The increasing need for high-quality data services puts special demands on delay and round-trip delay. The specific design requirement is directly related to the specific service; therefore it is hard to give any concrete guidelines here.

Backhaul Issues

On UMTS and HSPA the data service required in the indoor cell will demand bandwidth on the backhaul transmission line from the base station to the network. This is a major issue for the mobile operators; it does not make sense if the indoor system can support 14 Mbps if the backhaul transmission is limited to 2 Mbps.

7

Noise

Noise performance and noise calculations are the most important aspects in radio planning. The quality of any radio link is not defined by the absolute signal level, but by the signal-to-noise ratio. The lower the noise is relative to the signal level, the better the radio performance. Radio quality and mobile data rates are dependent on a certain signal-to-noise ratio. The better the SNR, the better the quality, the higher the data rate on the radio link will be.

Many RF planners consider the issue regarding noise in DAS systems, on the UL of their link budget to be 'black magic'. However, if you do not overcomplicate the subject of noise, the concept is really pretty straight forward and easy to understand. Understanding noise and noise calculations is crucial for the radio planner in designing any RF system. Radio planners need to understand that passive loss in a DAS system will also increase the total system noise figure. Noise power generated by amplifiers and the base station also has a big impact on the performance of the radio link.

Over the following pages, I will present the basics of noise, by presenting examples of amplifier configurations and analyzing the performance step by step. All the examples on the next pages have been calculated on a standard scientific pocket calculator (the one on the cover on this book). You do not need advanced computer programs to be in control of the noise calculations when designing passive or active DAS solutions. However, the creation of spreadsheets in Excel for doing the trivial part of the work with these calculations is recommended. It saves a lot of time, and you can limit the mistakes.

7.1 Noise Fundamentals

Any object capable of allowing the flow of any electric current will generate noise. The noise is generated by random vibrations of electrons in the material. The vibrations, and hence the noise power, are proportional to the physical temperature in the material. This noise is referred to as 'thermal noise'.

Indoor Radio Planning: A Practical Guide for GSM, DCS, UMTS and HSPA Morten Tolstrup

7.1.1 Thermal Noise

Thermal noise is 'white noise'; white noise has its power distributed equally throughout the total RF spectrum, from the lowest frequency all the way to the highest microwave frequency.

Spectral Noise Density

In other words, the power spectral density of white noise is constant over the RF frequency spectrum; hence the noise power is proportional to the bandwidth. If the bandwidth of the RF channel is doubled, the thermal noise power will also double (+3 dB).

Thermal Noise Level

In order to be able to calculate the noise power of a given bandwidth, you need to establish the 'base noise', that is, the noise in 1 Hz bandwidth. Knowing this level, you can simply multiply the noise power in 1Hz by the bandwidth in Hz. Thermal noise power is defined as

$$P = 10\log(KTB)$$

Where P is the noise power at the output of the thermal noise source (dB W), K is Boltzmann's constant = 1.380×10^{-23} (J/K), T is temperature (K) and B is the bandwidth (Hz).

Reference Noise Level per 1 Hz

At room temperature (17 °C/290 K), in a 1 Hz bandwidth we can calculate the power:

$$P = 10\log(1.380 \times 10^{-23} \times 290 \times 1 = -204\,(\text{dB W})$$

where dBm takes its reference as 1 mW. The relation between 1 W and 1 mW is $1/0.001 = 1000 = 30\,\text{dBm}$, i.e. $1\,\text{W} = 30\,\text{dBm}$. Therefore you can calculate that the thermal noise power at (17 °C/290 K) in a 1 Hz bandwidth in dBm:

$$\text{thermal noise power} = -204 + 30 = -174\,\text{dBm/Hz}$$

where −174 dBm/Hz in noise power is the reference for any noise power calculation when designing radio systems. Relative to the bandwidth, you can use the reference level of the −174 dBm/Hz and simply multiply it by the actual bandwidth of the radio channel. It does not matter if it is a radio service that operates on 450 MHz or at 2150 MHz, the noise power/Hz will be the same if the radio channel bandwidth is the same.

Reference Noise Level for GSM and DCS/UMTS

The noise power density is frequency-independent: the noise power remains the same no matter what the radio frequency. The noise power is equally distributed throughout the spectrum and is related to the bandwidth of the RF channel.

GSM and DCS

Using −174 dBm/Hz, we can calculate the thermal noise floor for the 200 kHz channel as used for GSM/DCS. We just calculate the thermal noise in that bandwidth:

$$KTB \text{ for GSM } (200\,\text{kHz}) = -174\,\text{dBm/Hz} + 10\log(200.000\,\text{Hz}) = -121\,\text{dBm}$$

The −121 dBm is therefore the absolute lowest noise power we will have in a 200 kHz GSM/DCS channel.

UMTS

The UMTS radio channel is 3.84 MHz. This makes the thermal noise floor for UMTS:

$$KTB \text{ for UMTS}(3.84\,\text{MHz}) = -174\,\text{dBm/Hz} + 10\log(3.840.000\,\text{Hz}) = -108\,\text{dBm}$$

7.1.2 Noise Factor

The noise factor (F) is defined as the input signal-to-noise ratio (S/N or SNR) divided by the output signal-to-noise ratio. In other words the noise factor is the amount of noise introduced by the amplifier itself, on top of the input noise. For a passive system such as a cable, the SNR would also be degraded, by attenuating the signal nearer to the thermal noise floor would equally degrade the SNR.

The noise factor is a linear value:

$$\text{noise factor } (F) = \frac{\text{SNR}_{(\text{input})}}{\text{SNR}_{(\text{output})}}$$

7.1.3 Noise Figure

The noise figure (NF) is noise factor described in dB, and is the most important figure to note on the uplink of any DAS or amplifier system. The NF will affect the DAS sensitivity on the uplink, and will determine the performance of the uplink. This will be the limiting factor for the highest possible data rate the radio link can carry, no matter whether it is GPRS, EDGE, HSUPA or any other service. The lower the NF, the better the performance.

$$\text{noise figure (NF)} = 10\log(F)$$

7.1.4 Noise Floor

The noise floor is the noise power at a given noise figure at a given bandwidth; you can calculate the noise power or noise level for a receiver, amplifier or any other active component.

$$\text{noise power (noise floor)} = KTB + \text{NF} + \text{gain of the device}$$

A typical case for a GSM BTS is shown in Figure 7.1.

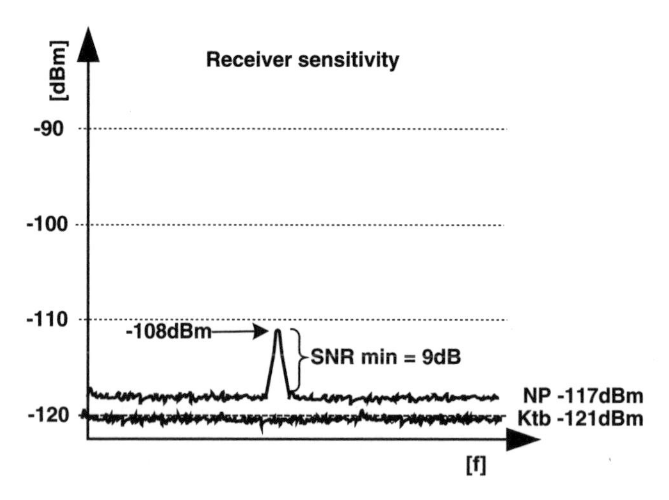

Figure 7.1 GSM receiver with 4 dB NF

7.1.5 The Receiver Sensitivity

Knowing the NF and bandwidth of the receiver in the BTS, you can calculate the receiver sensitivity. However, first you need to calculate the noise floor of the receiver.

Noise Floor

As an example, we can take a GSM receiver as shown in Figure 7.1 with an NF of 4 dB, then the noise floor for that receiver can be calculated as:

$$\text{receiver noise floor} = KTB + \text{NF}$$
$$\text{receiver noise floor} = -121\,\text{dBm} + 4\,\text{dB} = -117\,\text{dBm}$$

This is the base noise power, the absolute lowest signal level present at the receiver input, a noise power of -117 dBm, a noise level generated by the receiver itself.

Sensitivity
With reference to the noise floor, you can calculate the minimum signal level for any given service that the receiver is able to detect; you just need to reference the required SNR for the service, and make sure the service is the required level above the noise floor of the receiver.

If, for example, you want to provide a GSM voice service, you need a 9 dB SNR. You can calculate the minimum required level the receiver will be able to detect to produce the quality needed for voice (without adding any fading margins):

$$\text{receiver sensitivity} = \text{receiver noise floor} + \text{service SNR requirement}$$
$$\text{receiver sensitivity} = -117 + 9 = -108\,\text{dBm}$$

Thus the lowest detectable signal for a voice call will be -108 dBm. Therefore the receiver sensitivity for GSM voice is -108 dBm in this example, excluding any fading margins.

7.1.6 Noise Figure of Amplifiers

As in the example with the receiver above, you can calculate the noise increase in a single amplifier, knowing the bandwidth of the signal, the NF and the gain of the amplifier.

Example: 30 dB Amplifier with 10 dB NF

Using an amplifier for GSM with a 200 kHz bandwidth with 30 dB gain and an NF of 10 dB (as shown in Figure 7.2), you can calculate what happens if you feed the amplifier a signal of −90 dBm (as shown in Figure 7.3). Some would consider this a relatively low signal, but the quality, the SNR of the input signal, is actually good because in this example there is no interference from any source present at the input. The only noise power present on the input of the amplifier is the *KTB* (−121 dBm/200 kHz). Therefore we can calculate that the signal has an SNR of 31 dB [−90 dBm − (−121 dBm)] – this is in fact a really good quality signal.

Nevertheless, you use the amplifier to boost the signal level, and you can now calculate the consequences for the signal, the noise and the SNR at the output of the amplifier.

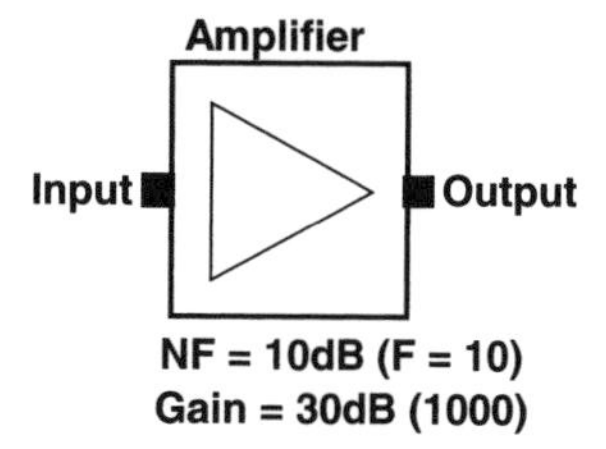

Figure 7.2 GSM 30 dB/10 dB NF amplifier

New Noise Floor

The input signal from Figure 7.3 is fed into the amplifier shown in Figure 7.2 with the specified 30 dB gain and 10 dB NF. Naturally the amplifier will amplify the signal from −90 to −60 dBm, but it will also amplify the noise at the input (*KTB* in this example) from −121 to −91 dBm. In addition to the 30 dB amplification of the noise, the amplifier will add an additional 10 dB to the noise power due to the 10 dB NF (as shown in Figure 7.3), so the output noise level from the amplifier is:

$$\text{NFloorOut} = \text{NFloorIn} + \text{Gain} + \text{NF}$$

$$\text{NFloorOut} = -121\,\text{dBm} + 30\,\text{dB} + 10\,\text{dB} = -81\,\text{dBm}$$

The Noise Figure Degrades the SNR

The result is amplification of the wanted signal by 30 dB, from −90 dBm at the input to −60 dBm at the output – so we have a higher signal level. However, what happened to the quality of the signal, the SNR?

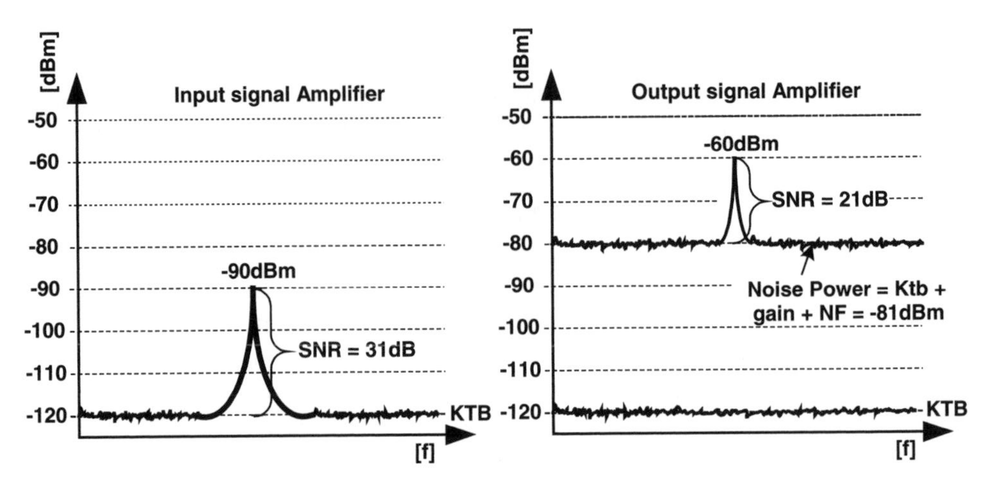

Figure 7.3 GSM 30 dB/10 dB NF amplifier

We actually degraded the SNR of the signal from 31 to 21 dB. In fact, we degraded the performance and data rate for the base station on the uplink due to the 10 dB NF degradation of the SNR. We can conclude that an amplifier will always degrade the SNR, if you consider the amplifier as a single block.

Why Do We Use Amplifiers?

We just calculated that the amplifier will degrade the SNR, due to the NF of the amplifier. Then why do we use amplifiers at all, if they degrade the SNR of the signal; after all, the performance of the signal is not the level but the SNR? To understand this we need to analyze the impact of loss of passive cable in the system (Sections 7.2.2 and 7.2.3) and the impact of loss in cascaded amplifier systems (Section 7.2.4).

7.1.7 Noise Factor of Coax Cables

To understand how and when you can use amplifiers, to maximize the performance of the DAS system, you need to understand the impact of the attenuation of a passive coax cable or component on the total system noise figure performance. You need to realize that any loss of a passive cable or any other passive component prior to the base station will also impact and degrade the SNR, like the path loss on the radio link or the NF of an amplifier will.

This may sound a bit odd; after all the passive cable or component will not produce any noise power, but it will degrade the SNR performance of the system just the same. The passive losses will diminish the signal, the 'S' in the SNR, whereas active elements will raise the 'N' in the SNR. Both will result in degraded SNR.

We know from the previous section that the noise factor (F) is defined as the input SNR divided by the output SNR:

$$\text{noise factor } (F) = \frac{\text{SNR}_{(\text{input})}}{\text{SNR}_{(\text{output})}}$$

Example of NF of a Passive Cable with 30 dB Loss

Let us analyze the noise performance of a passive coax cable with 30 dB loss, a typical attenuation for a passive DAS. In the example we feed the input of the cable a GSM signal level of −75 dBm, as shown in Figure 7.4. We can calculate the SNR at the input of the cable to be:

$$\text{SNR} = -75\,\text{dBm} - (-121\,\text{dBm}) = 46\,\text{dB}\ (\text{factor} = 39811)$$

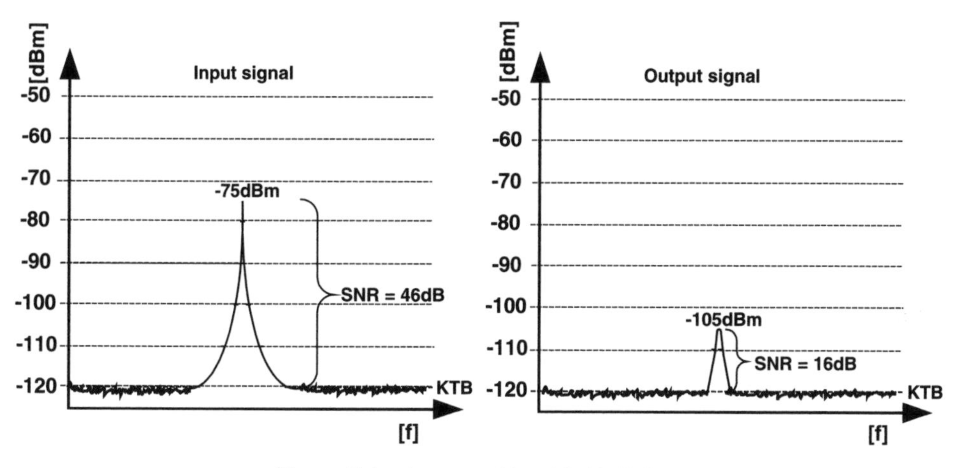

Figure 7.4 A coax cable with 30 dB loss

At the output of the cable the signal is now attenuated with the loss of the cable by 30 dB, and the signal level is now:

$$\text{output signal level} = \text{signal input} - \text{attenuation}$$
$$\text{output signal level} = -75\,\text{dBm} - 30\,\text{dB} = -105\,\text{dBm}$$

The SNR at the output of the cable is:

$$\text{noise is still the } KTB,\ \text{so the SNR} = -105\,\text{dBm} - (-121\,\text{dB}) = 16\,\text{dB}\ (\text{factor} = 39.8)$$

A value of 16 dB will be acceptable for most services, but might impact the performance on HSPA.

Now that we have the SNR for the input signal and the output signal we can calculate the noise factor:

$$\text{noise factor } (F) = \frac{39811}{39.8} = 1000$$

In dB:

$$\text{noise figure} = 10\log(F)$$
$$\text{noise figure} = 10\log(1000) = 30\,\text{dB}$$

The noise figure of the cable is 30 dB (noise factor 1000). We can then conclude that the noise figure of a passive component is the same as the loss or attenuation of the component.

The SNR of the signal is now degraded by 30 dB due to the loss of the cable. Like any NF degradation this will severely impact the performance of the RF link.

The Good SNR is Lost Forever

Amplifying the signal at the end of the cable would not help; we would also amplify the noise (the *KTB*) and degrade the SNR with the NF of the amplifier.

Conclusion, Loss on Passive Cables

Any loss or attenuation of a passive cable or passive component will degrade the NF with the attenuation in dB – dB per dB. Once the SNR is degraded cable, you are getting closer to the thermal noise floor, the *KTB*. There is no way back, not even with an amplifier at the output of the cable; this amplifier will only degrade the SNR even more – you simply cannot retrieve the SNR once it is lost.

7.2 Cascaded Noise

Components rarely work on a stand-alone basis. You will need to connect the base station to the DAS via cables. The passive DAS consists of many passive cables and many passive components, adding up the losses from the base station to the antenna and vice versa. Using active DAS you still need to use a passive cable to connect to the main unit, and the active DAS consists of amplifiers and other active components. Hybrid DAS is a mixture of a large portion of passive components and cables, with active amplifiers and distribution units.

When you chain amplifiers and passive components or cables in a system, it is called a 'cascaded' system. Most RF systems will be cascaded systems. The amplifier will amplify the signal and the noise of the preceding amplifier or passive component or cable, and therefore the NF builds up.

7.2.1 The Friis Formula

We can calculate the noise factor (F) of any number of chained amplifiers or passive components (stages) using the cascaded noise formula known as the Friis formula. The Friis formula can be used for any number of cascaded components, from 1 to n. F_1/G_1 is the noise factor and gain of stage 1, F_2/G_2 is the noise factor and gain of stage 2, etc.

$$F_S = F_1 + \left[\frac{F_2 - 1}{G_1}\right] + \left[\frac{F_3 - 1}{G_1 \times G_2}\right] + \cdots \left[\frac{F_n - 1}{G_1 \times G_2 \times G_3 \ldots G_{(n-1)}}\right]$$

The values for gain and NF have to be input as linear values (factors).

Linear Gain (G)

To calculate the linear gain to use in the Friis formula, or gain factor, we use this formula:

$$\text{linear gain} = 10^{[\text{gain (dB)}]/10}$$

Example: Linear Gain

$$6\,\text{dB gain will be } G = 3.98$$
$$6\,\text{dB attenuation } (-6\,\text{dB gain})\ G = 0.251$$

Linear Noise Factor (*F*)

We calculate the linear noise factor to use in the Friis formula, using the same formula:

$$\text{linear noise factor } (F) = 10^{[\text{NF (dB)}]/10}$$

Example

$$20\,\text{dB NF (amplifier) will be } F = 100$$
$$20\,\text{dB loss (cable NF} = 20\,\text{dB)} F = 100$$

Calculation of Cascaded Noise

Knowing the gain factor and noise factor of the elements in a cascaded system, we are now able to calculate the performance of the whole system.

Include the Passive Losses
We need to include NF and loss (attenuation = negative gain) of the cables interconnecting the amplifiers in the system as stages in the cascaded system, in order to accommodate the noise factor and attenuation of these losses.

The gain factor of a passive cable is of course less than zero due to the attenuation of the cable. When a passive component or cable is one of the stages, we need to inset the noise and factor of the cable, as just calculated in the example above.

7.2.2 Amplifier After the Cable Loss

In Section 7.1 we calculated the degradation of the signal level and SNR caused by a passive cable or DAS, so why do we not just use an amplifier to boost the signal again? Let us have a look; in this case the amplifier is connected to the end of the coax cable to compensate for the attenuation of the cable loss of the signal, as shown in Figure 7.5. However, is this the best approach? Let us analyze the system performance end-to-end.

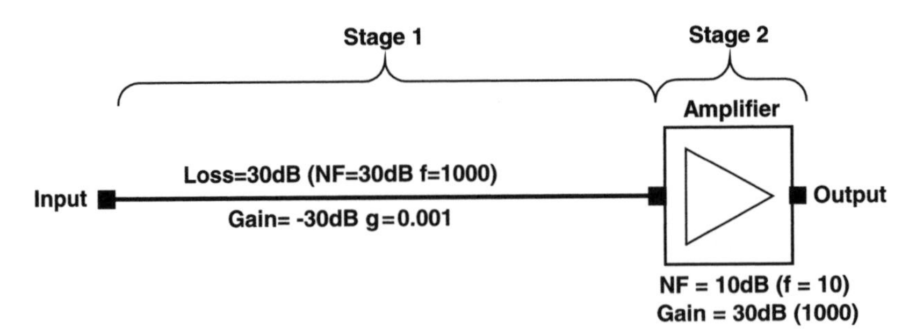

Figure 7.5 Cascaded system, with the passive cable as the first stage

Example
This is a two-stage system. The passive cable is stage 1 and the amplifier stage 2.

We can use the Friis formula to calculate the cascaded performance of the system: the first stage, as shown in Figure 7.5 (always counting the stages from the input of the system), is the coax cable; the second stage is the active amplifier. Inserting the values in the Friis formula of this two-stage system we get the cascaded noise calculation

$$F_S = F_1 + \left[\frac{F_{2-1}}{G_1}\right]$$
$$F_S = 1000 + \left[\frac{10-1}{0.001}\right]$$
$$F_S = 1000 + [9000]$$
$$F_S = 10\,000$$
$$\text{NF} = 10\log(10\,000) = 40\,\text{dB}$$

We can clearly see in the formula that stage 2 is highly affected by the attenuation of stage 1, with a noise factor of 1000 and a gain factor of 0.001

Signal Power and Signal-to-noise Performance

We know now that the cascaded NF of the system is 40 dB, but what happens with the signal and the SNR?

Input Signal
We input −70 dBm at the input of the cable.

Signal Power
We can calculate the signal level throughout the system:

$$\text{Signal input} = -70\,\text{dBm}$$
$$\text{Signal at the output of the cable} = -100\,\text{dBm}$$
$$\text{Gain of amplifier} = 30\,\text{dB}$$
$$\text{Signal at the output of the amplifier} = -70\,\text{dBm}$$

The amplifier actually compensates for the loss of the cable. We have restored the signal level, but what about the quality of the signal, what about the SNR? We can start by calculating the noise power at the amplifier output. We already know the signal level to be −70 dBm.

Noise Power

After we have verified that the system maintains the signal level, we must have a look at the impact of the different stages on the noise power: noise input = *KTB* = −121 dBm; noise at the output of the cable = −121 dBm (it cannot be lower than *KTB*, and the cable does not generate noise power); noise at the input of the amplifier = −121 dBm; noise output from the amplifier = noise input + gain + NF, = −121 dBm + 30 dB + 10 dB = −81 dBm. The cascaded noise formula gave an NF of 40 dB for the system.

We could also have used this formula for the system:

$$\text{NFloorOut} = \text{NFloorIn} + \text{gain} + \text{NF}$$
$$\text{NFloorOut} = -121 + 0 + 40 = -81\,\text{dBm}$$

Signal-to-noise Calculation

Now that we have calculated the signal level as well as the noise level on the output, the SNR is then easy to calculate:

$$\text{SNR} = \text{signal power} - \text{noise power} = -70\,\text{dBm} - (-81\,\text{dBm}) = 11\,\text{dB}$$

Knowing the cascaded NF of the system, we could also have calculated the SNR on the output more easily using this formula:

$$\text{SNR output} = \text{SNR input} - \text{NF}$$
$$\text{SNR input} = -70\,\text{dBm} - KTB = -70\,\text{dBm} - (-121\,\text{dBm}) = 51\,\text{dB}$$
$$\text{SNR output} = 51 - 40\,\text{dB} = 11\,\text{dB}$$

Conclusion: Passive Loss will Degrade the System

According to the cascaded noise formula, it is clear that the first stage of the system is the dominant factor on the system performance. In this case, the passive coax cable has the predominant role. The 30 dB factor of the cable plays a dominant role in the cascaded noise performance, owing to the high NF caused by the loss of the cable.

The example clearly shows that all passive loss (loss = NF) prior to the first amplifier can be added to the amplifier's NF to calculate the system NF. We can conclude that, to obtain the best performance in a cascaded system, we need to have a low NF in stage 1, and a gain better than 0 dB.

7.2.3 Amplifier Prior to the Cable Loss

As we just documented in Section 7.1.1, having the amplifier placed after the cable did not perform that well. We concluded that this was mainly due to the high NF of stage 1, the passive coax cable. We need as low an NF as possible in stage 1 to optimize the NF of the total cascaded system. Let us try to swap the cable and the amplifier, making the amplifier stage 1 and the cable stage 2, and evaluate the performance to see if we can confirm this strategy.

Example
The first stage in Figure 7.6 (always counting from the input) is now the amplifier, and the second stage is the passive cable. Inserting the values of this two-stage system in the cascaded noise formula we get the cascaded noise calculation:

$$F_S = F_1 + \left[\frac{F_2 - 1}{G_1}\right]$$

$$F_S = 10 + \left[\frac{1000 - 1}{1000}\right]$$

$$F_S = 10 + [0.99]$$

$$F_S = 10.99$$

$$\mathrm{NF} = 10\log(10.99) = 10.41\,\mathrm{dB}$$

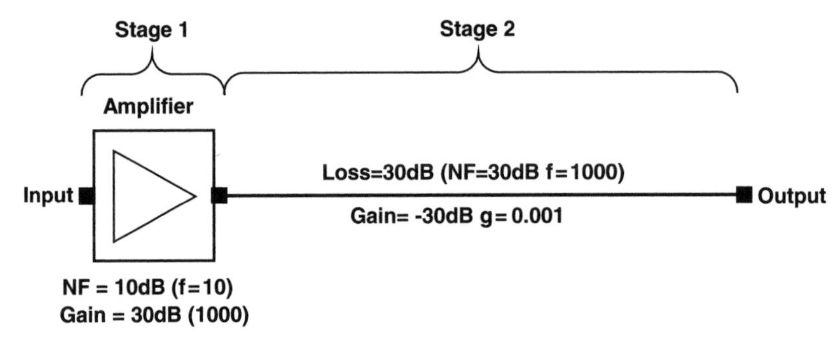

Figure 7.6 Cascaded system, with the passive cable as the last stage

Better NF

We can see that now the NF is improved close to 30 dB, and the first configuration with the amplifier after the cable the NF is 40 dB. The NF in the new configuration is only 10.41 dB. This is almost the only contribution of the 10 dB NF of the amplifier in stage 1. It is evident in the formula that the passive cable, now stage 2, has only a minor impact on the total NF of the system.

Let us have a look at what happens with the signal and noise level through the system, by applying the same signal at the input as in the previous example.

Input Signal
We input −70 dBm at the input of the cable.

Signal Power

We can calculate the signal level throughout the system:

$$\text{signal input} = -70\,\text{dBm}$$

$$\text{gain of amplifier} = 30\,\text{dB}$$

$$\text{signal at the output of the amplifier} = -40\,\text{dBm}$$

$$\text{loss of cable} = 30\,\text{dB}$$

$$\text{signal at the output of the cable} - 70\,\text{dBm}$$

As in the example from Section 7.1.1, the amplifier still compensates for the loss of the cable, with regards to the signal – the system is 'gain transparent': −70 dBm input and −70 dBm output.

Noise Power

We have verified that the system maintains the signal level; we need to analyze the noise performance and the noise power throughout the system:

$$\text{noise input} = KTB = -121\,\text{dBm}$$

$$\text{noise output amplifier} = KTB + \text{gain} + \text{NF} = -121\,\text{dBm} + 30\,\text{dB} + 10\,\text{dB} = -81\,\text{dBm}$$

$$\text{noise output cable} = \text{noise input cable} - \text{attenuation} = -81\,\text{dBm} - 30\,\text{dB} = -111\,\text{dBm}$$

Signal-to-noise Calculation

The SNR is easy to calculate:

$$\text{SNR} = \text{signal power} - \text{noise power} = -70\,\text{dBm} - (-111\,\text{dBm}) = 41\,\text{dB}$$

The SNR is now 30 dB better (compared with Figure 7.5), simply due to the amplifier being stage 1, prior to the passive cable (stage 2).

Conclusions: Amplifiers Prior to Passive Loss can Improve Performance

From the two examples in Sections 7.1.1 and 7.1.2 it is evident that, by placing the amplifier at the input of the system and preferably having minimum a gain equal to the subsequent cable loss, you can improve the link. In this example this is an improvement of 30 dB. It is evident in the cascaded noise formula that F1 plays the major role.

In the example the noise figure of the link is improved from 40 to 10.41 dB; just by moving the amplifier to the first stage of the signal chain you gain 29.58 dB in NF. This will 'clean' the UL signal and improve the SNR of the link dramatically.

7.2.4 Problems with Passive Cables and Passive DAS

In a passive DAS the first amplifier in the system is in the base station receiver. The loss of the passive DAS will impact the NF of the system as we just calculated from Section 7.2.2. The loss of the passive DAS will severely impact the NF and the SNR on the UL, limiting the UL data service performance. One decibel of passive loss prior to the base station will degrade the system NF by 1 dB; in passive systems the attenuation is typically in the range 25–40 dB. Every decibel lost on the passive cable is to be subtracted from the maximum allowed path loss for the given link budget. The passive losses of the coax cable do not have to be a problem, but they will degrade the uplink service range of the DAS antenna, as with any other loss on the radio link.

Doing the link budget calculation for a passive DAS, you will need to calculate every antenna individually if there is different loss to each antenna from the base station. However we also must have a pragmatic approach: if the radio service we are designing the DAS to requires an SNR of 14 dB, we should not worry much about whether the link is performing at 20 or 34 dB in SNR.

This is How a Low Noise Amplifier Works

This is the exact reason for deploying mast-mounted amplifiers (LNAs) in mobile systems, which are frequently used in UMTS deployment. A small low noise amplifier is mounted close to the antenna, in order to compensate for the loss of the feeder cable running to the base station. Often the local LNA inside the base station will be turned off, and the tower mounted LNA will compensate for the gain in the system. By doing so, the loss of the feeder is compensated and the gain in the system is applied where it should be, as close to the receive antenna as possible, before any passive losses degrade the SNR.

The LNA boosts the performance and raises the data rate on the uplink. The UL is often the limiting factor on UMTS, especially at the higher data rates; this will impact the more demanding uplink data services like HSUPA most. The LNA will improve the cell range on the uplink, and thus save on roll-out cost of the network operator, or obtain better performance for the same roll-out cost.

This Also Applies for Active DAS Systems

The exact same principle is used by some manufacturers of active DAS systems; the best performing active DAS will actually have the remote unit placed close to the antenna. Inside this remote unit is the first UL amplifier, the LNA. However, this only applies for pure active systems; hybrid active DAS will, as with any other cascaded system, suffer on the NF from the passive losses prior to the remote unit. For DAS systems where the UL data performance plays a major role, especially UMTS and HSPA systems, you need to be very careful. Do a detailed analysis to see if the uplink data service you are designing can cope with the NF of the DAS.

Uniform Performance

The pure active DAS system in Figure 7.7 will have the same NF at each antenna. Therefore there will be the same uniform uplink performance throughout the DAS system, making link budget calculation and planning easy.

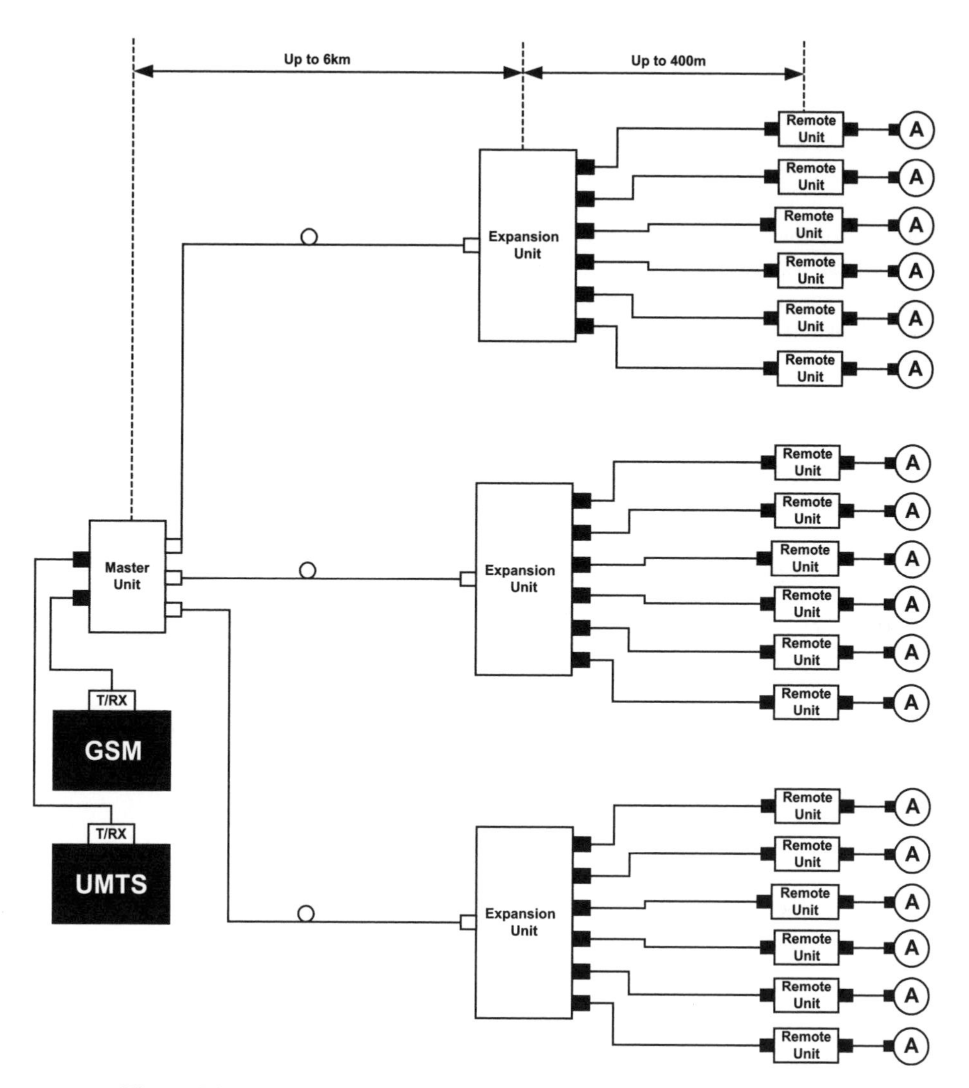

Figure 7.7 The pure active DAS: amplifiers located close to the antennas

For hybrid systems the loss of the passive prior to the remote unit will impact the performance of the NF, and might limit the uplink data speed. Do check with noise calculations of the total system from antenna to base station in order to be sure.

Conclusion: Higher Data Rates with Low NF

The impact from the NF on the UL is evident in the link budget, especially when the radio channel is servicing higher data rates, for UMTS and for HSPA. Refer to Chapter 8 for more details of the link budget.

7.3 Noise Power

From the example in Section 7.2, it is clear that the NF could be improved by installing the amplifier prior to the coax cable. Place the amplifier as near as possible to the signal, the antenna. However, it is also a matter of having control of the noise power, especially in noise-sensitive systems like UMTS/HSPA. In these systems the SNR of the system plays a major role for the high data rate services.

Let us have a closer look at what happens with the input signal as shown in Figure 7.9 (−90 dBm) and the noise in the system. Backing up the calculations with a 'spectrum analyzer look', it all becomes more obvious, compared with 'just' looking at formulas.

7.3.1 Calculating the Noise Power of a System

An example of a pre-amplifier 30 dB gain with 10 dB NF and 30 dB loss on cable is shown in Figure 7.8. The spectrum in Figure 7.9 looks at the input signal.

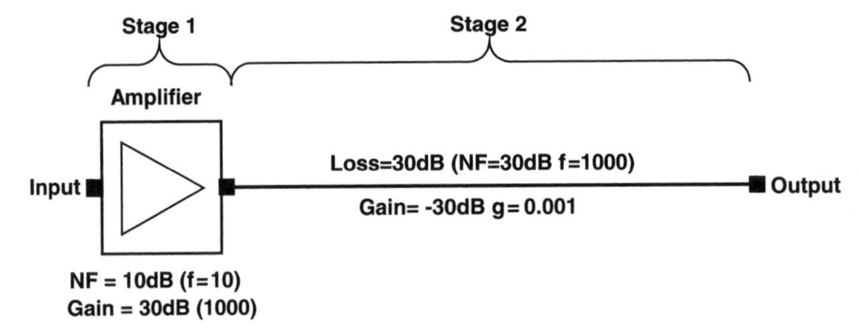

Figure 7.8 Pre-amplifier solution, −90 dBm input signal

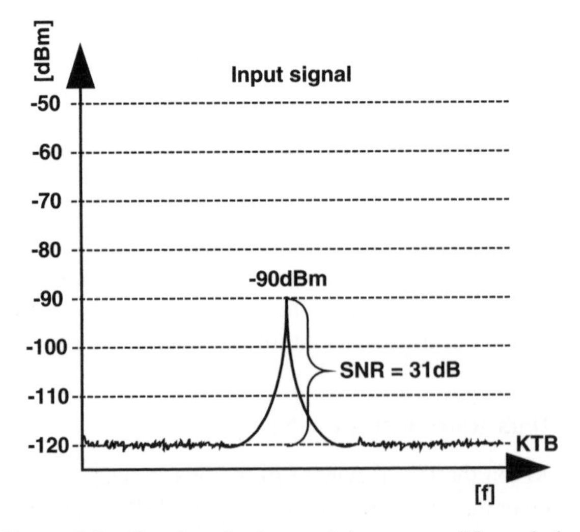

Figure 7.9 Signal at the input of the pre-amplifier solution

Signal-to-noise Ratio at the Input of the Amplifier

This is a GSM system operating at 200 kHz bandwidth; the *KTB* is −121 dBm (noise floor) and we can calculate the signal-to-noise ratio:

$$\text{SNR} = \text{signal/noise (linear)}$$

or in dB,

$$\text{SNR} = \text{signal power} - \text{noise (dB)}$$
$$\text{SNR} = -90\,\text{dBm} - (-121\,\text{dBm}) = 31\,\text{dB}$$

Output Signal from Amplifier

Analyzing the output of the amplifier (Figure 7.8) as shown in Figure 7.10, we can see the gain of the signal, as well as the noise increase caused by the gain and the NF.

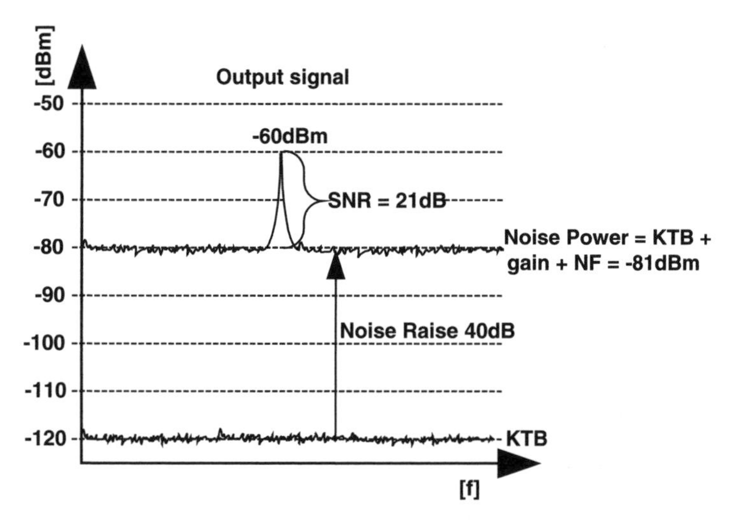

Figure 7.10 The output of the amplifier, signal and raised noise floor

Amplifier Output, Signal Level

The amplifier has 30 dB gain and a NF of 10 dB, so the signal is now:

$$\text{output signal} = \text{input signal} + \text{gain}$$
$$\text{output signal} = -90\,\text{dBm} + 30\,\text{dB} = -60\,\text{dBm}$$

Amplifier Output, Noise Level

The amplifier also amplifies the input noise, the *KTB* (−121 dBm) and adds 10 dB of noise due to the NF of the amplifier; we can calculate the new noise power:

$$\text{NFloorOut} = \text{NFloorIn} + \text{gain} + \text{NF}$$

$$\text{output noise floor} = \text{input noise floor} + \text{NF} + \text{gain}$$

$$\text{output noise floor} = -121\,\text{dBm} + 10\,\text{dB} + 30\,\text{dB} = -81\,\text{dBm}$$

Amplifier, SNR Degradation

The NF of the amplifier will degrade the signal-to-noise ratio:

$$\text{SNR output} = \text{SNR input} - \text{NF}$$

$$\text{SNR output} = 31\,\text{dB} - 10\,\text{dB} = 21\,\text{dB}$$

Output Signal from the Coax

Present at the output of the passive cable will be the attenuated signal level from the amplifier, as well as the noise power out of the amplifier. Both are attenuated by the cable loss (30 dB), as shown in Figure 7.11.

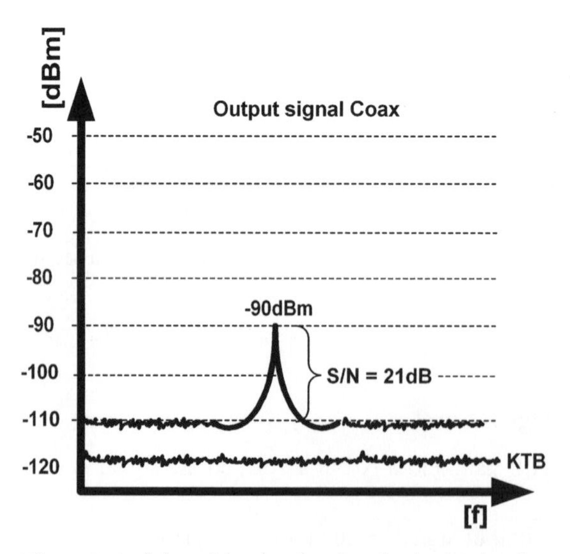

Figure 7.11 Signal at the output of the cable; the signal and raised noise floor are attenuated by the loss of the cable

Output, Coax Cable

The coax cable has 30 dB of attenuation, the same as the 30 dB gain of the amplifier. We can calculate the output of both the signal and noise power of the cable:

$$\begin{aligned}
\text{output signal} &= \text{input signal} - \text{cable loss} \\
\text{output signal} &= -60\,\text{dBm} - 30\,\text{dB} = -90\,\text{dBm} \\
\text{output noise floor} &= \text{system input noise floor} + \text{system NF} + \text{system gain} \\
\text{output noise floor} &= -121\,\text{dBm} + 11\,\text{dB} + 0\,\text{dB} \\
\text{output noise floor} &= -111\,\text{dBm} \\
\text{SNR} &= -90\,\text{dBm} - (-111\,\text{dBm}) = 21\,\text{dB}
\end{aligned}$$

SNR is Maintained

Obviously the cable will attenuate both the wanted signal and the noise, but as long as we make sure that the signals are not attenuated down to a level where the noise power from the amplifier will reach the *KTB* (-121 dBm) (no other noise is present than the *KTB* in this example), the SNR of the system will remain intact. In this example, -81 dBm $-$ (-121 dBm) $=$ 40 dB, we can keep the impact of the SNR minimal. However, if the attenuation of the cable is higher than the input noise power – *KTB*, the SNR will be degraded by the difference between the attenuation and the margin to the *KTB*. This is due to the fact that the *KTB*, -121 dBm (200 kHz GSM), is the absolute lowest level. Once the noise power 'hits' the *KTB* the SNR will be degraded and we will never be able to restore the lost SNR.

High SNR Means Higher Data Rates

The performance on any radio link is not related to the absolute signal power, but to the SNR of the radio channel. This is demonstrated in Figure 7.12; signal 1 is 'only' -90 dBm,

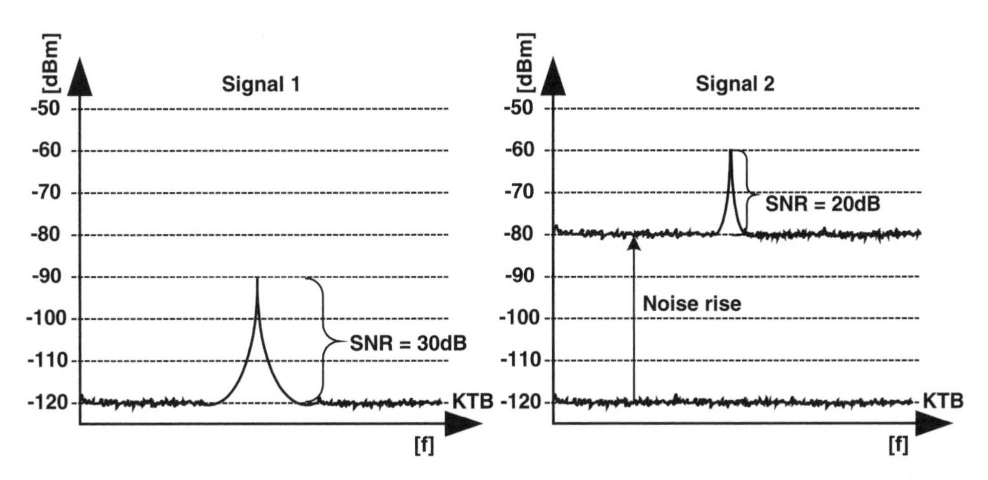

Figure 7.12 It is the SNR of the signal that is important for the performance, not the absolute signal level

whereas signal 2 is 30 dB higher. However, the SNR in signal 1 is 10 dB better; therefore this link will perform much better even if the signal is 30 dB lower.

Far too often all focus is on level, but keep in mind that the quality of the radio service is not defined by the signal level, but the SNR. The better the SNR, the higher the data rates on the radio link.

When designing 3G DAS solutions, the loss of passive cables has a big impact, due to the higher losses on higher frequencies. The stricter SNR requirements for high data rates will also demand good radio link quality, especially when designing HSPA solutions.

Conclusion

From the examples just covered, we can conclude that, to some extent, we can compensate for the loss of a coax cable with the use of pre-amplifiers. For maximum performance we must have the lowest possible NF of the amplifier, and a gain that can compensate for the loss of the passive cable following the amplifier.

7.4 Noise Power from Parallel Systems

Often there will be more than one noise signal (noise power sources) present at the input of a system. This could be two parts of an active DAS, using two repeaters, that needs to be combined to a base station receiver, like the GSM (200 kHz) example in Figure 7.13.

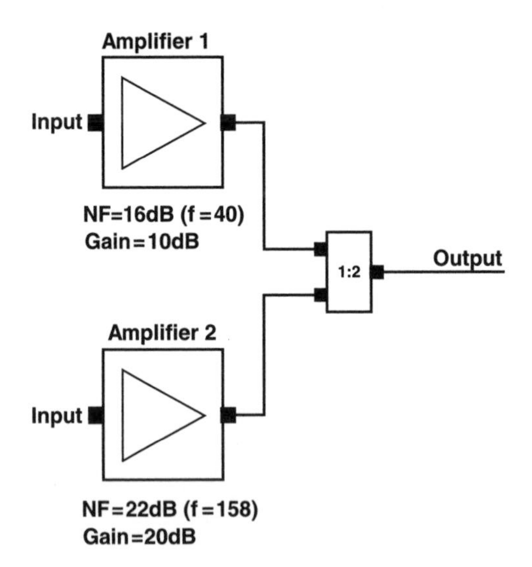

Figure 7.13 Two parallel noise sources combined into one output

7.4.1 Calculating Noise Power from Parallel Sources

Random Power Signals

Noise is random (white noise), and two noise signals will be noncorrelated. Thus the noise powers (in Watts) of two sources can be added in order to calculate the total noise power from the system.

First we need calculate the noise floor for each of the two amplifiers (GSM):

$$\text{NFloorOut} = \text{NFloorIn} + \text{gain} + \text{NF}$$
$$\text{NFloorOut} = KTB + \text{gain} + \text{NF}$$
$$\text{NFloor Out amplifier 1} = -121\,\text{dBm} + 10\,\text{dB} + 16\,\text{dB} = -95\,\text{dBm}$$
$$\text{NFloor Out amplifier 2} = -121\,\text{dBm} + 20\,\text{dB} + 22\,\text{dB} = -79\,\text{dBm}$$

Watts

We need to sum up the noise, by adding the power contribution from each source.

$$\text{power total} = \text{power 1} + \text{power 2} \ldots \text{power } (n)$$

Now we convert the noise powers from dBm to mW:

$$P\ (\text{mW}) = 10^{\text{dBm}/10}$$
$$-95\,\text{dBm} = 136 \times 10^{-12}\ (\text{mW})$$
$$-79\,\text{dBm} = 12.58 \times 10^{-9}\ (\text{mW})$$

Then we can sum up the two powers:
combined noise power $= 136 \times 10^{-12} + 12.58 \times 10^{-9} = 12.89 \times 10^{-9}$ (mW)

Convert to dBm

We convert the noise power back to dBm:

$$\text{noise floor (dBm)} = 10\log(\text{mW})$$
$$\text{noise floor (dBm)} = 10\log(12.89 \times 10^{-9}) = -78.89\,\text{dBm}$$

The Combiner

The noise power is combined in a splitter with 3 dB loss ($F = 0.5$); the output noise floor will be:

$$\text{noise floor output} = -78.89\,\text{dBm} - 3 = -81.89\,\text{dBm}$$

OBS!

Note that for the ease of the example, the loss of the cables and the insertion loss of the splitter have not been accounted for.

7.5 Noise Control

Noise control is important, especially in cascaded systems with multiple amplifiers and interconnecting cable. Gains have to be adjusted according to noise figures and losses of all the individual system components in order to optimize the performance of the whole system, end to end.

7.5.1 Noise Load on Base Stations

When an amplifier or an active DAS injects a noise power signal on the uplink port of the base station, it may potentially cause problems that will affect the performance of the base station. The main issue is that the injected noise will desensitize the receiver in the base station, limiting the performance of the uplink. This issue needs special attention in applications where an active DAS is connected to a macro cell, or if the base station is also serving a passive DAS. In these applications we need to insure that we configure the active part of the DAS with regards to gain and attenuation to minimize the effect of the uplink degradation. To select the correct UL attenuation in order to minimize the noise power, refer to Chapter 10. Furthermore, the raised noise floor has additional side effects for GSM and UMTS systems.

7.5.2 Noise and GSM Base Stations

Injected Noise Might give Interference Alarms

GSM base stations will monitor the noise level in idle mode time slots to estimate any potential interference problems from mobiles in traffic using the same frequency, in nearby or distant cells, PIM products, etc. This enables the base station to provide warnings about potential interference issues that might degrade the performance.

When the noise floor on the GSM base station increases due to any external noise source, like the active DAS, which might inject noise power on the uplink of the base station, this will cause the BSS system to generate an 'idle mode interference alarm'. The trigger level for this alarm can be changed, but the solution is to attenuate the noise to a level below the alarm value (see Chapter 10 for more details).

7.5.3 Noise and UMTS Base Stations

UMTS Systems: Noise will Affect Admission Control

Noise control is paramount for the performance of any WCDMA system, like UMTS. This is the motivation for very strict power and noise control in UMTS. UMTS base stations will evaluate (measure) the noise rise on the uplink in order to evaluate the load in the cell (in UMTS traffic increase is equal to the noise increase).

The UMTS base station will have a preset reference for the base level of the noise floor with no traffic, i.e. the NF of the base station itself, and with this as a reference the base station can calculate the current load (noise) in the cell. The purpose of this noise control is not to admit new traffic if the cell exceeds the pre-set maximum noise increase or load rate. A load rate of 50% in a UMTS cell is equal to a noise rise of 3 dB.

If we inject noise into a UMTS base station that increases the noise floor by 3 dB, then the base station might be tricked into evaluating the current traffic load as 50%, even with no traffic. The base station will then only admit another 10% traffic if the maximum load is preset to 60%.

The result will be that this base station will only be capable of picking up 10% traffic and will be heavily underutilized if we do not minimize the noise load from the external source.

Solution

First of all, the uplink gain of any in-line amplifier in a DAS system should only be adjusted to a level where it compensates for the losses of the subsequent passive network. This will insure that the noise build-up is kept to a minimum, and the noise power injected into the base station or amplifier is at the level of the *KTB*, minimizing the ramped-up noise of the system.

The solution is to attenuate the noise (and signal), in order to lower the noise load of the base station. As we have just clarified, we can do that without affecting the SNR as long as we keep clear of the *KTB* (see Section 7.3.1).

It is impossible to completely remove the noise power on the input without degrading the SNR of the system. Luckily for UMTS, where the noise control is crucial for the performance of admission control, you can offset the base station noise reference level. For further details refer to Chapter 10.

7.6 Updating a Passive DAS from 2G to 3G

In recent years RF designers have tried to prepare their passive DAS designs for UMTS to some extent. Often the planning of these passive DAS was done years prior to the launch of their UMTS service. Thus, it lacks a UMTS link budget, guidelines and knowledge as to how the NF caused by the attenuation of the passive system would impact the 3G uplink performance once implemented.

Many of these GSM systems were 'future proofed' to support UMTS, normally by using passive components that covered the 2100 MHz band, and sometimes a slightly heavier coax cable.

For some of these projects it has become clear that, in some areas of the building, especially in the areas serviced by the most remote antennas, the antennas with the highest loss have problems with UL data speed on UMTS, and in particularly with HSUPA.

7.6.1 The 3G/HSPA Challenge

The reason for this is often the increased loss, the higher noise figure on the passive DAS when the frequencies get higher, but also the much stricter SNR requirements for the high data rate on 3G/HSPA. These requirements are covered in more detail in Chapter 8.

Solution

As we have just explored in the first sections of Chapter 7, we can use amplifiers to improve the performance in the part of the passive DAS that is having difficulties with the loss/noise figure.

Often antennas close to the base station and those with relatively low loss will perform acceptable. Typically it will be just the more distant part of the passive DAS that needs an upgrade to perform on UMTS/HSPA.

Example

This is a real-life practical example on why and how we can upgrade a passive DAS from 2G to 2G + 3G. This passive DAS, shown in Figure 7.14, was originally designed for GSM900. All components, splitters, tappers, etc., were prepared for 1800 and 2100 MHz, so the system

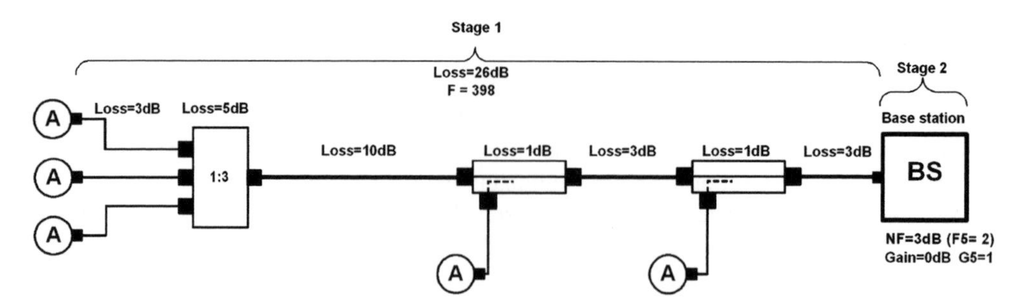

Figure 7.14 GSM passive DAS with 26 dB loss to the three most remote antennas

was to some extent prepared for the UMTS upgrade, which was expected within 5–7 years after installing the original DAS system.

7.6.2 The UMTS Problem

However, after the deployment of UMTS, it turns out that there are issues for the performance on the uplink data service in the areas covered by the three most remote antennas. The radio planner clarified that the problem was due to the high loss on the three remote antennas, as shown in Figure 7.15.

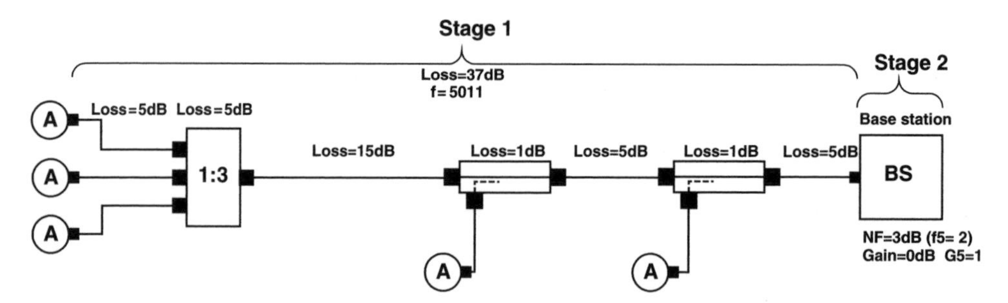

Figure 7.15 UMTS on the passive DAS designed for GSM, now with excessive (37 dB) loss to the three most remote antennas

Cascaded noise calculations showed that the losses on the UMTS band (2100 MHz) were now 37 dB, and that this increased the noise figure for the system performance for the three antennas to 40 dB and degraded the data service in that area of the building.

The UL Problem on UMTS

The problem is related to the high NF that impacts the UL data performance. As we can see in the calculation, stage 1 plays a major role in the degradation, being the passive DAS.

Cascaded Noise Calculation

We can calculate the NF of this two-stage system:

$$F_S = F_1 + \left[\frac{F_2 - 1}{G_1}\right]$$

$$F_S = 5011 + \left[\frac{2 - 1}{0.0001995}\right]$$

$$F_S = 10\,022$$

$$\text{NF} = 10\log(10\,012) = 40\,\text{dB}$$

Conclusion on the Existing Passive DAS

The problem is related to the high NF, and impacts the UL data performance. As we can see in the calculation, stage 1 plays a major role in the degradation, stage 1 being the passive DAS.

UMTS UL Data Coverage at 40 dB NF

When analyzing the link budget (Chapter 8), it becomes clear that the coverage range for 384 kbps is only about 11 m in the dense office environment. The practical experience with degraded uplink 384 kbps data performance is confirmed by the link budget calculation. Even the data services with lower data rates and less strict requirements are limited. For UMTS UL Data service:

- 384 kbps = 11 m.
- 128 kbps = 14 m.
- 64 kbps = 16 m.

Several solutions were considered in order to solve the problem.

7.6.3 Solution 1, In-line BDA

The initial idea was to install one amplifier after the last tapper, in order to boost all three antennas as shown in Figure 7.16, thus only needing to install one active element in the system, and thereby saving on hardware and installation costs.

Cascaded Noise Calculation

We can calculate the NF of this, now four-stage, system:

$$F_S = F_1 + \left[\frac{F_2 - 1}{G_1}\right] + \left[\frac{F_3 - 1}{G_1 \times G_2}\right] + \left[\frac{F_4 - 1}{G_1 \times G_2 \times G_3}\right]$$

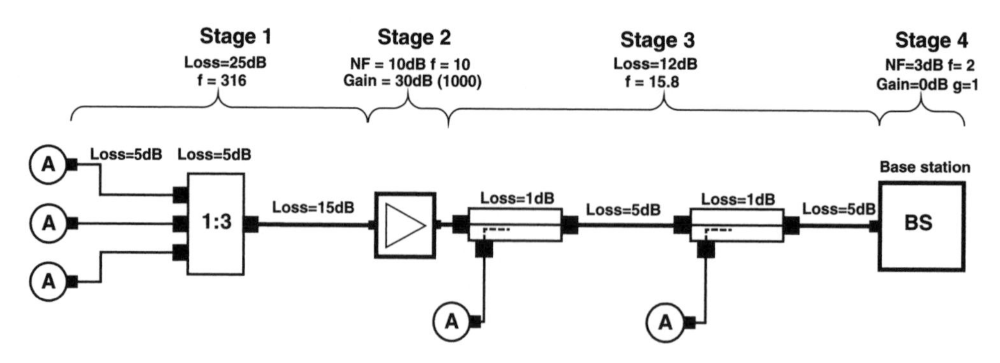

Figure 7.16 UMTS upgrade by deploying an amplifier at a central point in the system

$$F_S = 316 + \left[\frac{10-1}{0.00316}\right] + \left[\frac{15.8-1}{0.00316 \times 1000}\right] + \left[\frac{2-1}{0.00316 \times 1000 \times 0.63}\right]$$

$$F_S = 316 + [2848] + [4.68] + [0.05]$$

$$F_S = 3168$$

$$\mathrm{NF} = 10\log(3168) = 35\,\mathrm{dB}$$

Conclusion on Upgrade with Central BDA

This system with one in-line bi-directional amplifier BDA and repeater is only a 5 dB improvement on the uplink performance. The link budget calculation shows that this is clearly not enough to get the noise figure down to a level that will improve the performance to the required service level on higher data rates. The NF was only improved by 5 dB.

From the cascaded noise formula we can see that stage 2 takes a big 'hit' from the 25 dB loss of the passive stage 1. Thus we conclude that, in order to increase the uplink performance of the system to a desired level, we must try to limit the impact of the passive loss prior to stage 1 of the system. We must deploy the pre-amplifier as close to the antennas as possible to minimize the impact of the coax loss.

Solution 2, Active DAS Overlay

In order to minimize the impact of any passive loss prior to the amplifier, we can now analyze the performance if we install the amplifier right next to each of the three antennas as shown in Figure 7.17. By using the strategy, we can eliminate the passive losses in stage 1 of the cascaded system. Note that the diagram only shows one amplifier, but since the three antennas have the exact same loss back to the base station, only one link is analyzed. The performance for all three antennas with three amplifiers will be the same.

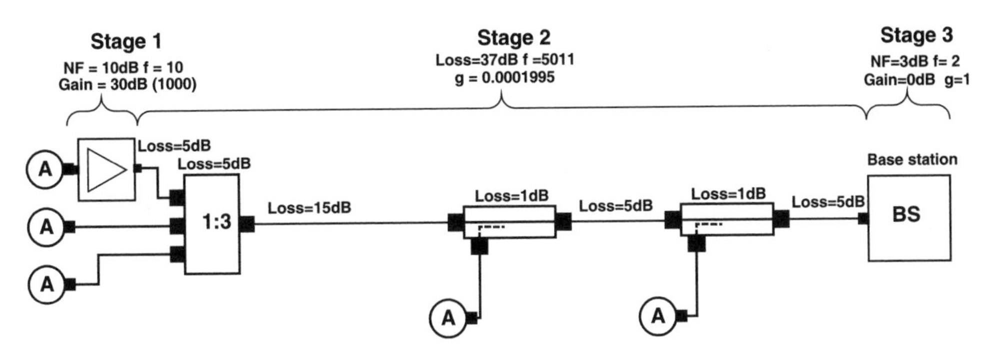

Figure 7.17 UMTS upgrade by deploying the amplifier at the antenna

Cascaded Noise Calculation

We can calculate the NF of this, now three-stage, system:

$$F_S = F_1 + \left[\frac{F_2 - 1}{G_1}\right] + \left[\frac{F_3 - 1}{G_1 \times G_2}\right]$$

$$F_S = 10 + \left[\frac{5011 - 1}{1000}\right] + \left[\frac{2 - 1}{1000 \times 0.0001995}\right]$$

$$F_S = 20.02$$

$$\mathrm{NF} = 10\log(20.02) = 13.01\,\mathrm{dB}$$

Conclusion with BDAs Installed Close to the Antennas

The cascaded noise analysis shows the result; there is a stunning 27 dB improvement of the noise figure. This will boost the uplink performance on the desired UMTS data rates, and even perform well on HSUPA. Furthermore, we can actually reuse most of the old GSM installation by upgrading it with amplifiers close to the antennas. This will insure a fast deployment and limited interruption of the mobile service in the building during the installation of the upgrade.

The part of the existing passive system with relative low loss we can leave intact; here the performance is acceptable. We only need to pinpoint the problem areas, and upgrade these antennas.

Be careful about the noise power injected in the base stations uplink. In this case the attenuation of the passive DAS components, from the BDA to the base station, will attenuate the noise power, and minimize the noise load on the base station. See Chapter 10 for more details on how to minimize the noise load on UMTS base stations using attenuators.

UMTS UL Data Coverage at 13 dB NF

The link budget analysis shows the boost in the uplink performance. The cure is clear: install the amplifier as close to the antenna as possible. This is actually how pure active DAS works (see Section 4.4.2):

- 384 kbps = 39 m.
- 128 kbps = 51 m.
- 64 kbps = 59 m.

Note that ideally we should have calculated the combined NF of the three parallel BDAs, but it turns out that the DL is now the limiting factor, and we must use another approach.

The Downlink

We just analyzed how to boost the performance on the uplink UMTS data performance, on a passive DAS designed for GSM, but what about the performance of the downlink coverage? After all we have 37 dB of loss on the DL from the base station room in the basement to the topmost floors with the problem area.

In this case we are using a +40 dBm base station (CPICH at 10% = +30 dBm), and the CPICH power into the antenna will only be −3 dBm.

The Problem is in the Top of the Building

The problem area is actually located in 'zone C' of the building (see Section 3.5.6), on the highest floors (the executive management floor), where we need a good dominant signal from the in-building DAS system and preferably HSDPA performance.

Lack of DL Signal, CPICH Dominance

According to the free space loss formula (see Section 5.2.8), we have a path loss from the antenna of about 67 dB at 25 m from the antenna, giving us a CPICH level of about −3 dBm − 67 dB = −70 dBm.

The design level for this floor is actually −65 dBm CPICH, so we need to boost the DL power at the antenna in order to get dominance and good SNR and maintain pilot dominance inside the building.

When analyzing the link budget, we see that for the data service for DL, we have less than 12 m of service radius.

UL/DL Service Balance

Often the highest data load is on the DL, and we have just improved the UL with the pre-amplifier to these service radiuses:

UL UMTS Service (Dense Environnent)

- 384 kbps = 39 m.
- 128 kbps = 51 m.
- 64 kbps = 59 m.

There is an imbalance between the uplink and downlink performance, making the system downlink-limited. We will need to boost the downlink level in order to balance the performance and coverage on UL/DL.

Boost both UL and DL on UMTS

We need to boost the DL as well as the UL in order to balance the links; the solution would be to install two-way (UL/DL) amplifiers close to all the distant antennas. These types of amplifiers are often referred to as bi-directional amplifies, but by installing remote BDAs in the building, other considerations arise,

Installation and Performance Issues with Remote BDAs
Often it will be a major challenge to install all the required in-line UMTS BDAs, in these typical, large indoor projects where a passive DAS designed for GSM needs an update to perform on UMTS. The challenge is more than installing the BDAs.

Power Supply
You are frequently not allowed to tap into the existing AC power supply but will need to install a new 'AC group' for the BDA cluster. Installing new AC power groups for this throughout a large building is time-consuming, expensive and has a heavy installation impact on the building. You also need to consider if power supply back-up will be needed for these distributed BDAs. This distribution no-break power will further add to the cost and complication of the system.

GSM Coverage
Make sure that the selected BDAs will not degrade the performance on the existing GSM system. The BDAs must have a 'bypass' function that makes sure that any GSM (and UMTS) signals will be bypassed through the amplifier in case of malfunction or AC power loss.

Commissioning
The distributed BDAs frequently do not have a centralized control point, so setting and tuning them must be done locally for each individual BDA. They are also often installed in locations that are not easily accessible, such as above ceilings (they have to be close to the antennas to perform).

Operations and Maintenance and Surveillance
In case of a fault on the system, it is challenging to find the cause. A cluster of BDAs often lacks a common alarm structure, and a common external alarm interface to the base station.

7.6.4 Solution 2: Active DAS Overlay

There is actually a way to upgrade the passive DAS with remote BDAs located at the antenna, having a common alarm infrastructure with the alarm interface to the base station and a common centralized power supply, if we use the system from Section 4.4.3 (shown in Figure 7.18). If you combine this small active DAS with the passive DAS in the problem areas, the problem will be solved.

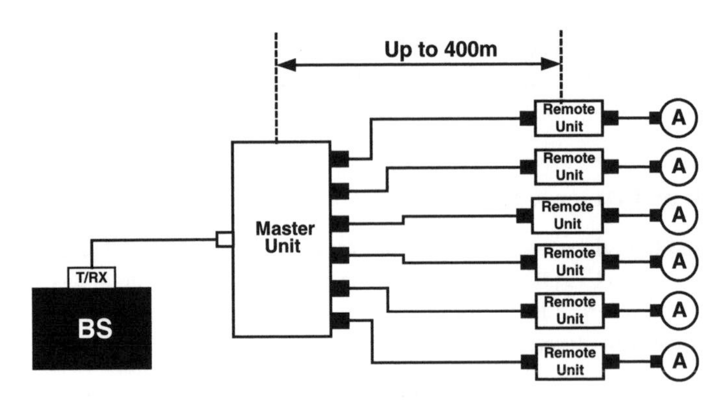

Figure 7.18 Small active DAS, with amplifiers (RU) close to the antennas and no loss

The Existing Passive DAS, Designed for GSM

This 14-floor office building is designed for GSM (the passive DAS is shown in Figure 7.19), and the performance on GSM is perfect. After several years of good GSM operation on the DAS, UMTS was deployed. Initially the UMTS base station was connected to the DAS, expecting full UMTS performance on the existing DAS. However it turned out that there were problems similar to those described in the previous section: lack of pilot dominance and high noise figure on floors 13 and 14. The users on these executive floors suffered from degraded UMTS data performance, dropped calls and no HSPA performance. Measurements on-site confirmed the problem as lack of dominance in 'zone C' the most distant from the base station, and with high signal levels from the outside network present inside the building. As in the previous example, the cascaded noise figure of the antennas covering the topmost floors was calculated, and link budget calculations confirmed the problem.

The users on these executive floors needed an upgrade fast; the problem was solved by deploying small local uplink and downlink amplifiers close to the antennas in the problem, as described below.

The Upgrade to 3G

By deploying the small active system from Figure 7.18, you can increase the DL power and improve the noise figure of the DAS on the two problem floors 13 and 14 (see Figure 7.20). The downlink power from the active DAS is 15 dBm power per carrier (PPC), so the DL could be boosted; 15 dB uplink gain and an NF of 16 dB could improve the performance of the uplink.

You use filters installed in the base station room (S), and at the antenna locations on the two topmost floors 13 and 14 (F) to inject the UMTS signal and to separate the GSM.

Noise Control

When connecting the active DAS system to the existing passive DAS, you must be aware that the noise power from the active DAS can desensitize the base station and degrade the

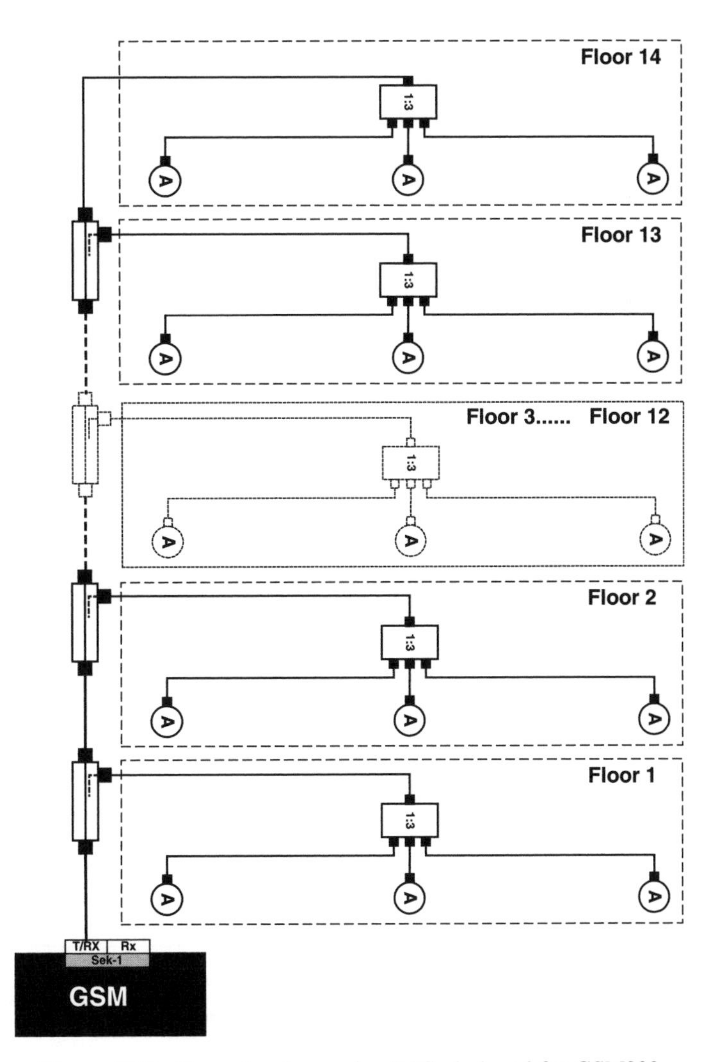

Figure 7.19 Example of a passive DAS, designed for GSM900

uplink performance on the passive section of the DAS. In order to avoid this, we need to insert an appropriate attenuator to attenuate the injected noise power. Refer to Chapter 10 for calculation of the optimum attenuator.

The Performance

The link budget calculation shows that, after the upgrade, the DL will have more power, producing higher downlink levels, and the uplink will perform much better due to the lower noise figure of the system.

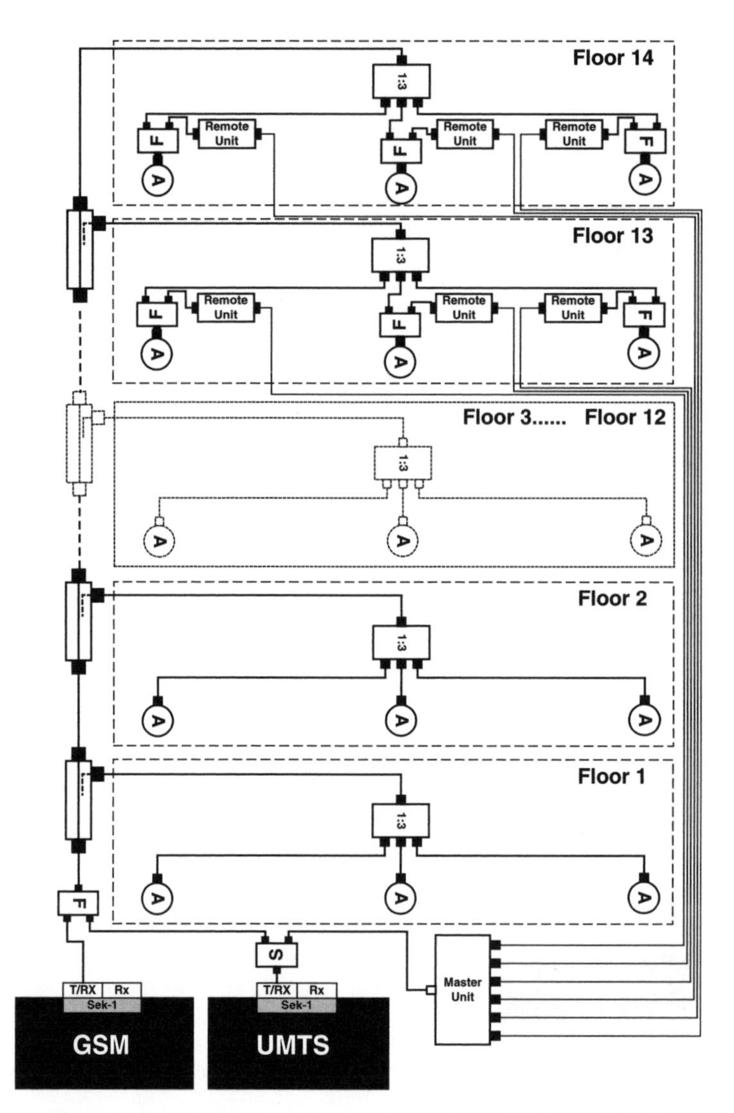

Figure 7.20 Updated DAS servicing for both GSM and UMTS

The system (in Figure 7.20) was implemented, and the link budget calculation was confirmed by measurements. A summary of real-life measurement results from office buildings using this upgrade approach can be seen in Table 7.1.

Table 7.1 UMTS UL/DL performance improvement after upgrade of the passive DAS

Passive DAS loss	NF improvement	DL gain	E_c/I_0 gain
30 dB	13 dB	5 dB	2 dB
35 dB	14 dB	12 dB	8 dB
40 dB	15 dB	17 dB	14 dB

These measurements confirm the theory, and results from other real-life upgrades and implementations also show similar results. This approach is applicable on existing passive DAS, boosting the DL signal and raising the UL data speed.

Still Full GSM Service

The GSM service is still maintained. The only 'cost' on GSM is the insertion loss of the filters, less than 1 dB. In practice there will be no performance degradation, and the GSM service in the building is maintained at the current level.

Conclusion: Upgrading Passive DAS to UMTS Performance

The use of remote located BDAs close to the existing antennas can increase the UL and DL performance on an existing DAS originally designed for GSM. Using the small version of an active DAS is applicable in practice, and use of thin IT-type cabling between the master unit and the remote units ensures a rapid deployment of the system. In some cases you might be able to reuse some of the existing IT cables, minimizing the need to install new ones. The small active DAS uses central power, distributed to the remote units from the master unit, so there will be no need for local power supply at the antenna locations. The master unit has external alarm outputs that can be connected to the base station in order to maintain full visibility of the status of the active elements (the MU and RUs).

7.6.5 Conclusions on Noise and Noise Control

All electronic components generate noise. The density of the noise is constant over the spectrum, and the power is −174 dBm/Hz. This base noise is referred to as the *KTB*. By multiplying the −174 dBm/Hz by the bandwidth of the radio channel, we can calculate the absolute minimum noise level for GSM (200 kHz) as −121 dBm, and UMTS (3.84 MHz) = −108 dBm.

The performance of a radio link is defined as the signal-to-noise ratio, not the absolute signal level. Any noise figure of an amplifier or loss of a passive cable will degrade the radio link. The loss or attenuation of a passive cable or passive component will degrade the NF with the attenuation in decibels. Once the SNR is degraded by the passive cable, you are getting closer to the thermal noise floor, the *KTB*. Then there is no way back, not even with an amplifier at the output of the cable. This amplifier would only degrade the SNR even more – you simply cannot retrieve the SNR once it is lost. Placing an uplink amplifier at the end of the passive DAS, close to the base station, will not improve the SNR, but degrade the SNR with the noise figure of the amplifier.

From the two examples in Sections 7.2.2 and 7.2.3 it is evident that you must place the amplifier at the input of the system. It is preferable to have a low noise figure in the amplifier and minimum gain as the subsequent cable has loss. With this strategy you can improve the link, and compensate for most of the impact of the NF from the passive loss in the following cable. In the example the link was improved by almost 30 dB. If you make sure you keep below the maximum loss, SNR is maintained, and the performance on the RF link is maintained.

When using amplifiers prior to the base station, you must be careful with the noise you inject on the uplink of the base station. The uplink gain of, any in-line amplifier in a DAS system should only be adjusted to a level where it compensates for the losses of the following passive network. This will insure that the noise build-up is kept to a minimum, and the noise power injected into the base station or amplifier is at a level of the *KTB*, minimizing the ramped-up noise of the system. You can decrease the noise power down close to the *KTB* with only minimum noise impact of the link.

It was also demonstrated how this basic knowledge about noise and noise control enables us to upgrade an existing passive DAS designed for GSM to perform high data rates on UMTS. By deploying an active DAS overlay on the part of passive DAS that are designed for GSM, you could boost the performance on the uplink to service for UMTS/HSPA. This strategy will boost the uplink as well as the downlink. Understanding noise and noise performance and calculations is the basis of all radio planning.

8

The Link Budget

The link budget (LB) is the fundamental calculation for planning of any RF link between a transmitter (Tx) and a receiver (Rx). In two-way calculations we actually have two LB calculations, one for the DL and one for the UL.

The result of the link budget calculations is the maximum allowable path loss (APL) from the base station to the mobile in the downlink and the maximum allowable link loss on the reverse link, from the mobile to the base station the uplink. You need to include all the attenuation and gains of the end-to-end signal path from the Tx to the Rx, the attenuation due to the distance, adding the clutter loss of the environment, cable attenuation and antenna gains. You will also need to include a safety margin to provide a given probability of the desired signal, accounting for fading margins and body losses.

Depending on the type of distribution system you design, there are various parameters to take into account when calculating the link budget. Based on these parameters for the DAS, the radio service requirement, and the impact of noise from existing signal sources operating on the same frequency or channel, you must calculate the link budget for both links, the uplink and the downlink, to determine the service range of the system in both directions. The link budget for the particular radio service range can be either UL- or DL-limited; this will depend on the parameters affecting the links. One cell can be downlink-limited for one type of service, and the same cell can be uplink-limited for another service (service profile).

8.1 The Components and Calculations of the RF Link

Let us try to break down the link budget and have a look on the most important parameters in the calculation. The simplest LB calculation looks, in principle, like this:

$$\text{Rx level (dBm)} = \text{Tx power (dBm)} - \text{cable attenuation (dB)} - \text{propagation losses} + \text{antenna gain (dB)}$$

However, the real-life GSM and UMTS LB are more complex, so you will need to break down the link budget calculation into more detail.

Indoor Radio Planning: A Practical Guide for GSM, DCS, UMTS and HSPA Morten Tolstrup

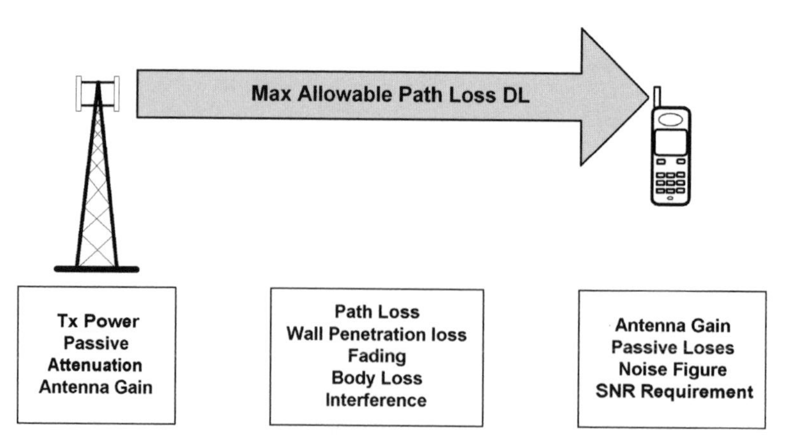

Figure 8.1 Principles of the link budget (DL)

8.1.1 The Maximum Allowable Path Loss

The essential purpose of making the link budget is to calculate the maximum allowable path loss, as shown in Figure 8.1. Once you have calculated the APL, you can calculate the service area and service radius to use for antenna placement in the building.

Note that the example in Figure 8.1 is rather simplified, and merely shows a few of the factors you need to take into account when calculating the link budget. This simplified example shows only the downlink, but both the downlink and the uplink APL need to be calculated, for all services on the DAS.

8.1.2 The Components in the Link Budget

In order to understand the details in the link budget for indoor DAS design, you need to break it down to the all various components of gains and losses. Then you can understand how the different parameters interact and affect the link budget calculation.

The components of the basic link budget are shown in Figure 8.2. To be able to do a detailed analysis, you need to take a look at the different components of the link budget calculation shown in basic form in Table 8.1, one by one (the symbols from Table 8.1 are defined below).

- *a*, BS power (dBm): this is the generated RF power from the base station, at the antenna connector of the output of the base station rack.
- *b*, Feeder loss (dB): this is the attenuation of the coax cable from the BS to the antenna. The loss is symmetrical for the UL and DL.
- *c*, BS antenna gain (dBi): this is the antenna gain (directivity) of the BS antenna.
- *d*, EiRP (dBm): effective isotropic radiated power – this is the radiated power from the base station antenna. It is BS power − feeder loss + antenna gain.
- *e*, MS antenna gain (dBi): the mobile terminal antenna has a gain that we need to include in the LB. This antenna gain may in fact be negative! There are many measurements

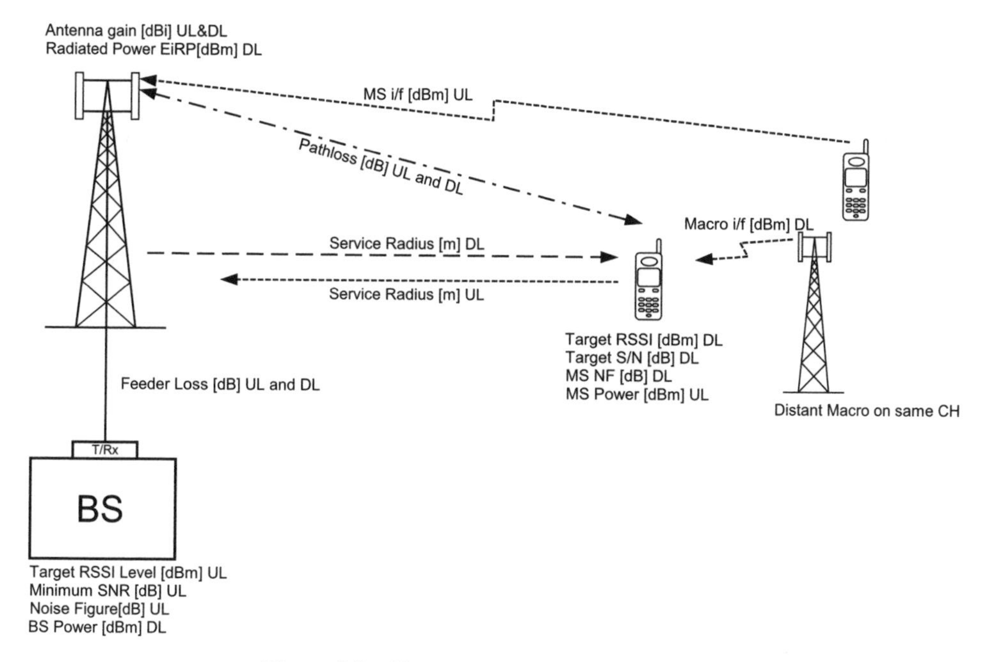

Figure 8.2 The components of the link budget

Table 8.1 Link budget example, GSM downlink (GSM900)

BS transmitter				
BS Tx power	40	dBm	*a*	Input
Feeder loss	35	dB	*b*	Input
BS antenna gain	2	dBi	*c*	Input
EiRP	**7**	dBm	*d*	$a - b + c$
MS receiver				
MS antenna gain	0	dBi	*e*	
MS noise figure	8	dB	*f*	GSM type, 8 dB; UMTS type, 7 dB
MS noise floor	**−113**	dBm	*g*	
Thermal noise floor	−121	dBm	*h*	GSM = −121 dBm, UMTS = −108 dBm
Interference	−120	dBm	*i*	
Service SNR requirement	9	dB	*j*	Signal-to-noise demand
Mobile sensitivity	**−103.2**	dBm	*k*	
The RF channel				
Log–normal shadow fading	10	dB	*l*	
Multipath fading margin	6	dB	*m*	
Body loss	3	dB	*n*	
Total margin	**19**	dB	*o*	$l + m + n$
Rx minimum level	**−84.2**	dBm	*p*	$k + o$
Maximum allowable path loss	**91.2**	dB	*q*	$d - p$
Service radius from antenna	**35**	m	*r*	Example based on PLS 38.5

For explanation of *a* to *r*, see text.

available that show the gain for various types of mobiles. Some of these measurements show mobiles with antenna gain down below −7 dBi. That is important to realize this when doing the LB calculation.

- *f*, Mobile station noise figure, dB: the amplifiers and electronics inside a receiver will generate noise. The relative power of that noise is defined as the noise figure. This NF together with the thermal noise floor will define the reference for the noise floor at the MS; refer to Chapter 7 for more details.
- *g*, MS noise floor: as an example consider a typical GSM MS with an NF of 8 dB. You simply add the NF to the thermal noise to calculate the noise floor in the MS

$$\text{noise floor} = \text{noise power at } 1\,\text{Hz} + \text{NF} + 10\log(\text{bandwidth})$$

$$\text{noise floor} = -174\,\text{dBm/Hz} + \text{NF} + 10\log(\text{bandwidth})$$

$$\text{noise floor} = -174\,\text{dBm/Hz} + 8\,\text{dB} + 10\log(200\,000\,\text{Hz}) = -113\,\text{dBm}$$

- *h*, Thermal noise floor (dBm): depending on the operating bandwidth of the radio channel, there will be certain thermal noise floor. This is a physical constant. At room temperature (17°C), the thermal noise floor is defined as:

$$\text{noise floor} = -174 + 10\log(\text{bandwidth})$$

So for GSM(200 kHz) the thermal noise floor will be:

$$\text{GSM} = -174\,\text{dBm/Hz} + 10\log(200\,000\,\text{Hz}) = -121\,\text{dBm}$$

For UMTS (3.84Mc) it will be

$$\text{UMTS} = -174\,\text{dBm/Hz} + 10\log(3\,840\,000\,\text{Hz}) = -108\,\text{dBm}$$

- *i*, Interference: interfering base stations transmitting on the same carrier must be taken into account. This interference will add to the noise floor, increasing the signal requirement to fulfill the SNR. It is very important to take this interference into account when doing the LB. It is highly recommended always to perform a measurement (see Section 5.2.2) of the interference level present in the building, and to use this input when calculating the link budget. In this example the interference is set to −120 dBm (very low).

UMTS Noise Increase

On UMTS we also need to take into account the noise increase due to the traffic load in the cell. Therefore, with 50% of load in the cell we need to add 3 dB in noise increase; refer to Section 2.4.3 for more detail on how to calculate the noise increase.

- *j*, Service SNR (Signal to Noise Ratio) requirements (dB): for the specific service we need to support over the radio link there will be a quality requirement, in terms of SNR. This is a definition of how strong the signal receive level has to be above the noise floor in the channel in order for the RF service to work. If we have a service that needs 9 dB SNR (GSM voice) in order to work, the lowest acceptable signal would be:

$$\begin{aligned}\text{Required signal level} &= -113\,\text{dBm (MS noise floor)} + 9\,\text{dB (service requirement)}\\ &= -104\,\text{dBm}\end{aligned}$$

The SNR Requirement is Related to the Data Rate

The critical parameter that affects the service level on the UL and DL is not the absolute signal level on the link but the quality of the RF link, the SNR. For different services there will be different demands on the quality of the RF link. The higher the service demands (data rates), the better the RF link needs to be. This is why the higher data rates on UMTS and HSPA are more sensitive to any degradation of the indoor DAS noise figure or attenuation. Therefore we must insure the lowest possible noise figure and attenuation of the DAS, in order to perform at the higher data rates, and cater for future data services.

UMTS

For UMTS design we need to define the desired E_b/N_o and also apply the processing gain of the radio service.

Table 8.2 Processing gain

User rate	Processing gain
12.2 kbps	25 dB
64 kbps	18 dB
128 kbps	15 dB
384 kbps	10 dB

$$\text{Processing gain} = \text{chip rate/user data rate}$$

$$\text{Processing gain} = 3.84\,\text{M/user rate (linear)}$$

$$\text{Processing gain} = 10\log(3.84\,\text{M/user rate})\ \text{(dB)}$$

Using these formulas we can calculate the processing gain of the specific UMTS data service as shown in Table 8.2.

Examples

$$\text{Processing gain for high-speed data, 384 kbps: } 10\log(3.84\,\text{Mcps}/384\,\text{kb}) = 10\,\text{dB}$$

$$\text{Processing gain for speech 12.2 kbps: } 10\log(3.84\,\text{Mcps}/12.2\,\text{kbps}) = 25\,\text{dB}$$

The use of orthogonal spreading codes and processing gain is the main feature of UMTS/WCDMA, giving the system robustness against self-interference. The main reason is the frequency reuse factor of 1, due to the rejection of noise from other cells/users.

When we know the required bit power density, E_b/N_o(energy per bit/noise) for the specific service (voice 12.2 kbps + 5 dB, data + 20 dB), we are able to calculate the required signal-to-interference ratio.

Voice at 12.2 kbps needs approximately 5 dB wideband signal-to-interference ratio minus the processing gain, $5 - 25\,\text{dB} = -20\,\text{dB}$. In practice, the processing gain means that the signal for the voice call can be 20 dB lower than the interference/noise, but still be decoded.

Soft Handover Gain (UMTS)

In traditional macro link budgets you will include a soft handover gain, typical using 3 dB. This is because the users on the edge of the cell are in soft handover, and using macro diversity improves the link. However, for indoor UMTS planning I recommend not using the soft handover gain in the link budget. You should minimize the areas of soft handover, to limit the load on the node B and the network.

GSM Example

The impact on the data service offerings in a typical indoor GSM900 system can be seen in Table 8.3. This example shows first and foremost the demand on the SNR and the *C/I* (channel/interference).

For GSM voice the demand is a minimum of 9 dB *C/I*. In this example (the downlink budget in Table 8.1), the maximum allowable link loss on the DL in order to provide voice service is 97 dB. Therefore, the total loss from the base station to the mobile must not exceed 97 dB. In this environment (moderate dense office) that corresponds to about 55 m (note that other parameters and margins also play a major role – the sensitivity of the mobile, the noise figure of the mobile, the fading margin, co-channel interference etc.; all these parameters will be elaborated later on in this chapter).

Table 8.3 Examples of GSM coverage indoors on different service requirements; note that the signal level in this example is with no interference

LB example, GSM900 at 12 dBm EiRP, moderate dense office				
Service	Minimum *C/I*	Maximum APL	Radius	DL level
GSM voice	9 dB	97 dB	55 m	–85.0 dBm
EDGE-MCS1 8.8 kbps	9.5 dB	96 dB	53 m	–84.5 dBm
EDGE-MCS2 11.2 kbps	12 dB	94 dB	46 m	–82.0 dBm
EDGE-MCS3 14.8 kbps	16.5 dB	89 dB	35 m	–77.5 dBm
EDGE-MCS4 17.6 kbps	21.5 dB	84 dB	26 m	–72.5 dBm
EDGE-MCS5 22.4 kbps	14.5 dB	91 dB	39 m	–79.5 dBm
EDGE-MCS6 29.6 kbps	17 dB	89 dB	34 m	–77.0 dBm
EDGE-MCS7 44.8 kbps	23.5 dB	82 dB	23 m	–70.5 dBm
EDGE-MCS8 54.4 kbps	29 dB	77 dB	16 m	–65.0 dBm
EDGE-MCS9 59.2 kbps	32 dB	74 dB	14 m	–62.0 dBm

In this case the DL level needed to provide sufficient signal level is −85 dBm. Whereas one can see that the service requirement for EDGE-MCS9 (59.2 kbps) is 23 dB stricter on *C/I*, with a maximum allowable path loss of 74 dB. This results in a service level of −62 dBm, or a reduction of the service range from the indoor DAS antenna from 55 m to support voice to only 14 m to support EDGE-MCS9.

- *k*, Mobile sensitivity: interference, i.e. signals transmitting on the same frequency as the supported service, will desensitize the receiver in the mobile.

Knowing the interference coming from other base stations using the same DL frequency (measured with a test receiver as described in Section 5.2.2), and knowing the mobile noise figure, the mobile antenna gain and the SNR requirement for the specific service, you can then calculate the mobile receiver sensitivity for the specific service:

$$\text{Rx sensitivity} = 10\log(10^{i/f+\text{THnoise}}) + 10^{(\text{MSNF}+\text{THnoise})/10} - \text{MS antenna gain} + \text{service Req.}$$

In this example, shown in Table 8.1, the MS sensitivity is calculated to −103.2 dBm.

The Radio Channel

- $l + m$, Fading margins: in any RF environment there will be fading due to reflections and diffractions of the RF signal. Several books and reports have been written about fading margins and RF planning; this is beyond the scope of this book. Typically, in indoor environments you should use a total fading margin of around 16–18 dB to obtain 95% area probability of the desired coverage (Refer to [6]).

RF DL Planning Level

If we continue to calculate using the example from Figure 8.1, we can calculate the planning level for the RF signal:

- *n*, Body loss (dB): the MS will be affected by the user, who will act as 'clutter' between the MS and the BS antenna. To take this into account in the LB, you will need to apply 'body loss' to the calculation. A typical number used for body loss is 3 dB.
- *o*, Total design margin: in this example, for the total design margin *i*,

$$\text{total fading margin} = \text{log} - \text{normal shadow fading} + \text{multipath fading} + \text{body loss}$$
$$\text{total fading margin} = 10\,\text{dB} + 6\,\text{dB} + 3\,\text{dB} = 19\,\text{dB}$$

- *p*, Minimum level at cell edge (RxMin): knowing the mobile sensitivity and the total fading margin, we can now calculate the design goal at the cell edge:

$$\text{RxMin} = \text{mobile sensitivity} + \text{design margin}$$
$$\text{RxMin} = -103.2\,\text{dBm} + 19\,\text{dB} = -84.2\,\text{dBm}$$

This is 0.8 dB higher than the signal requirement shown in Table 8.3, and is due to the −120 dBm interference on the channel, as shown in Table 8.1.

- *q*, Maximum allowable path loss: finally, we can calculate the maximum link loss that can fulfill the design criteria. We know the radiated power EiRP and the minimum signal level at the cell edge; therefore we can calculate the APL:

$$\text{APL} = \text{EiRP} - \text{RxMin}$$

$$\text{APL} = 7\,\text{dBm} - (-84.2\,\text{dBm}) = 91.2\,\text{dB}$$

- *r*, Calculating the antenna service radius: now we have established the APL, we can calculate the service range of the antenna in the particular environment:

$$\text{coverage radius (m)} = 10^{(\text{APL}-\text{PL1m})/\text{PLS}}$$

Remember the Uplink!

We have only done half the link budget at this point. We need to perform the exact same calculation, in reverse, for the uplink (as shown in Table 8.4). For GSM indoor planning, the LB will often be limited by the DL power. However, it is important to confirm this in the LB,

Table 8.4 Link budget example, GSM uplink

MS transmitter				
MS Tx power	33	dBm	*a*	Input
MS antenna gain	0	dBi	*b*	Input (can be negative!)
EiRP	**33**	dBm	*c*	$a + b$
DAS receiver				
BS noise figure	3	dB	*d*	GSM type, 3 dB; UMTS type, 4 dB
DAS noise figure	0	dB	*e*	The NF of the active DAS
DAS passive loss	35	dB	*f*	The loss of the passive DAS
System noise figure	**38**	dB	*g*	$(d + f)$ or $(d + e)$
Thermal noise floor	−121	dBm	*h*	GSM = −121 dBmUMTS = −108 dBm
Interference	−120	dBm	*i*	
Service SNR requirement	9	dB	*j*	Signal-to-noise demand
DAS antenna gain	2	dBi	*k*	
BS sensitivity	**−76**	dBm	*l*	
The RF channel				
Log–normal shadow fading	10	dB	*m*	
Multipath fading margin	6	dB	*n*	
Body loss	3	dB	*o*	
Total margin	**19**	dB	*p*	$m + n + o$
Rx minimum level	**−57.0**	dBm	*q*	$l + p$
Maximum allowable path loss	**91.2**	dB	*r*	$c - q$
Service radius from antenna	**23**	m	*s*	Example based on PLS 38.5

so always perform a two-way link budget calculation for GSM as well. For UMTS/HSPA you could well be limited by the UL, depending on the load profile of the cell.

Balance the Link

You will have different service distance on the DL compared with the UL. Try to balance your LB, maybe with parameter adjustments on the cell, once it is in service. It makes no sense to blast to high power on any link UL or DL if you are dependent on the other link limiting the service radius of the cell. Then you might as well power down the dominant link and balance the LB, thus limiting the interference from that particular link to adjacent cells or mobiles.

8.1.3 Link Budgets for Indoor Systems

When designing indoor systems, there are some additional parameters you need to take into account in the link budget calculation. Different parameters need to be considered, depending on whether it is a passive or an active DAS you are designing (see Figure 8.3).

8.1.4 Passive DAS Link Budget

The passive DAS is based on coaxial cables and splitters, as described in Section 4.3. From a link budget calculation it is fairly easy to design a passive DAS. The attenuation of the DAS

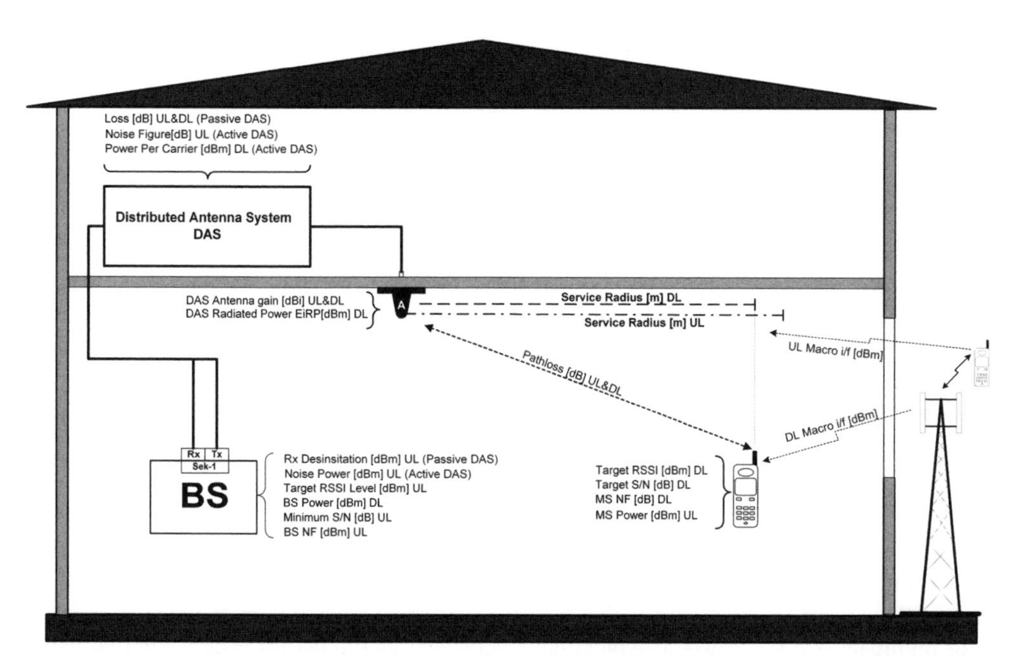

Figure 8.3 Components of the indoor link budget

will degrade both UL and DL with the same loss, corresponding to added link loss due to distance. The challenge is to get the exact information regarding the attenuation to each antenna in the system. This depends on the exact installation and cable route. Therefore detailed information on the actual installation is needed, and this is upfront before any system is installed. This can be a challenge for the radio planner, often making passive DAS design more installation planning exercise than radio planning.

The Downlink Calculation

The passive DAS will attenuate the power from the signal source, base station or repeater, so be sure to offset the DL power with the loss of the passive DAS when calculating the link budget. Be aware that you need to evaluate the DL link budget for all antennas in the system, owing to the fact that the individual antennas in the system will have different losses, and hence different radiated power levels and service ranges. Be very careful not to overdrive the passive DAS, especially when operating a multicarrier or multioperator system. It is often forgotten how high the power on the passive DAS actually is, especially on the components close to the base stations. One example is a system with four GSM base stations, each feeding 43 dBm to eight carriers, a total of 24 carriers; 43 dBm is 20 W, and the composite power will be 58 dBm, or 630 W. Even using good PIM with specifications of 120 dBc, you might have severe intermodulation products (refer to Section 5.7.3 for more details on PIM).

The Uplink Calculation

The UL part of the link budget calculation of the passive DAS is also straightforward; the attenuation of the DAS will impact the NF of the base station, as described in Section 7.1.7. The different attenuation from the base station to each individual antenna will result in different UL coverage ranges from each antenna, so be sure to calculate all antenna locations.

8.1.5 Active DAS Link Budget

The active DAS is described in detail in Section 4.4. An active DAS will typically be 'transparent', so the downlink signal you feed into the DAS you will have out of the antennas, and vice versa for the uplink plus or minus the gain you set in the system. This makes active DAS very easy to design and implement since you do not have to take cable distances and losses into concern. However, there are other things you need to pay attention to when calculating the link budget for an active DAS.

The Downlink Calculation

The limitation on the downlink power in an active DAS is usually the composite power capabilities of the DL amplifier in the 'remote unit'. The RU is located close to the antenna and will have one shared amplifier that the entire spectrum supported by the DAS is sharing. For multiband DAS you would typical have dedicated amplifiers for each supported band. Typically you would have a power level at one carrier of, for example, 15 dBm, and then

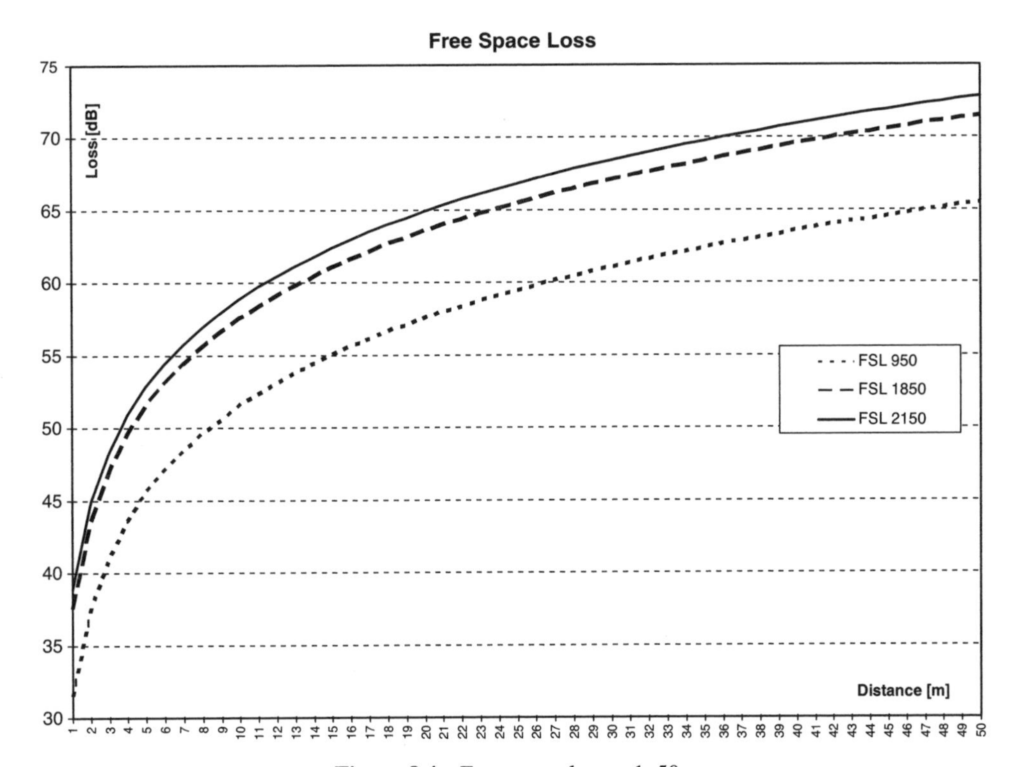

Figure 8.4 Free space losses 1–50 m

every time you doubled the number of carriers you would back off about 3–3.5 dB in order not to overdrive the system and cause distortion. Be very careful only to operate the active DAS system at the specified PPC (see Section 4.12.2 for more details). Otherwise you might cause interference problems for own and other services.

Hybrid DAS Link Budget

The hybrid DAS as described in Section 4.5 is a mix of passive and active DAS; you need to incorporate both strategies from above when doing the link budget calculation.

8.1.6 The Free Space Loss

The Path Loss

A major component in the LB is the path loss between the Tx and Rx antenna. This loss depends on the distance between the Tx and Rx antenna, and the environment that the RF signal has to pass. Free space loss (Figure 8.4) is a physical constant. This simple RF formula is valid up to about 50 m from the antenna when in line-of-sight indoors. The free space loss does not take into account any additional clutter loss or reflections, hence the name. The free space loss formula is:

$$\text{free space loss (dB)} = 32.44 + 20(\log F) + 20(\log D)$$

where f = frequency (MHz) and D = distance (km).

8.1.7 The Modified Indoor Model

In addition to the free space loss, we have the loss due to the environment. The main component from the indoor environment is penetration losses through walls and floor separations. The complete 'model' for indoor propagation loss could look like this:

$$\text{path loss} = \text{free space loss} + \text{wall loss}$$

This simple model gives a fair accuracy, provided that you have sufficient data for wall losses for the particular building (see Figure 8.5).

Indoor Calculator MTO. V.1.0.1

Inputs

	No. of Walls	Att-900	Att-1800	
Wall Att. Wall-Type 1	1	4	8	**[dB]**
Wall Att. Wall-Type 2	1	12	16	**[dB]**
Wall Att. Wall-Type 3	0	14	18	**[dB]**
Distance to antenna			25	**[m]**

	900MHz	1800MHz	
BTS RX Sensitivity	-107	-104	**[dBm]**
BTS Max Power	33	33	**[dBm]**
MS RX Sensitivity	-102	-100	**[dBm]**
MS Max Power	33	33	**[dBm]**
DAS Coax System loss	35	35	**[dB]**

Results

	900MHz	1800MHz	
Pathloss	75.46	89.56	**[dB]**
RX signal UL @ BTS	-77.46	-91.56	**[dBm]**
RX signal DL @ MS	-77.46	-91.56	**[dBm]**
C/N Margin DL @ MS excl. Fading	-13.46	-29.56	**[dB]**
C/N Margin UL @ BTS excl. Fading	-8.46	-25.56	**[dB]**

IB-Model
900 :L(dB) = 91.5 + 20log d + p*W(k)
1800:L(dB) = 97.6 + 20log d + p*W(k)

L(dB) = 32.5 + 20log f + 20log d + k * F(k) + p*W(k) + D(d-db)
L = path loss (dB)
f = frequency (MHz)
d = Distance MS / DAS Antenna
k = No. of floor separations, the signal has to pass
F = floor separation loss.
P = No. of walls the signal has to cross
W = wall attenuation
D = linear attenuations factor dB/m
db =50m "breakpoint" additional 0.2 dB/m

Figure 8.5 Indoor 'model' calculator in Excel, the first homemade tool I ever made for indoor designs

It is straightforward to create your own model and also to calculate the free space loss, and add the losses of the various walls. Obviously this method is not 100% accurate – no prediction tools actually are. However, it gives you a fair estimate of the possible path. The tool shown in Figure 8.5 is my first indoor link calculator; I have done about 75 indoor GSM designs using this calculator.

8.1.8 The PLS Model

A widely used model for calculating the path loss relies on 'path loss slopes' (PLS). These PLS are different attenuation slopes for different types of environments and frequencies. A general model is based on empirical analysis of a vast number of measurement samples in these different types of environments.

How is the PLS Established?

The PLS is derived from measurement samples in different environments (see Figure 8.6). The path loss is measured at different distances from the antenna, and it is possible to average out these thousands of measurements and calculate a path loss slope.

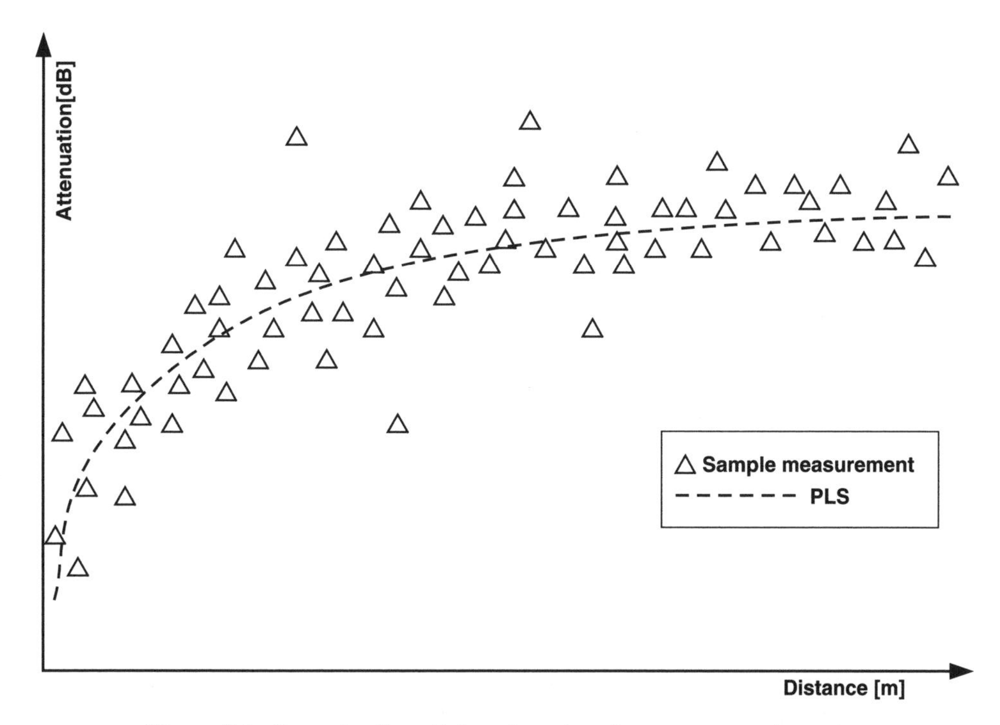

Figure 8.6 Example of a path loss slope, based on measurement samples

It is important to note that the PLS is an average value, and that there will be both positive and negative variations from the PLS. It is my experience that these variations will be covered by the fading margins suggested in Section 8.1.2.

Table 8.5 PLS constants for different environments

Type of environment	900 MHz PLS	1800/2100 MHz PLS
Open environment, few RF obstacles Parking garage, convention center	33.7	30.1
Moderately open environment, low to medium number of RF obstacles Factory, airport, warehouse	35	32
Slightly dense environment, medium to large number of RF obstacles Shopping mall, office that is 80% cubicle and 20% hard wall	36.1	33.1
Moderately dense environment, medium to large number of RF obstacles Office that is 50% cubicle and 50% hard wall	37.6	34.8
Dense environment, large number of RF obstacles Hospital, office that is 20% cubicle and 80% hard wall	39.4	38.1

It is very important to use the right PLS constant, according to the environment. The model based on the PLS values from Table 8.5 looks like this:

$$\text{path loss (dB)} = \text{PL at 1 m (dB)} + \text{PLS} \times \log(\text{distance, m})$$

You can calculate the free space loss at 1 m, using the free space formula from Section 8.1.6:

- 950 MHz = 32 dB.
- 1850 MHz = 38 dB.
- 2150 MHz = 39 dB.

You will find different PLS values from various sources; the internet is a good place to start. You might even develop your own PLS based on indoor RF survey measurements (see Section 5.2.5) in the various environment types and on the frequencies you use. Or you can use Table 8.5 as a start, and fine tune these PLS to match your needs. Do understand that these types of RF models are just guidelines, and can never be considered 100% accurate. Use these models accordingly, and apply some common sense based on your experience. If in doubt, then always do an RF survey measurement to verify the model used (see Section 5.2.5). Thus it would also be possible to establish the correct fading margin for the particular environment measured.

In Figure 8.7 you can see an example of the suggested PLS in Table 8.5 for a dense office environment, for GSM, DCS and UMTS frequencies. There is considerable difference between the PLS attenuation and the free space loss, as shown in Figure 8.8. Note that these PLS models assume that you use common sense when placing the antennas, so only use visible antenna placement below the ceiling (refer to Section 5.3 for more details).

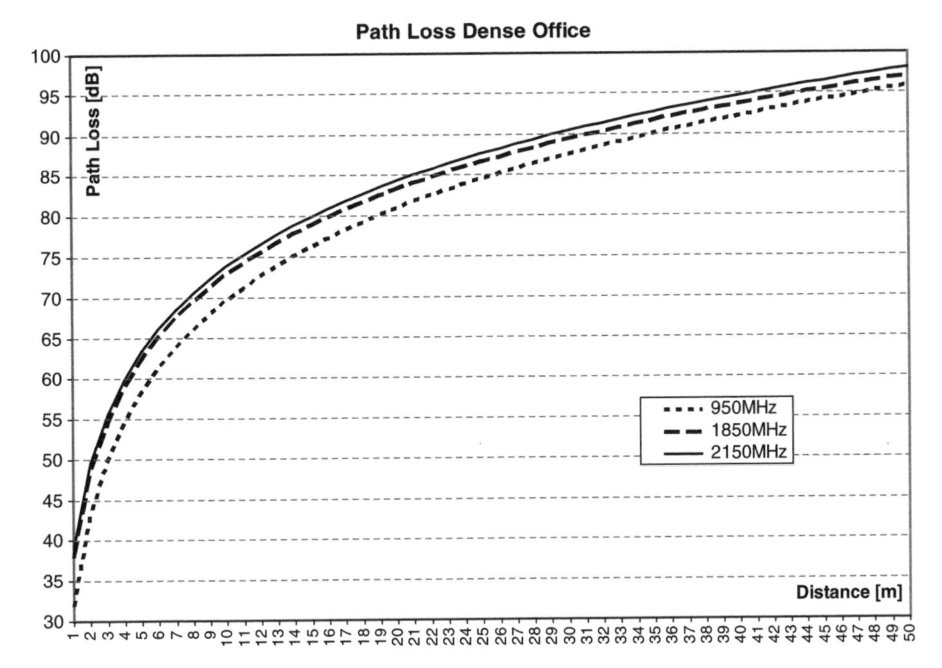

Figure 8.7 Path loss based on PLS for a dense office, 1–50 m

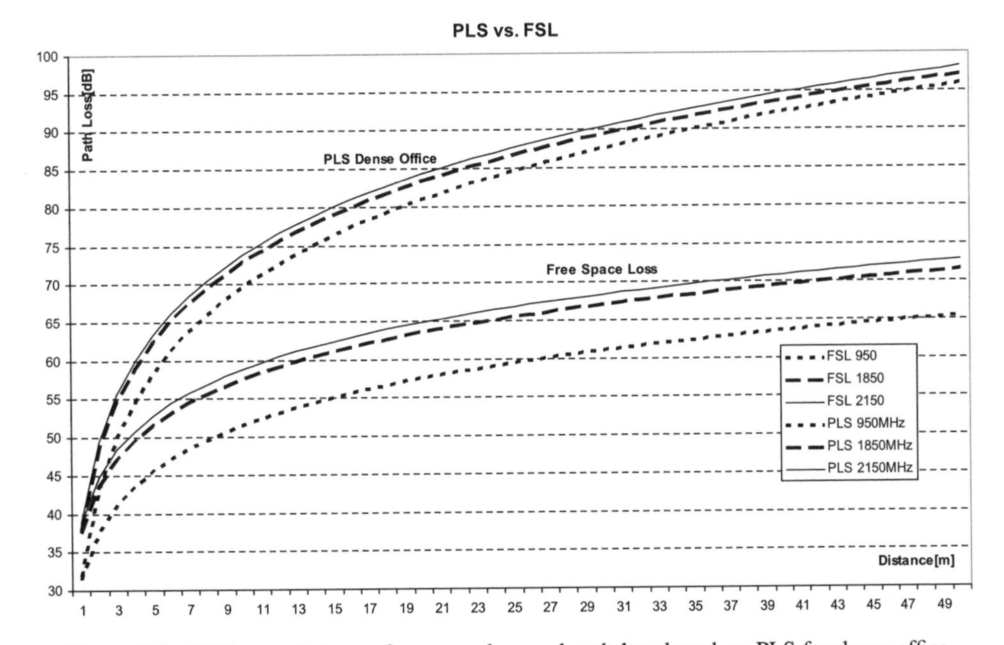

Figure 8.8 Difference between free space loss and path loss based on PLS for dense office

Table 8.6 GSM 1800 example of DL service range from 2 dBi omni antenna

GSM 1800 service radius at four different DL levels	DL EiRP from antenna							
	7 dBm		12 dBm		17 dBm		22 dBm	
DL target level	*−65*	*−70*	*−65*	*−70*	*−65*	*−70*	*−65*	*−70*
	−75	*−80*	*−75*	*−80*	*−75*	*−80*	*−75*	*−80*
Open	14 m	20 m	20 m	30 m	30 m	44 m	44 m	64 m
Coverage radius	30 m	44 m	44 m	65 m	65 m	94 m	94 m	138 m
Moderately open	12 m	17 m	17 m	25 m	25 m	35 m	35 m	50 m
Coverage radius	25 m	35 m	35 m	50 m	50 m	72 m	72 m	103 m
Slightly dense	11 m	16 m	16 m	22 m	22 m	31 m	31 m	44 m
Coverage radius	22 m	31 m	31 m	44 m	44 m	63 m	63 m	89 m
Moderately dense	10 m	14 m	14 m	19 m	19 m	26 m	26 m	37 m
Coverage radius	19 m	26 m	26 m	37 m	37 m	51 m	51 m	71 m
Dense	8 m	11 m	11 m	15 m	15 m	20 m	20 m	27 m
Coverage radius	15 m	20 m	20 m	27 m	27 m	36 m	36 m	50 m

Log–normal shadowing fading, 10 dB; multipath fading margin, 6 dB; body loss, 3 dB; no interference; 8 dB MS NF.

8.1.9 Calculating the Antenna Service Radius

When you have completed the link budget calculation and have established the APL, you can calculate the service range of the antenna in the particular environment:

$$\text{coverage radius (m)} = 10^{(\text{APL}-\text{PL1m})/\text{PLS}}$$

Examples of DL Service Ranges for Indoor Antennas

This is just an example on the downlink service ranges from an indoor DAS antenna (Table 8.6). Make your own link budget calculation to establish what the actual service range is for both uplink and downlink for your particular design.

9

Tools for Indoor Radio Planning

The trade of indoor radio planning relies on several types of tools to aid the radio planner with the process. These tools help in accurately predicting and verifying radio coverage, capacity and quality. More logistical tools may be also needed to calculate the component count, equipment lists and project costs. Obviously tools for design documentation, diagrams, etc., are needed as well. There are many different types of tools, and each operator has a tool box that relies on different types of tools for the various aspects of the process.

The choice of the tools you should select is must be relevant to your individual needs. If you only design three to five indoor projects per year, it might be alright to use standard spreadsheets for link budget calculations and simple diagram drawing programs for documentation. However, if you are dedicated to indoor radio planning on a full-time basis and design 20 or more indoor systems per year, it would be worthwhile investing in automatic tools that can make your design process more cost-efficient and improve the quality of your work. Remember that the design documentation should also be used by other parties in the project, so all documentation must be clear and precise.

In the following, examples of tools are presented. These are just examples, and many different tools from various manufacturers are available, each with their own pros and cons.

9.1 Live and Learn

Whatever nice tools you might have on your PC, it is highly recommended to get out there in the building, walk the turf, conduct measurements, take part in the optimization campaign and try to participate in a couple of late-night installations of one of your designs. That will enable you to get a feel for the whole process. Indoor planning is much more than 'just' desktop work on the PC.

An essential part of the learning process for the radio planner is also to gain experience from the indoor solutions and installations that are live with real traffic. Conduct the post installation RF verification surveys yourself; do these measurements enough times to gain experience so that you can trust your 'RF sense'. Real-life feedback from the designs you have done is vital, and the launch pad for improvement.

Indoor Radio Planning: A Practical Guide for GSM, DCS, UMTS and HSPA Morten Tolstrup

Another valuable input for the radio planner is feedback on network performance statistics of the indoor cells implemented. Monitor the call drop rate, quality, traffic RF performance and handovers to learn where the weak point in the design is.

9.2 Diagram Tools

All indoor radio designs need to be well documented, including diagrams. It is essential to use an electronic tool that enables you to store and update the diagram documentation in the future.

Always leave a set of diagrams of the DAS on-site, preferably with all components and their location in the building. It can be very helpful if you draw the diagram, using the floor plans as the backdrop.

9.2.1 Simple or Advanced?

The simplest tools can take you a long way. Most diagrams and an illustration in this book were done using Microsoft Visio (as the example in Figure 9.1). Visio is very easy to use, and you can easily create your own stencil containing the DAS components you use. You can also import a picture format of the floor plan you are working on, and draw components on top. This is a very easy approach and can be useful if you do only a few DAS designs a month. If you are designing DAS on a larger scale, it is defiantly worthwhile considering a more automatic tool that can help you with the total process. Many designers are also using AutoCAD for DAS design; this can be advantageous if you can get the floor plans in AutoCAD format.

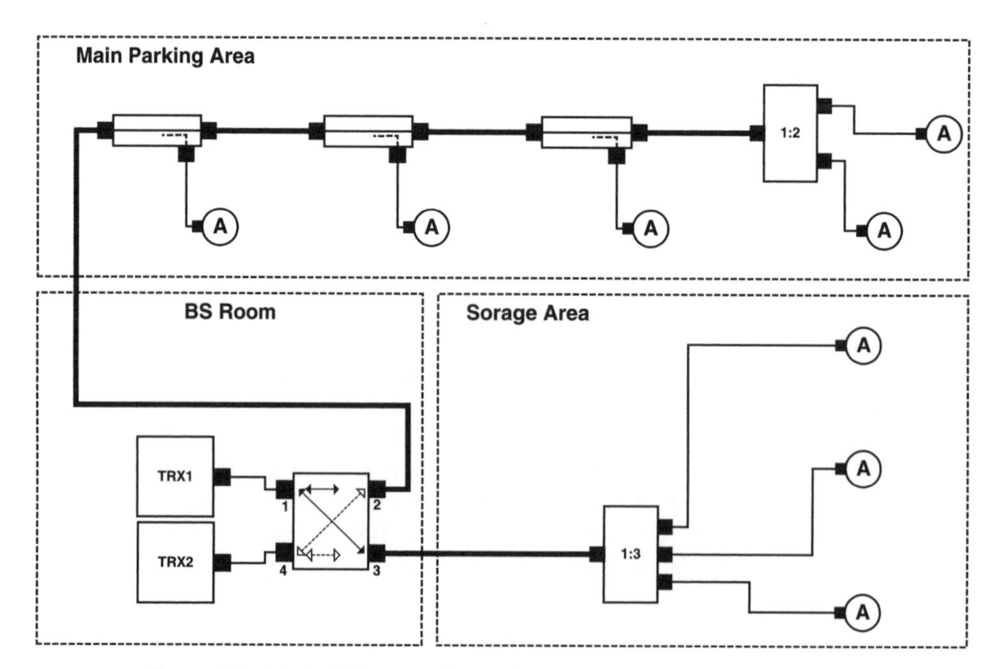

Figure 9.1 Typical diagram of a small passive DAS, done with MS Visio

You are welcome to contact me if you want a copy of my Visio stencil I used for this book. If you are doing a lot of indoor designs, the optimum solution would be to get a tool similar to the one described in Section 9.7, which is what I use on a daily basis.

9.3 Radio Survey Tools

In some cases it is advisable to perform an RF survey in the building, to verify the antenna locations and the model you are using for simulation of the coverage in the building (refer to Section 5.2.5).

9.3.1 Use Only Calibrated Equipment

You will need a calibrated transmitter, preferably battery operated, and a pole or tripod that enables you to place the survey transmit antenna at the same location as the planned antenna. The transmitter needs to be able to have adjustable output power so you can simulate the same power as the final antenna.

Log the Measurement Result

You will also need calibrated measurement tools that are able to log the measurement, and preferably save the data on a PC. The best system enables you to place the measurement on a floor plan that is imported to the system (see Section 5.2.6). If you are using, or plan to use, a design program that can do indoor modeling and RF prediction, it is important to chose a radio survey tool that can export the measurement data to that tool, and thereby measurements from the actual building can be used to fine tune or calibrate the prediction model. Needless to say, a radio planner never leaves home without a 'test mobile' that is able to display, sometimes even log, the basic radio levels and parameters.

9.4 The Simple Tools and Tips

I am a big fan of simple, straightforward tools. In radio planning, as in so many other cases, some of the best tool and tips are often the simplest ones.

9.4.1 Use a Digital Camera

A digital camera is a must for the radio survey, but also for installation documentation. It is recommended to document all antenna and equipment locations with digital pictures. Name the files accordingly, so the picture of the proposed installation location of antenna 'A1' is saved as 'A-1.jpeg', etc. It is also useful to use a laser pointer to pinpoint the exact antenna location when recording the picture; then there will be no debate with the installer as to where the agreed antenna location is.

9.4.2 Use the World Wide Web

Prior to the survey it can be useful to visit the homepage of the company who owns the building where you are implementing the DAS. Often you can find site plans, photos and other useful information that can help you prepare the draft design concept.

9.4.3 *Traffic Calculations*

Obviously the Erlang table is one of the main tools when calculating voice capacity (see Section 6), and you will find a copy in the Appendix. Excel is a useful tool for the basic traffic calculations. It is very easy to produce a dedicated calculator for a specific need. The example in Table 9.1 is a typical easily made tool for calculating the traffic profile of an indoor cell, by adding the various users.

Table 9.1 Traffic profile calculator for calculating the traffic load of an indoor cell

User profile/type	mE/user	Number of users	Total E
Normal private user	15	25	0.375
Business user, not 100% WO	30	200	6.000
SOHO, 100% WO	50	0	0.000
Normal business user, 100% WO	70	40	2.800
Office user, 100% WO	100	100	10.000
Heavy business/telecomm, 100% WO	140	5	0.700
User def.-1	0	0	0.000
User def.-2	0	0	0.000
Total			*19.875*

9.5 Tools for Link Budget Calculations

Any science pocket calculator will do fine, but it is preferable to us a spreadsheet type of link budget calculator, such as the example in Table 9.3 and 9.4. Most of my tools are done using standard Excel for windows. Dedicated tools for link budget calculations are available, but the flexibility of a standard Excel spreadsheet makes it very easy to upgrade and maintain. You can easily integrate the results of other calculations, such as cascaded noise (as shown in Table 9.2) or power per carrier, into your link budget tool if it is all based on Excel.

Table 9.2 Simple cascaded noise calculator using Excel

Cascaded Noise

Total NF	Stage 3	Stage 2	Stage 1	Passive Loss
17.38 *[dB]*	BS-NF 6 [dB]	UL-Att 10 [dB]	Active Gain 0 [dB] Active NF 12.0 [dB]	Loss 0.0 [dB]
Calculated total Noise Figure	*The Noise Figure of the Basestation*	*The UL attenator value*	*Gain and Noise Figure of the Active system*	*Attenuation of passive losses, prior to the first stage (RAU)*

Input all the RED fields

NF = NF1 + (NF2 - 1) / G1 + (NF3 - 1) / (G1 x G2) + P

NF = 15.85 + (10 - 1) / 1 + (3.98 - 1) / (1 x 0.1) + 0.0

NF = 54.7 = 17.38 dB + 0.0 = *17.38 dB*

Table 9.3 Link budget tool in Excel, GSM uplink

MS transmitter		
MS Tx power	33	dBm
MS antenna gain	0	dBi}
EiRP	**33**	**dBm**
DAS receiver		
BS noise figure	3	dB
DAS noise figure	0	dB
DAS passive loss	35	dB
System noise figure	**38**	**dB**
Thermal noise floor	−121	dBm
Interference	−120	dBm
Service SNR requirement	9	dB
DAS antenna gain	2	dBi
BS sensitivity	**−76**	dBm
The RF channel		
Log–normal shadow fading	10	dB
Multipath fading margin	6	dB
Body loss	3	dB
Total margin	**19**	dB
Rx minimum level	**−57.0**	dBm
Maximum allowable path loss	**91.2**	dB
Service radius from antenna	**23**	m

9.6 Tools for Indoor Predictions

It is true that once you have done 20 or 30 indoor design, radio surveys and post installation measurements, you will have a really good feeling, and even a sixth 'RF sense'. It is important to recognize and trust that feeling and experience, but *always* do a link budget, and do use a link calculation, at least in the expected worst case areas of the building.

If you only use your experience, you might make expensive mistakes, or expensive overdesigns of the DAS solutions. If you want to utilize your experience, the best way is to first plot the antennas on the floor plan, then re-check with link budget and propagation simulations, maybe even RF survey measurement with a test transmitter in the building, to confirm you are right. This will make your 'gut feeling' even better.

9.6.1 Spreadsheets Can Do Most of the Job

Like the link budget calculator, you can use a standard Excel spreadsheet to predict the coverage in the building. This approach is limited to calculating a specific level or distance in a specific area or place in the building. Therefore you will simulate the worst case in the building, and also several location samples that represent the building.

Table 9.4 Indoor 'model' calculator, the first homemade tool I ever made for indoor designs

Indoor Calculator MTO. V.1.0.1

Inputs

	# of Walls	Att-900	Att-1800	
Wall Att. Wall-Type 1	1	4	8	[dB]
Wall Att. Wall-Type 2	1	12	16	[dB]
Wall Att. Wall-Type 3	0	14	18	[dB]
Distance to antenna			25	[m]

	900MHz	1800MHz	
BTS RX Sensitivity	-107	-104	[dBm]
BTS Max Power	33	33	[dBm]
MS RX Sensitivity	-102	-100	[dBm]
MS Max Power	33	33	[dBm]
DAS Coax System loss	35	35	[dB]

Results

	900MHz	1800MHz	
Pathloss	75.46	89.56	[dB]
RX signal UL @ BTS	-77.46	-91.56	[dBm]
RX signal DL @ MS	-77.46	-91.56	[dBm]
C/N Margin DL @ MS excl. Fading	-13.46	-29.56	[dB]
C/N Margin UL @ BTS excl. Fading	-8.46	-25.56	[dB]

IB-Model

$$900: L(dB) = 91.5 + 20\log d + p^{*}W(k)$$

$$1800: L(dB) = 97.6 + 20\log d + p^{*}W(k)$$

$$L(dB) = 32.5 + 20\log f + 20\log d + k * F(k) + p^{*}W(k) + D(d-db)$$

L = path loss (dB)
f = frequency (MHz)
d = Distance MS / DAS Antenna
k = # of floor speerations, the signal has to pass
F = floor separation loss.
P = # of walls the signal has to cross
W = wall attenuation
D = linear attenuationsfactor dB/m
db =50m "breakpoint" additional 0.2 dB/meter

9.6.2 The More Advanced RF Prediction Models

Some tools are able to import floor plans and to define the loss of the individual wall and floor separations. The most advanced of these prediction tools can even calculate 'ray-tracing', estimating the signals being reflected throughout the building in three dimensions.

It is very useful to use these types of tools, where you also can include the nearby macro sites in your simulation. You can produce *C/I* plots or best server plots and save a lot of investment, while keeping the planning mistakes to a minimum.

Any tool will only be good if you use it correctly; no tool is better than the quality of the input you provide. Performing an accurate RF simulation of a building can be time-consuming; you need all the details of penetration loss through walls and floor separations, reflectivity of the surfaces, etc. It is evident that these tools cannot be 100% accurate, but they allow you to simulate the environment and to experiment by moving antennas around, thereby getting a good feeling on how and where to place the antennas. If in doubt, always do a verification survey with a test transmitter and receiver.

9.7 The Advanced Toolkit (RF-*vu* from iBwave.com)

There are many types of tools for the various processes with regards to indoor radio planning. In many cases you end up with separate tools for each task: one tool for drawing the diagram, one tool for the link budget, one tool for calculating the losses and powers at the antennas, one tool for inter-modulation calculations, one tool to produce equipment lists, etc. It can be challenging to handle all the different input and outputs of data from these tools so that you are sure that each tool is updated with accurate project data. However there is one tool that combines all the aspects of the indoor design process into one (the tool I use on a daily basis). This is not a sales pitch for that specific tool, but I am sure you will find the software very useful and cost-saving if you do indoor radio planning on a frequent base.

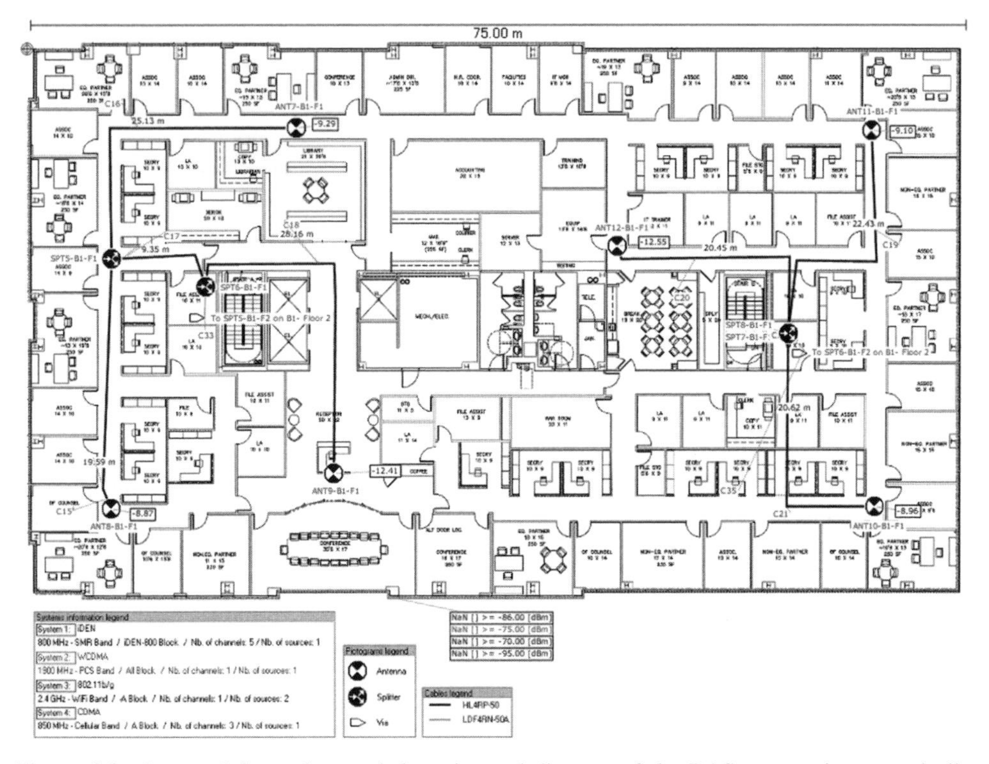

Figure 9.2 Imported floor plan, scaled to size and diagram of the DAS on top; it automatically calculates cable distance, losses and installation costs

9.7.1 *Save Time, Keep Costs and Mistakes to a Minimum*

This tool is not only easy to use, but it saves a lot of planning costs, and keeps mistakes to a minimum, because all data is linked, so if you update a component on the diagram, the floor plan, simulation and equipment lists are updated simultaneously.

9.7.2 *Import Floor Plans*

You can easily import any standard type of floor plan, e.g. *.gif, *.jpeg, *.pdf, and even AutoCAD. When you have imported the floor plan, you scale it to match the actual size of the building (as shown in Figure 9.2). All floor plan can be aligned together in order to have a perfect three-dimensional modeling of the building. The cable lengths will then be automatically measured horizontally and vertically based on the floor plan scales.

9.7.3 *Diagram and Floor Plan*

When you place components such as antenna, coupler, amplifier, cable and others on the floor plan, the schematic diagram is automatically drawn for the user, so the diagram will always be accurate and match the floor plan, and vice versa.

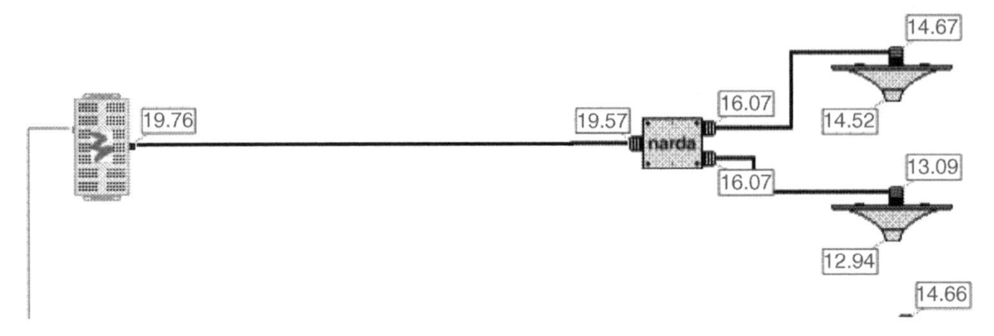

Figure 9.3 A small section of the DAS diagram – power levels at the input and output of each component are calculated

In Figure 9.3 you can see a screen shot of a simple floor plan with two antennas. The software automatically calculates the actual cable distance due to the scaled floor plan. The cable distance and the predicted cell border are updated in real time when moving the components on the floor plan.

You can choose what type of supplementary information you would like the diagram to present; in the example in Figure 9.3 it is the power level in dBm at the input and output of each component.

9.7.4 *Schematic Diagram*

The schematic diagram is where the user can see all the components connected together and analyze the RF power at each single connector (as shown in Figure 9.3). Each component is drawn for better identification but also contains all its technical specifications in order for the tool to simulate its action as it would perform in real life. Most of the new in-building projects are now multiband and multitechnology so people are using the same infrastructure, as a big RF pipeline, to carry multiple types of signals at different frequencies. The schematic view gives the ability to the user to analyze the interaction of all the signals going through a component and perform the specific RF calculations based on the particularity of the technology. The tool performs many downlink and uplink calculations, which is useful to increase speed and accuracy of the design. Large quantities of projects are made using passive DAS, which is hard to balance without a tool, especially when there are many components. The uplink portion is even worse since the cascaded noise figure of the system is crucial but very hard to calculate manually. When modifications are required due to a site visit or other factors, the tool will save a lot of time by re-doing all the calculations in a second. The link budget, noise calculations, power and inter-modulation calculations are also done within the tool, and it supports the various mobile and communication protocols from around the world.

9.7.5 Error Detection

The software has detailed warnings on many different topics and will prevent the user from making errors on the projects. The software will check for errors such as connecting two outputs together, incorrect connectors (gender and type), composite RF power going through a component, overdriving an amplifier, and many more.

9.7.6 Component Database

The tool has a large database of components from the different manufacturers, so the design calculation and documentation will show and use the real data, updated online from the manufacturer. This also includes the antenna radiation diagram, so you can make accurate predictions, using the actual antenna you plan to implement. It is easy to create your own components if needed, or change the existing component data.

9.7.7 Equipment List and Project Cost Report

You can define the cost per component, the installation cost per component or meter of cable, and with the push of a button you will get a complete shopping list and project cost. This enables the radio planner to fairly accurately estimate the project cost and the business case.

9.7.8 RF and Installation Report

Many other reports are available to help the user to quickly analyze and optimize the project. Other reports let the user look at the link budget of each antenna at each band, analyze the average output power per antenna to better balance its system and give a detailed cable routing report to installers in order to speed up the installation process.

9.7.9 Multisystem or Multioperator

The software is capable of planning dual band or multioperator systems, and will show the different bands with different color coding.

9.7.10 Importing an RF Survey

You can import a results file from the RF survey measurement (as shown in Figure 9.4), and compare data with the prediction model. It is very easy to find installation problems when comparing the post installation survey with the prediction.

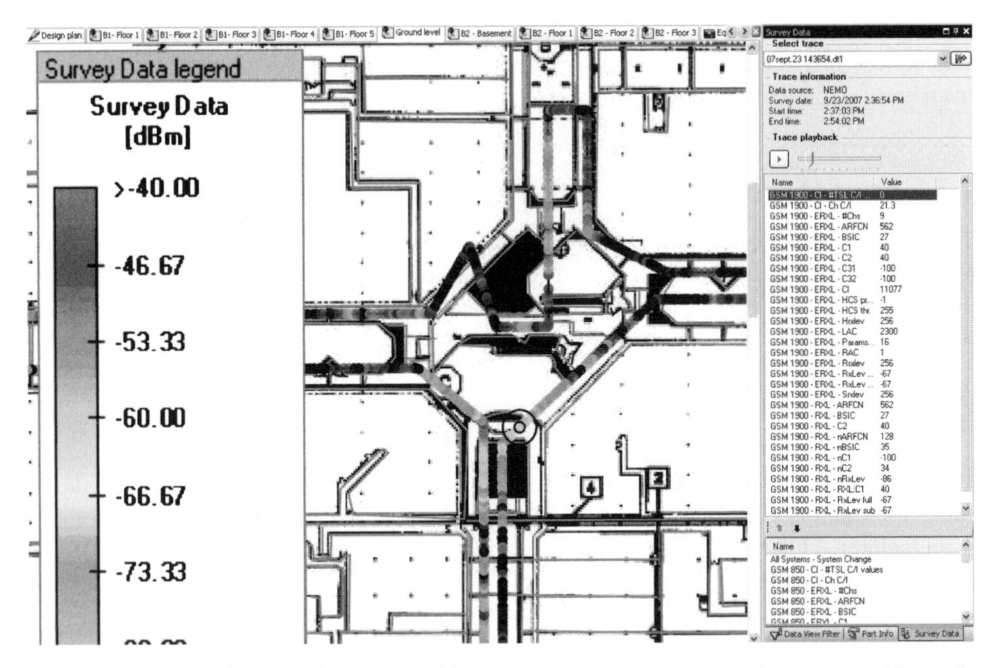

Figure 9.4 Imported survey plot on top of the RF design floor plan; it can be used to calibrate the prediction model used

9.7.11 Site Documentation

It is possible to include site documentation (as shown in Figure 9.5), as build pictures and mock-ups in order to help the installers avoid mistakes and help the building owner to understand the installation.

9.7.12 RF Propagation

You can select various prediction models and modes (as shown in Figures 9.6 and 9.7), from COST231 to 3D Dominant Path, which is an optimized ray tracing model. The database contains multiple types of materials in order to define as well as possible the building characteristics and the environment. Three-dimensional prediction can be performed and analyzed for all the bands at the same time and displayed in different formats, such as Best Server, Link Loss or Signal Strength. The threshold can also be defined by the technology to identify which band will meet the technology requirement and which one will need to be improved.

Figure 9.5 Example of installation mock-up, which is very useful for the installation process

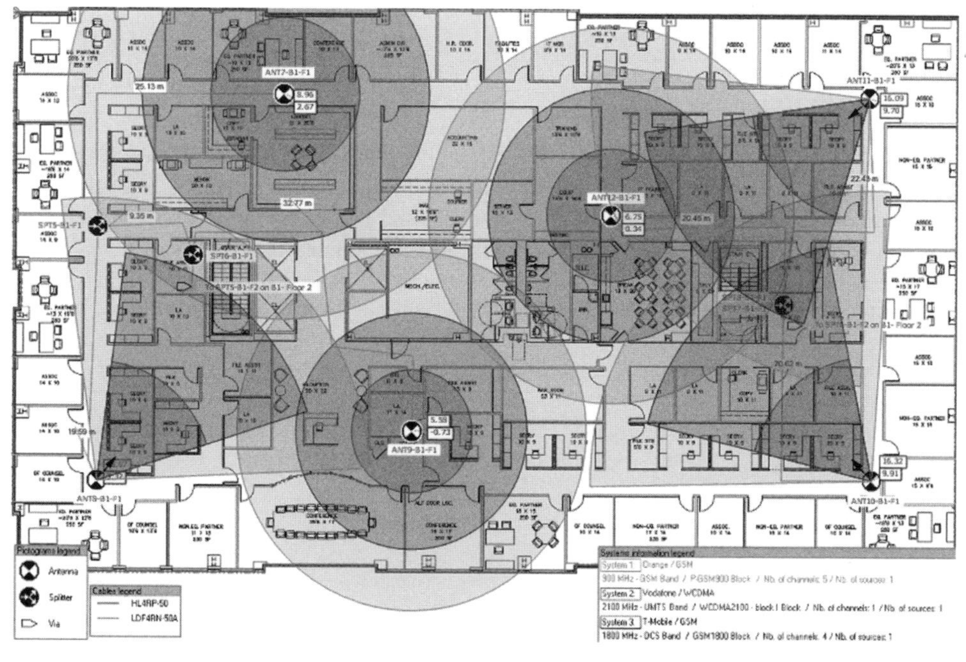

Figure 9.6 It is possible to plot the PLS-based coverage range for the different services on the DAS, called the floor plan

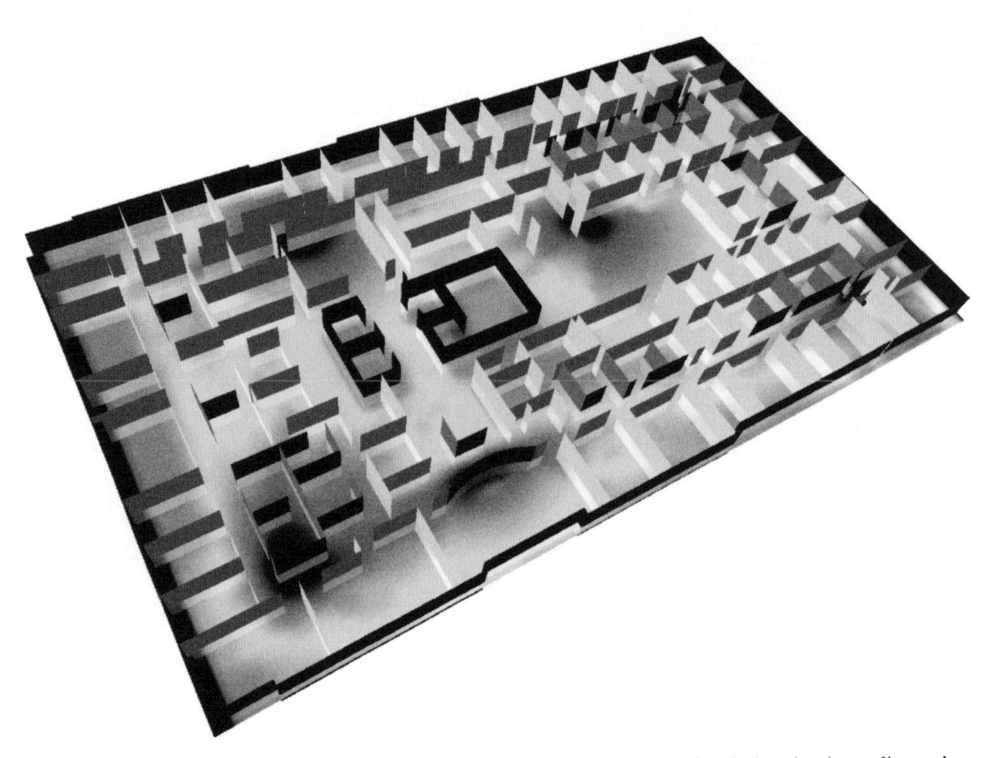

Figure 9.7 The same DAS project as shown in Figure 9.6; this is simulation in three dimensions (obviously in color in real life)

9.7.13 Fully Integrated

The best thing about the tool is that all the data is integrated into the same platform, so it is easy to update the project. The documentation, the calculations, reports, diagrams, floor plans and all the project info is contain in one file and can easily be shared across the in-building value chain. Replacing single or multiple components is a matter of a few clicks with the mouse as all data is linked. This can save a lot of project time and limit the number of mistakes.

10

Optimizing the Radio Resource Management Parameters on Node B When Interfacing to an Active DAS, BDA, LNA or TMA

When connecting any external uplink or downlink amplifiers to node B, you will offset the UL and DL. It will be necessary to evaluate and adjust the basic power control, system-info, noise and timing parameters in the BSS system in order to optimize the performance of an indoor active DAS system. If the parameters in the radio access network (RAN) (node B and RNC) are not tuned to cope with the offsets in DL power, noise load and timing offset, there will be a serious impact on the performance of the indoor system.

10.1 Introduction

UMTS is an advanced system. As it is noise- and power-limited, enhancement of the UL and DL performance can be obtained by the use of UL and DL remote amplifiers to overcome the passive losses from the base station (node B) to the antenna. This is the basic principle of an active DAS, which functions as distributed amplifiers in the system, offsetting the DL power and to some extend loading the base station with noise power on the uplink.

10.1.1 UMTS Radio Performance is All About Noise and Power Control

It is a fact that, in order to have optimum performance on a UMTS system, we need to insure:

- *Strict power control*: all mobiles use the same UL frequency, so all mobiles have to be controlled so that the received signal strength from them on the UL of the node B is kept at

Indoor Radio Planning: A Practical Guide for GSM, DCS, UMTS and HSPA Morten Tolstrup

the same receive level (see Section 2.4.5). The offset of the node B DL power caused by a BDA, TMA or active DAS has to be considered in the node B setting.

- *Noise load is controlled in the system*: node B will measure the total UL noise (traffic) continuously, and use that measurement for admission control, to make sure that the overall system noise increase on the UL is kept below the desired threshold. The added UL noise from the active DAS, BDA or remote LNA has to be considered in the node B evaluation of noise/traffic increase.
- *Delay of the DAS (synchronization window)*: delay in the external amplifier system has to be incorporated in the timing and cell size.

10.1.2 UMTS RF Parameter Reference is Different from GSM

In GSM the reference for the radio parameters is the antenna connectors on top of the base station rack. However, for UMTS it was foreseen that external equipment between the base station and the antenna would offset the downlink power, as well as the uplink noise load. The 3GPP-UMTS specification foresees that external equipment will be connected to the node B antenna connector (3GPP no's 25.215 and 25.104). This equipment will typically be an LNA (to enhance the uplink) or TMA (tower-mounted amplifier, to enhance the downlink).

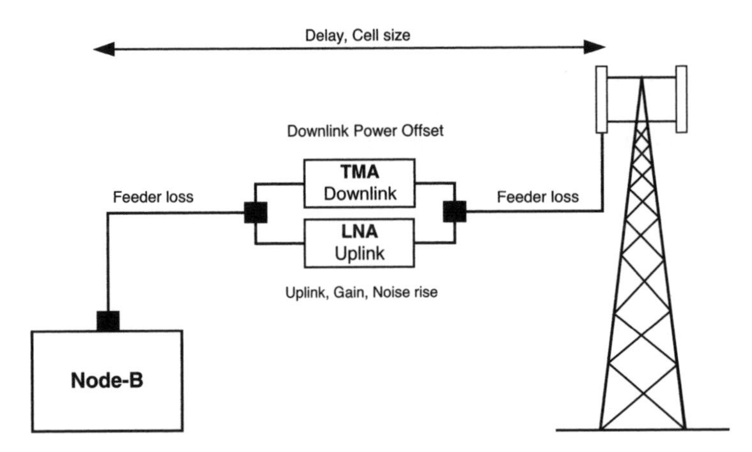

Figure 10.1 UMTS base station with external TMA or LNA; the noise offset on the UL and the power offset will affect the performance of the noise and power control

10.1.3 Adjust the Parameters

The 3GPP specification takes the performance of this external equipment (see Figure 10.1) into account in the parameter set, by having the BS antenna as the reference. This is different from the GSM specification where the antenna connector of the base station is the reference. This change of reference for the node B parameter is motivated by the fact that UMTS is a power- and noise-limited system, so it is crucial that the performance of any external equipment is taken into account the radio resource management parameter evaluations.

10.1.4 How to Adjust this in the RAN

Typically, there will be some offset adjustments to consider in the parameter set of the specific RAN vendor. The actual parameter name varies from vendor to vendor, but the principle is the same. In order for node B to be 3GPP-compliant, these offsets must be tunable, in order to offset the reference for the radio resource management in the BSS.

In the typical RAN you use the 'external TMA/LNA' function; here you define the DL power at the antenna, the noise figure and gain on the uplink, and then the RAN tunes all the control parameters accordingly.

10.1.5 Switch Off the LNA in Node B when Using Active DAS

When connecting an active DAS to node B, it is pretty much like connecting a TMA and a LNA to node B. Some vendors to have an easy way of taking the offset off the DL power and UL noise increase into account, by setting node B into 'external TMA' and 'external LNA' mode. Then you just define the NF and DL power of the DAS, and everything is taken into account. Note that the pure active system has a UL amplifier close to the antenna, and by using the UL gain in the pure active system, and switching off the internal LNA in node B, the UL performance increases because to the basics of cascaded noise; it is always best to have the gain closest to the source (the antenna).

10.2 Impact of DL Power Offset

When node B is connected to an active DAS, e.g. a 40 dBm, the CPICH power will typically be 30 dBm (10%/−10 dB). With one UMTS carrier on the active system, the maximum power will be, for example, 15 dBm, so the CPICH power of the remote access unit will be 6–10 dB lower. However the system information transmitted by node B will (if left unchanged) still inform the mobiles that the transmitted CPICH power is 30 dBm from node B. These 25 dB in offset between the system information and the actual CPICH power will cause problems during access burst, if left unchanged.

10.2.1 Access Burst

When a mobile is in idle mode (open loop power control), it will monitor and decode the system information broadcast by node B (transmitted CPICH power). In order to enable the mobile to access the cell with the correct initial access burst level (not to overshoot the UL), node B broadcasts the CPICH Tx Power as a reference to the mobile. This CPICH Tx power level is used by the mobile to calculate the link loss. Therefore, the mobiles are able to start the initial access sequence with a power level that will insure that this access burst does not overpower all the other traffic on the UL (as shown in Figure 10.2).

Example

A mobile is decoding a cell where the BS reports that the CPICH is transmitted at +33 dBm; the mobile receives the CPICH signal at −85 dBm. Thus the mobile will calculate the link

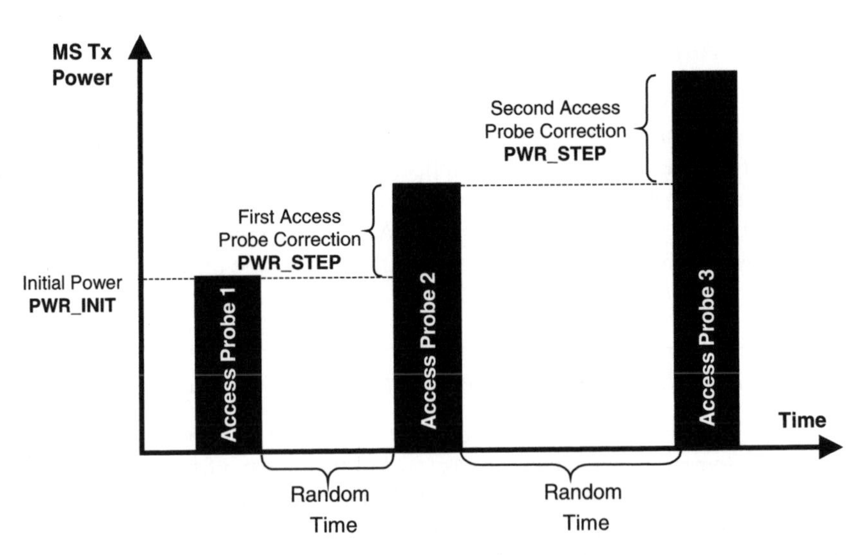

Figure 10.2 The UMTS access burst principle: PWR_INIT = CPICH_Tx_Power − CPICH_RSCP + UL_Interference + UL_Required_CI, where PWR_Init = calculated initial MS power, for the first access burst; CPICH_Tx_Power = the BS broadcasts the transmitted CPICH power, so the MS can estimate the path loss; CPICH_RSCP = received CPICH level at the MS; UL_Interfeerence = UL interference level at the BS, which the MS UL signal has to overcome; UL_Required_CI = the required CI on the UL

loss to be:

$$\text{link loss} = \text{CPICH received} - \text{CPICH transmitted}$$

$$\text{link loss} = -85\,\text{dBm} - 33\,\text{dBm} = 118\,\text{dB}$$

Using this 118 dB of link loss, the mobile can estimate the initial power needed to overcome the link loss, for the first access burst.

10.2.2 Power Offset Between Node B and the Active DAS

The active DAS has a remote unit with an integrated DL power amplifier. The actual DL power will therefore depend on the composite power resource of that specific DL amplifier. In a typical active DAS a RU would typically transmit the CPICH at +5 dBm (one UMTS carrier at full power is +15 dBm, and with the CPICH set to −10 dB = 5 dBm CPICH).

If the signal source to the active system is a +43 dBm macro node B, the CPICH is +33 dBm, and the system information broadcasts +33 dBm as the CPICH reference power (not the actual +5 dBm radiated from the RU), there is an obvious problem: using the same example as above, where the mobile is receiving the CPICH at −85 dBm, the mobile will still decode the information broadcast by node B, which the CPICH is transmitted at +33 dBm by the active DAS. The mobile will once again calculate the link loss to be

118 dBm:

$$\text{link loss} = -85\,\text{dBm} - 33\,\text{dBm} = 118\,\text{dB}$$

The mobile will then calculate the needed UL power to overcome the 118 dB link loss when accessing node B. However, in reality the CPICH is not transmitted at +33 dBm, but at +5 dBm due to the power amplifier (PA) in the RU, so the real link loss is actually:

$$\text{link loss} = -85\,\text{dBm} - 5\,\text{dBm} = 90\,\text{dB}$$

The result is that the mobile will use to high power in the access bursts. This erroneous CPICH power information will cause the mobile to access the cell with a level 28 dB too high (118–90 dB).

This overshoot on the UL will cause:

- Excessive noise increase on the serving cell and potentially in the neighboring cells.
- Degraded service for other mobiles in the serving cell and potentially in the neighboring cell.
- Potential for dropped calls in serving cell and neighboring cell.
- UL blocking of the RU (trigging the UL automatic level control).

10.2.3 Solution

Thanks to 3GPP, the solution is easy. The broadcast information on the CPICH power has to be changed to the actual level transmitted level by the RU. It is important to note that the CPICH level itself (−10 dB) is not changed, but only the information on the absolute transmitted level is adjusted to the CPICH level of the RU.

10.2.4 Impact on the UL of Node B

The noise load of the UL is a limiting factor on the UL performance, so it is crucial for UMTS to control the noise increase in the cells.

10.2.5 Admission Control

Node B will constantly measure the overall noise power on the UL to evaluate the UL noise increase in the cell. The total noise power on the UL of the node B will be a result of:

$$\text{noise}_{\text{total}} = \text{own}_{\text{traffic noise}} + \text{other}_{\text{cells' noise}} + \text{node} \quad \text{B}_{\text{noise power}}$$

So in addition to the noise generated by the traffic in the cell, there will be a contribution of noise from users in other cells and finally the NF of the node B hardware. Using this noise measurement and adding the hardware noise of node B makes it possible to evaluate if new admitted traffic will cause a noise increase that exceeds the predefined limit for the maximum UL noise increase in the cell (as shown in Figure 10.3).

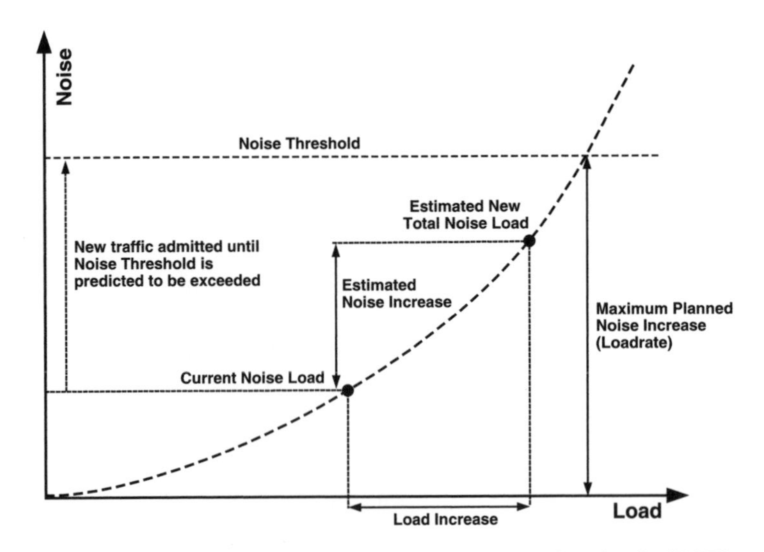

Figure 10.3 The principle of the admission control function in UMTS

10.3 Impact of Noise Power

An active DAS consists of amplifiers and, depending of the actual configuration, will have a given noise figure that adds to the noise on the UL of node B.

10.3.1 The UL Noise Increase on Node B

The active DAS needs to be configured so that the active DAS and UL attenuators are set at optimum values, minimizing the resulting noise increase on node B – without degrading the performance of the UL. However there will be a noise increase on the UL of node B. Therefore, it is crucial to tune the admission control parameter, with regards to the added noise power on the UL of the node B

Solution

$$\text{Tune the: node } B_{noise\ power}$$

This includes the noise increase caused by the active DAS (minimized by a correct UL attenuator). If this is not tuned, the result will be that the admission control assumes too high a UL traffic (noise level), limiting the UL capacity. In extreme cases (if no UL attenuator is used), this results in no call set-up in the cell.

10.4 Delay of the Active DAS

Connecting external equipment between node B and the antenna will cause an offset of the timing in the cell. In reality this delay will result in an increase in cell size (timing window).

An active DAS connected directly to the node B will typical add 2–3 km to the cell size. This delay or offset depends on the delay in the active DAS amplifiers, but also on the length of fibers and CAT5 cables. The longer the fiber and cable runs, the longer the delay will be, in some cases exceeding 15–20 km offset of the cell size. Some operators will set a given cell size according to the environment the cell is serving.

Typically, a node B serving an indoor solution would be set to a cell size of 2 km, so that node B does not pick up any traffic distant to the cell. Therefore if one does not consider the delay on an active indoor DAS, and sets the cell size to 2 km, the cell might not be able to set up calls on the indoor system, due to the delay of the DAS.

10.4.1 Solution

Set the cell size to cope with the offset caused by the delay in the active DAS.

10.5 Impact of External Noise Power

It is crucial to select the correct value for the UL attenuator between node B and the active system. The purpose of this attenuator is to minimize the UL noise power. However the correct value of the attenuator needs to be selected in order to have the minimum noise impact and not affect the UL (SNR) performance.

The ideal attenuator will be the difference between the noise power from the DAS and the *KTB*. The 'noise power normograph' in Figure 10.4 will help you to calculate the noise power in any given bandwidth, and thereby enable you to select the optimum UL attenuator. It is the standard noise calculations from Chapter 7, although the noise normograph might be easier to use. This is how to use the normograph.

- *The left graph* – this is the reference point for the noise figure from the DAS plus the UL gain set in the system, so if your NF from the system is 12 dB, and you use 3 dB UL gain, you should select 15 dB as the reference.
- *The right graph* – this is the reference point for the RF bandwidth for the noise calculation, so for GSM you select 200 kHz for UMTS 3.84Mc (Figure 10.4).

10.5.1 To Calculate the Noise Power

You select the RF bandwidth and the NF and draw a line between the two points; then the noise power is given by the point where this line crosses the center graph. Too calculate the noise power of a UMTS node B with an NF of 4 dB, the two reference points are the UMTS bandwidth (3.84Mc) on the right graph and the NF (4 dB) on the left graph. The line connecting these two points crosses the noise power graph at −104 dBm [4 dB NF + (−108 dBm)], so this is the noise power loading the Rx port of node B, and this new RAN admission reference for zero traffic on node B.

10.5.2 To Calculate the UL Attenuator

The example shows a UMTS (3.84Mc) system, connected to an active DAS with 20 dB NF. As we can see, the noise power is −89 dBm, loading the input of the node B, if we do not use

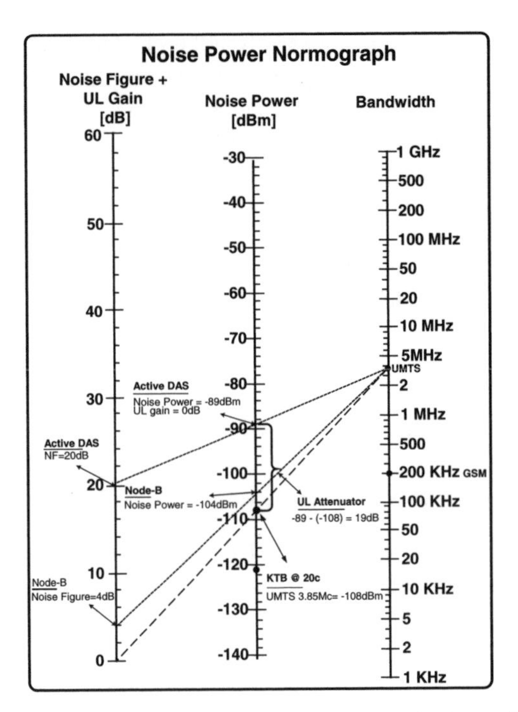

Figure 10.4 Noise normograph, an easy way to calculate noise power

any attenuator. The ideal attenuator is calculated as the difference between this noise power and the *KTB* (the *KTB* is the noise power at 0 dB NF). In this example the ideal attenuator is 19 dB, so in practice you would select a 20 dB attenuator and adjust the UL gain of the DAS to +1 dB. Note that selecting a higher attenuator will degrade the UL dynamic range and sensitivity of the node B. However, there will still be a slight impact of the noise power from the DAS when node B has a noise power of −106 dBm and the DAS with attenuator has −108 dBm of noise power, and we will need to sum up these two noise powers by adding the powers in Watts.

First we convert the dBm to W:

$$P\ (\text{mw}) = 10^{\text{dBm}/10}$$

The idle noise power on the base station $= -106\,\text{dBm} = 1.58 \times 10^{-11}\,\text{mW}$. The noise power from DAS $= -108\,\text{dBm} = 2.51 \times 10^{-11}\,\text{mW}$. Now we add the two noise powers:

$$\text{noise power} = \text{P1} + \text{P2} = 1.58 \times 10^{-11}\,\text{mW} + 2.51 \times 10^{-11}\,\text{mW} = 4.09 \times 10^{-11}\,\text{mW}$$

Then we convert back to dBm:

$$P\ (\text{dBm}) = 10 \log(\text{mW})$$
$$P\ (\text{dBm}) = 10 \log(4.09 \times 10^{-11}) = -103.88\,\text{dBm}$$

10.5.3 Affect on Admission Control

The resulting noise power load would be about −104 dBm, and the admission control should offset to take this into account. It would be a noise load increase of +2 dB on the UL of the base station, corresponding to a UL load of 37%. If the base station admission control was set to a maximum UL load of 50% (3 dB noise increase), this 'phantom load' would have the effect that the base station would only be able to admit 13% of traffic load (50% max UL load −37% phantom load). If no UL attenuator was used, the noise power injected to the base station would saturate admission control with phantom noise, and no traffic would be admitted to the cell.

References

This book is a result of more than 15 years' experience in indoor radio planning, working for a mobile operator as an RF-planner and as technical director for one of the leading suppliers of DAS equipment. The subjects, examples and approaches in this book are based on practical experience and indoor implementations I have been responsible for and directly involved in.

My interest is the RF part of these radio systems and not all the deep background core network or protocols. There are no direct transcripts or copies from other books, but I have to recommend some of the daily reference material I use myself. These are the books that helped me understand the basics, and that I use as my reference.

[1] Michel Mouly and Marie-Bernadette Pautet. *The GSM System for Mobile Communications*. Published by the authors (ISBN 2-9507190-0-7).

[2] P. E. Clint Smith and Curt Gervelis. *Cellular System Design & Optimization*. McGraw-Hill: New York.

[3] Harri Holma and Antti Toskala (eds). *WCDMA for UMTS, Radio Access for Third Generation Mobile Communications*. Wiley: Chichester.

[4] Jaana Laiho, Achim Wacker and Tomáš Novosad (eds). *Radio Network Planning and Optimisation for UMTS*, 2nd edn. Wiley: Chichester.

[5] Harri Holma and Antti Toskala (eds). *HSDPA/HSUPA for UMTS, High Speed Radio Access for Mobile Communications*. Wiley: Chichester.

[6] S. R. Saunders and A. Aragon-Zavala. *Antennas and Propagation for Wireless Communication Systems*, 2nd edn. Wiley: Chichester.

Appendix

Reference Material

Conversions of Watt to dBm, dBm to Watt

$$\text{dBm} = 10 \log\left(\frac{\text{power (W)}}{1\,\text{mW}}\right)$$

$$\text{P mW} = 10^{(\text{dBm}/10)}$$

Table A.1

Watt	dBm	Watt	dBm	Watt	dBm	Watt	dBm	Watt	dBm	Watt	dBm
0.001	0.00	**2.1**	33.22	**5.90**	37.71	**9.70**	39.87	**13.50**	41.30	**17.30**	42.38
0.002	3.01	**2.2**	33.42	**6.00**	37.78	**9.80**	39.91	**13.60**	41.34	**17.40**	42.41
0.003	4.77	**2.3**	33.62	**6.10**	37.85	**9.90**	39.96	**13.70**	41.37	**17.50**	42.43
0.004	6.02	**2.4**	33.80	**6.20**	37.92	**10.00**	40.00	**13.80**	41.40	**17.60**	42.46
0.005	6.99	**2.5**	33.98	**6.30**	37.99	**10.10**	40.04	**13.90**	41.43	**17.70**	42.48
0.006	7.78	**2.6**	34.15	**6.40**	38.06	**10.20**	40.09	**14.00**	41.46	**17.80**	42.50
0.007	8.45	**2.7**	34.31	**6.50**	38.13	**10.30**	40.13	**14.10**	41.49	**17.90**	42.53
0.008	9.03	**2.8**	34.47	**6.60**	38.20	**10.40**	40.17	**14.20**	41.52	**18.00**	42.55
0.009	9.54	**2.9**	34.62	**6.70**	38.26	**10.50**	40.21	**14.30**	41.55	**18.10**	42.58
0.010	10.00	**3**	34.77	**6.80**	38.33	**10.60**	40.25	**14.40**	41.58	**18.20**	42.60
0.020	13.01	**3.1**	34.91	**6.90**	38.39	**10.70**	40.29	**14.50**	41.61	**18.30**	42.62
0.030	14.77	**3.2**	35.05	**7.00**	38.45	**10.80**	40.33	**14.60**	41.64	**18.40**	42.65
0.040	16.02	**3.3**	35.19	**7.10**	38.51	**10.90**	40.37	**14.70**	41.67	**18.50**	42.67
0.050	16.99	**3.4**	35.31	**7.20**	38.57	**11.00**	40.41	**14.80**	41.70	**18.60**	42.70
0.060	17.78	**3.5**	35.44	**7.30**	38.63	**11.10**	40.45	**14.90**	41.73	**18.70**	42.72
0.070	18.45	**3.6**	35.56	**7.40**	38.69	**11.20**	40.49	**15.00**	41.76	**18.80**	42.74
0.080	19.03	**3.7**	35.68	**7.50**	38.75	**11.30**	40.53	**15.10**	41.79	**18.90**	42.76
0.090	19.54	**3.80**	35.80	**7.60**	38.81	**11.40**	40.57	**15.20**	41.82	**19.00**	42.79
0.100	20.00	**3.90**	35.91	**7.70**	38.86	**11.50**	40.61	**15.30**	41.85	**19.10**	42.81
0.200	23.01	**4.00**	36.02	**7.80**	38.92	**11.60**	40.64	**15.40**	41.88	**19.20**	42.83
0.300	24.77	**4.10**	36.13	**7.90**	38.98	**11.70**	40.68	**15.50**	41.90	**19.30**	42.86

(*continued*)

Indoor Radio Planning: A Practical Guide for GSM, DCS, UMTS and HSPA Morten Tolstrup

Table A.1 (*Continued*)

Watt	dBm	Watt	dBm	Watt	dBm	Watt	dBm	Watt	dBm	Watt	dBm
0.400	26.02	**4.20**	36.23	**8.00**	39.03	**11.80**	40.72	**15.60**	41.93	**19.40**	42.88
0.500	26.99	**4.30**	36.33	**8.10**	39.08	**11.90**	40.76	**15.70**	41.96	**19.50**	42.90
0.600	27.78	**4.40**	36.43	**8.20**	39.14	**12.00**	40.79	**15.80**	41.99	**19.60**	42.92
0.700	28.45	**4.50**	36.53	**8.30**	39.19	**12.10**	40.83	**15.90**	42.01	**19.70**	42.94
0.800	29.03	**4.60**	36.63	**8.40**	39.24	**12.20**	40.86	**16.00**	42.04	**19.80**	42.97
0.900	29.54	**4.70**	36.72	**8.50**	39.29	**12.30**	40.90	**16.10**	42.07	**19.90**	42.99
1.000	30.00	**4.80**	36.81	**8.60**	39.34	**12.40**	40.93	**16.20**	42.10	**20.00**	43.01
1.100	30.41	**4.90**	36.90	**8.70**	39.40	**12.50**	40.97	**16.30**	42.12	**25.00**	44.77
1.200	30.79	**5.00**	36.99	**8.80**	39.44	**12.60**	41.00	**16.40**	42.15	**30.00**	46.99
1.300	31.14	**5.10**	37.08	**8.90**	39.49	**12.70**	41.04	**16.50**	42.17	**35.00**	45.44
1.400	31.46	**5.20**	37.16	**9.00**	39.54	**12.80**	41.07	**16.60**	42.20	**40.00**	46.02
1.500	31.76	**5.30**	37.24	**9.10**	39.59	**12.90**	41.11	**16.70**	42.23	**50.00**	47.00
1.600	32.04	**5.40**	37.32	**9.20**	39.64	**13.00**	41.14	**16.80**	42.25	**60.00**	47.78
1.700	32.30	**5.50**	37.40	**9.30**	39.68	**13.10**	41.17	**16.90**	42.28	**70.00**	48.45
1.800	32.55	**5.60**	37.48	**9.40**	39.73	**13.20**	41.21	**17.00**	42.30	**80.00**	49.03
1.900	32.79	**5.70**	37.56	**9.50**	39.78	**13.30**	41.24	**17.10**	42.33	**90.00**	49.54
2.000	33.01	**5.80**	37.63	**9.60**	39.82	**13.40**	41.27	**17.20**	42.36	**100.00**	50.00

Noise Power Calculator

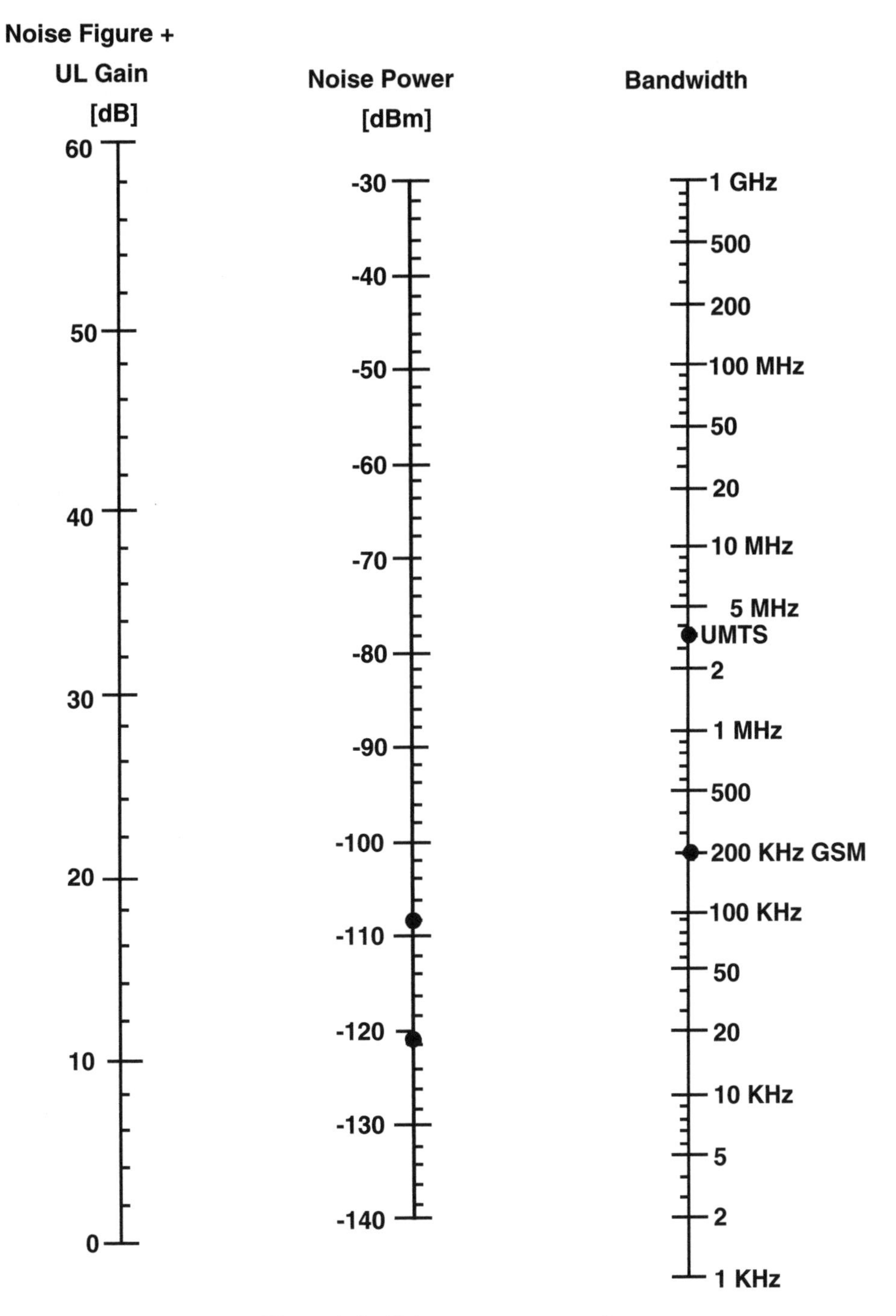

Figure A.1 Noise power normograph

Table A.2 Erlang B table, 1–50 channels, 0.01–5% grade of service

Number of channels	Blocking (GOS)					Number of channels
	0.1%	0.5%	1%	2%	5%	
1	0.0010	0.0050	0.0101	0.0204	0.0526	**1**
2	0.0458	0.1054	0.1526	0.2235	0.3813	**2**
3	0.1938	0.3490	0.4555	0.6022	0.8994	**3**
4	0.4393	0.7012	0.8694	1.0923	1.5246	**4**
5	0.7621	1.1320	1.3608	1.6571	2.2185	**5**
6	1.1459	1.6218	1.9090	2.2759	2.9603	**6**
7	1.5786	2.1575	2.5009	2.9354	3.7378	**7**
8	2.0513	2.7299	3.1276	3.6271	4.5430	**8**
9	2.5575	3.3326	3.7825	4.3447	5.3702	**9**
10	3.0920	3.9607	4.4612	5.0840	6.2157	**10**
11	3.6511	4.6104	5.1599	5.8415	7.0764	**11**
12	4.2314	5.2789	5.8760	6.6147	7.9501	**12**
13	4.8306	5.9638	6.6072	7.4015	8.8349	**13**
14	5.4464	6.6632	7.3517	8.2003	9.7295	**14**
15	6.0772	7.3755	8.1080	9.0096	10.633	**15**
16	6.7215	8.0995	8.8750	9.8284	11.544	**16**
17	7.3781	8.8340	9.6516	10.656	12.461	**17**
18	8.0459	9.5780	10.437	11.491	13.385	**18**
19	8.7239	10.331	11.230	12.333	14.315	**19**
20	9.4115	11.092	12.031	13.182	15.249	**20**
21	10.108	11.860	12.838	14.036	16.189	**21**
22	10.812	12.635	13.651	14.896	17.132	**22**
23	11.524	13.416	14.470	15.761	18.080	**23**
24	12.243	14.204	15.295	16.631	19.031	**24**
25	12.969	14.997	16.125	17.505	19.985	**25**
26	13.701	15.795	16.959	18.383	20.943	**26**
27	14.439	16.598	17.797	19.265	21.904	**27**
28	15.182	17.406	18.640	20.150	22.867	**28**
29	15.930	18.218	19.487	21.039	23.833	**29**
30	16.684	19.034	20.337	21.932	24.802	**30**
31	17.442	19.854	21.191	22.827	25.773	**31**
32	18.205	20.678	22.048	23.725	26.746	**32**
33	18.972	21.505	22.909	24.626	27.721	**33**
34	19.743	22.336	23.772	25.529	28.698	**34**
35	20.517	23.169	24.638	26.435	29.677	**35**
36	21.296	24.006	25.507	27.343	30.657	**36**
37	22.078	24.846	26.378	28.254	31.640	**37**
38	22.864	25.689	27.252	29.166	32.624	**38**
39	23.652	26.534	28.129	30.081	33.609	**39**
40	24.444	27.382	29.007	30.997	34.596	**40**
41	25.239	28.232	29.888	31.916	35.584	**41**
42	26.037	29.085	30.771	32.836	36.574	**42**
43	26.837	29.940	31.656	33.758	37.565	**43**
44	27.641	30.797	32.543	34.682	38.557	**44**
45	28.447	31.656	33.432	35.607	39.550	**45**
46	29.255	32.517	34.322	36.534	40.545	**46**
47	30.066	33.381	35.215	37.462	41.540	**47**
48	30.879	34.246	36.109	38.392	42.537	**48**
49	31.694	35.113	37.004	39.323	43.534	**49**
50	32.512	35.982	37.901	40.255	44.533	**50**

Unit Conversions

Decibel Conversion: Power

$$\mathrm{dB} = 10\log(P2/P1)$$

Decibels relative to power.

Decibel Conversion: Voltage

$$\mathrm{dB} = 20\log\,[V1/V2]$$

Decibels relative to voltage across the same resistance.

Decibel Conversion: Current

$$\mathrm{dB} = 20\log\,[I1/I2]$$

Decibels relative to current through same resistance.

Decibel Conversion: Milliwatts

$$\mathrm{dBm} = 10\log[\mathrm{signal\ (mW)}/1\ \mathrm{mW}]$$

Decibels relative to 1 mW.

Decibel Conversion: Microvolts

$$\mathrm{dB}\mu\mathrm{V} = 20\log[\mathrm{signal}\ (\mu\mathrm{V})/1\mu\mathrm{V}]$$

Decibels relative to 1 μV across the same resistance.

Decibel Conversion: Microamps

$$\mathrm{dB}\mu\mathrm{A} = 20\log[\mathrm{signal}\ (\mu\mathrm{A})/1\mu\mathrm{A}]$$

Decibels relative to one microamp through the same resistance.

Power Conversion: dBw to dBm

$$\mathrm{dBm} = \mathrm{dBw} + 30$$

Conversion from dBw to dBm.

Voltage Conversion: dBV to dBμV

$$\mathrm{dB}\mu\mathrm{V} = \mathrm{dBV} + 120$$

Conversion from dBV to dBμV.

Voltage to Power Conversion: dBμV to dBm

$$\mathrm{dBm} = \mathrm{dB}\mu\mathrm{V} - 107$$

where the constant 107 is as follows:
RF systems are matched to 50 Ω

$$P = V^2/R$$
$$10\log_{10}(P) = 20\log_{10}(V) - 10\log_{10}(50\Omega)$$
$$V = (PR)^{0.5} = 0.223\,\mathrm{V} = 223\,000\,\mu\mathrm{V}$$

For a resistance of 50 Ω and a power of 1 mW:

$$20\log_{10}[223\,000\,\mu\mathrm{V}] = 107\,\mathrm{dB}$$

Power Density

$$\mathrm{dBW/m^2} = 10\log_{10}(\mathrm{V/m} - \mathrm{A/m})$$

Decibel–Watts per square meter.

$$\mathrm{dBm/m^2} = \mathrm{dBW/m^2} + 30$$

where the constant 30 is the decibel equivalent of the factor 1000 used to convert between W and mW:

$$10\log_{10}[1000] = 30$$

Electric Field Voltage

$$\mathrm{V/m} = 10^{\{[(\mathrm{dB}\mu\mathrm{V/m})-120]/20\}}$$

Electric field voltage in Volts per meter.

Electric Field Current

$$\mathrm{dB}\mu\mathrm{A/m} = \mathrm{dB}\mu\mathrm{V/m} - 51.5$$

where the constant 51.5 is a conversion of the characteristic impedance of free space (120 Ω) into decibels: $20\log_{10}(120\,\Omega) = 51.5$.

$$\mathrm{A/m} = 10^{\{[(\mathrm{dB}\mu\mathrm{A/m})-120]/20\}}$$

Electric field current in Amps per meter.

Conversion Units

$$\text{dB} = \text{decibels (log 10)}$$
$$\mu = \text{micro} = 10 \times 10^{-6}$$
$$\text{m} = \text{milli} = 10 \times 10^{-3}$$
$$\text{dBw} = \text{decibels relative to 1 W}$$
$$\text{dBm} = \text{decibels relative to 1 mW}$$
$$\text{dBV} = \text{decibels relative to 1 V}$$
$$\text{dB}\mu\text{V} = \text{decibels relative to 1 }\mu\text{V}$$
$$\text{dB}\mu\text{A} = \text{decibels relative to 1 }\mu\text{A}$$

V = Volts, A = Amps, I = current, R = resistance Ω = Ohms (50), W = Watts, P = Power, m = meters

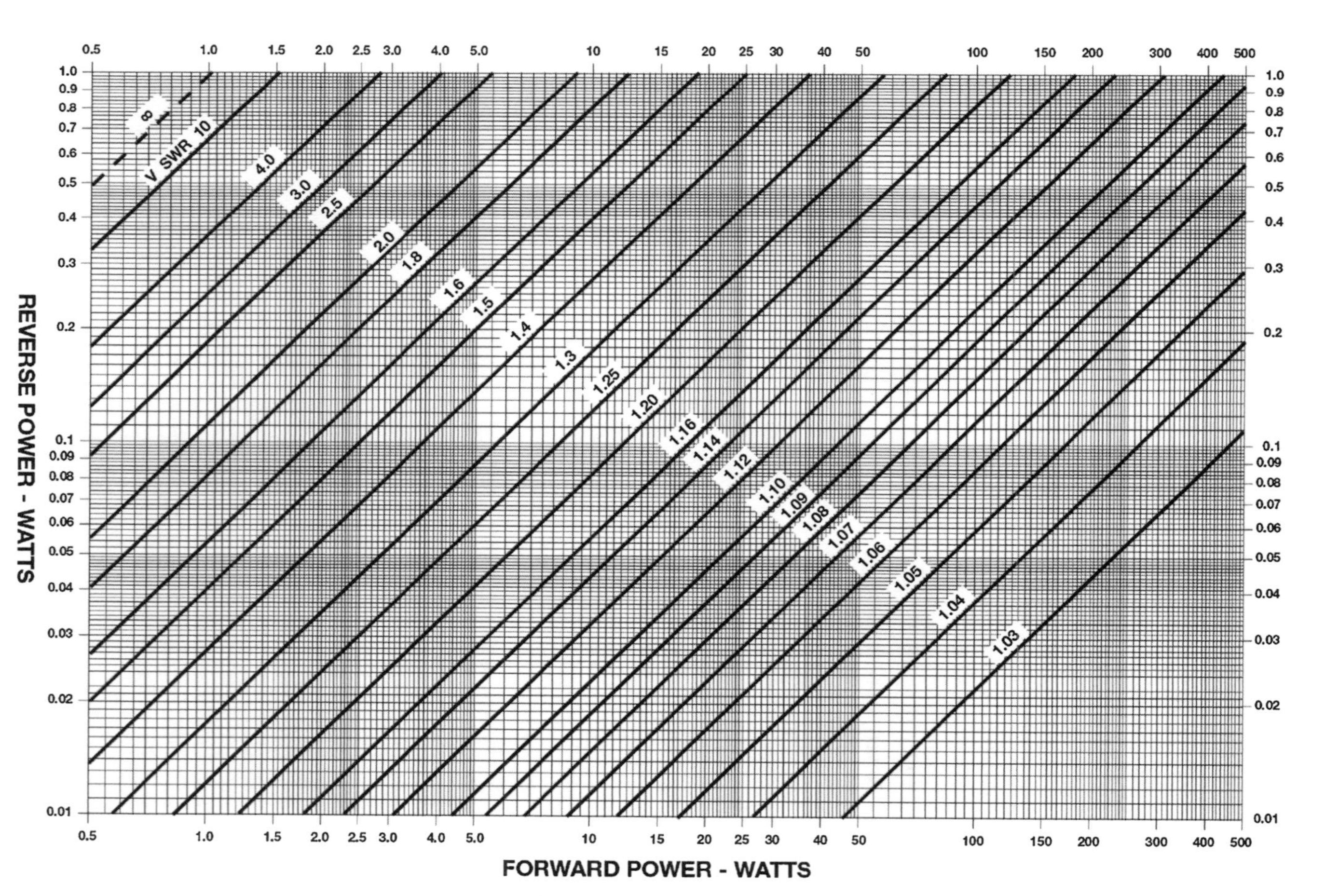

Figure A.2 VSWR chart

Table A.3 GSM 900 channels and frequencies

Channel	UL (MHz)	DL (MHz)	Channel	UL (MHz)	DL (MHz)	Channel	UL (MHz)	DL (MHz)
1	890 200	935 200	**41**	898 200	943 200	**81**	906 200	951 200
2	890 400	935 400	**42**	898 400	943 400	**82**	906 400	951 400
3	890 600	935 600	**43**	898 600	943 600	**83**	906 600	951 600
4	890 800	935 800	**44**	898 800	943 800	**84**	906 800	951 800
5	891 000	936 000	**45**	899 000	944 000	**85**	907 000	952 000
6	891 200	936 200	**46**	899 200	944 200	**86**	907 200	952 200
7	891 400	936 400	**47**	899 400	944 400	**87**	907 400	952 400
8	891 600	936 600	**48**	899 600	944 600	**88**	907 600	952 600
9	891 800	936 800	**49**	899 800	944 800	**89**	907 800	952 800
10	892 000	937 000	**50**	900 000	945 000	**90**	908 000	953 000
11	892 200	937 200	**51**	900 200	945 200	**91**	908 200	953 200
12	892 400	937 400	**52**	900 400	945 400	**92**	908 400	953 400
13	892 600	937 600	**53**	900 600	945 600	**93**	908 600	953 600
14	892 800	937 800	**54**	900 800	945 800	**94**	908 800	953 800
15	893 000	938 000	**55**	901 000	946 000	**95**	909 000	954 000
16	893 200	938 200	**56**	901 200	946 200	**96**	909 200	954 200
17	893 400	938 400	**57**	901 400	946 400	**97**	909 400	954 400
18	893 600	938 600	**58**	901 600	946 600	**98**	909 600	954 600
19	893 800	938 800	**59**	901 800	946 800	**99**	909 800	954 800
20	894 000	939 000	**60**	902 000	947 000	**100**	910 000	955 000
21	894 200	939 200	**61**	902 200	947 200	**101**	910 200	955 200
22	894 400	939 400	**62**	902 400	947 400	**102**	910 400	955 400
23	894 600	939 600	**63**	902 600	947 600	**103**	910 600	955 600
24	894 800	939 800	**64**	902 800	947 800	**104**	910 800	955 800
25	895 000	940 000	**65**	903 000	948 000	**105**	911 000	956 000
26	895 200	940 200	**66**	903 200	948 200	**106**	911 200	956 200
27	895 400	940 400	**67**	903 400	948 400	**107**	911 400	956 400
28	895 600	940 600	**68**	903 600	948 600	**108**	911 600	956 600
29	895 800	940 800	**69**	903 800	948 800	**109**	911 800	956 800
30	896 000	941 000	**70**	904 000	949 000	**110**	912 000	957 000
31	896 200	941 200	**71**	904 200	949 200	**111**	912 200	957 200
32	896 400	941 400	**72**	904 400	949 400	**112**	912 400	957 400
33	896 600	941 600	**73**	904 600	949 600	**113**	912 600	957 600
34	896 800	941 800	**74**	904 800	949 800	**114**	912 800	957 800
35	897 000	942 000	**75**	905 000	950 000	**115**	913 000	958 000
36	897 200	942 200	**76**	905 200	950 200	**116**	913 200	958 200
37	897 400	942 400	**77**	905 400	950 400	**117**	913 400	958 400
38	897 600	942 600	**78**	905 600	950 600	**118**	913 600	958 600
39	897 800	942 800	**79**	905 800	950 800	**119**	913 800	958 800
40	898 000	943 000	**80**	906 000	951 000	**120**	914 000	959 000

Table A.4 GSM 1800 channels and frequencies

Channel	UL (MHz)	DL (MHz)	Channel	UL (MHz)	DL (MHz)	Channel	UL (MHz)	DL (MHz)	Channel	UL (MHz)	DL (MHz)
			570	1 721.800	1 816.800	**630**	1 733.800	1 828.800	**690**	1 745.800	1 840.800
			571	1 722.000	1 817.000	**631**	1 734.000	1 829.000	**691**	1 746.000	1 841.000
512	1 710.200	1 805.200	**572**	1 722.200	1 817.200	**632**	1 734.200	1 829.200	**692**	1 746.200	1 841.200
513	1 710.400	1 805.400	**573**	1 722.400	1 817.400	**633**	1 734.400	1 829.400	**693**	1 746.400	1 841.400
514	1 710.600	1 805.600	**574**	1 722.600	1 817.600	**634**	1 734.600	1 829.600	**694**	1 746.600	1 841.600
515	1 710.800	1 805.800	**575**	1 722.800	1 817.800	**635**	1 734.800	1 829.800	**695**	1 746.800	1 841.800
516	1 711.000	1 806.000	**576**	1 723.000	1 818.000	**636**	1 735.000	1 830.000	**696**	1 747.000	1 842.000
517	1 711.200	1 806.200	**577**	1 723.200	1 818.200	**637**	1 735.200	1 830.200	**697**	1 747.200	1 842.200
518	1 711.400	1 806.400	**578**	1 723.400	1 818.400	**638**	1 735.400	1 830.400	**698**	1 747.400	1 842.400
519	1 711.600	1 806.600	**579**	1 723.600	1 818.600	**639**	1 735.600	1 830.600	**699**	1 747.600	1 842.600
520	1 711.800	1 806.800	**580**	1 723.800	1 818.800	**640**	1 735.800	1 830.800	**700**	1 747.800	1 842.800
521	1 712.000	1 807.000	**581**	1 724.000	1 819.000	**641**	1 736.000	1 831.000	**701**	1 748.000	1 843.000
522	1 712.200	1 807.200	**582**	1 724.200	1 819.200	**642**	1 736.200	1 831.200	**702**	1 748.200	1 843.200
523	1 712.400	1 807.400	**583**	1 724.400	1 819.400	**643**	1 736.400	1 831.400	**703**	1 748.400	1 843.400
524	1 712.600	1 807.600	**584**	1 724.600	1 819.600	**644**	1 736.600	1 831.600	**704**	1 748.600	1 843.600
525	1 712.800	1 807.800	**585**	1 724.800	1 819.800	**645**	1 736.800	1 831.800	**705**	1 748.800	1 843.800
526	1 713.000	1 808.000	**586**	1 725.000	1 820.000	**646**	1 737.000	1 832.000	**706**	1 749.000	1 844.000
527	1 713.200	1 808.200	**587**	1 725.200	1 820.200	**647**	1 737.200	1 832.200	**707**	1 749.200	1 844.200
528	1 713.400	1 808.400	**588**	1 725.400	1 820.400	**648**	1 737.400	1 832.400	**708**	1 749.400	1 844.400
529	1 713.600	1 808.600	**589**	1 725.600	1 820.600	**649**	1 737.600	1 832.600	**709**	1 749.600	1 844.600
530	1 713.800	1 808.800	**590**	1 725.800	1 820.800	**650**	1 737.800	1 832.800	**710**	1 749.800	1 844.800
531	1 714.000	1 809.000	**591**	1 726.000	1 821.000	**651**	1 738.000	1 833.000	**711**	1 750.000	1 845.000
532	1 714.200	1 809.200	**592**	1 726.200	1 821.200	**652**	1 738.200	1 833.200	**712**	1 750.200	1 845.200
533	1 714.400	1 809.400	**593**	1 726.400	1 821.400	**653**	1 738.400	1 833.400	**713**	1 750.400	1 845.400
534	1 714.600	1 809.600	**594**	1 726.600	1 821.600	**654**	1 738.600	1 833.600	**714**	1 750.600	1 845.600
535	1 714.800	1 809.800	**595**	1 726.800	1 821.800	**655**	1 738.800	1 833.800	**715**	1 750.800	1 845.800
536	1 715.000	1 810.000	**596**	1 727.000	1 822.000	**656**	1 739.000	1 834.000	**716**	1 751.000	1 846.000
537	1 715.200	1 810.200	**597**	1 727.200	1 822.200	**657**	1 739.200	1 834.200	**717**	1 751.200	1 846.200
538	1 715.400	1 810.400	**598**	1 727.400	1 822.400	**658**	1 739.400	1 834.400	**718**	1 751.400	1 846.400
539	1 715.600	1 810.600	**599**	1 727.600	1 822.600	**659**	1 739.600	1 834.600	**719**	1 751.600	1 846.600
540	1 715.800	1 810.800	**600**	1 727.800	1 822.800	**660**	1 739.800	1 834.800	**720**	1 751.800	1 846.800
541	1 716.000	1 811.000	**601**	1 728.000	1 823.000	**661**	1 740.000	1 835.000	**721**	1 752.000	1 847.000
542	1 716.200	1 811.200	**602**	1 728.200	1 823.200	**662**	1 740.200	1 835.200	**722**	1 752.200	1 847.200
543	1 716.400	1 811.400	**603**	1 728.400	1 823.400	**663**	1 740.400	1 835.400	**723**	1 752.400	1 847.400
544	1 716.600	1 811.600	**604**	1 728.600	1 823.600	**664**	1 740.600	1 835.600	**724**	1 752.600	1 847.600
545	1 716.800	1 811.800	**605**	1 728.800	1 823.800	**665**	1 740.800	1 835.800	**725**	1 752.800	1 847.800
546	1 717.000	1 812.000	**606**	1 729.000	1 824.000	**666**	1 741.000	1 836.000	**726**	1 753.000	1 848.000
547	1 717.200	1 812.200	**607**	1 729.200	1 824.200	**667**	1 741.200	1 836.200	**727**	1 753.200	1 848.200
548	1 717.400	1 812.400	**608**	1 729.400	1 824.400	**668**	1 741.400	1 836.400	**728**	1 753.400	1 848.400
549	1 717.600	1 812.600	**609**	1 729.600	1 824.600	**669**	1 741.600	1 836.600	**729**	1 753.600	1 848.600
550	1 717.800	1 812.800	**610**	1 729.800	1 824.800	**670**	1 741.800	1 836.800	**730**	1 753.800	1 848.800
551	1 718.000	1 813.000	**611**	1 730.000	1 825.000	**671**	1 742.000	1 837.000	**731**	1 754.000	1 849.000
552	1 718.200	1 813.200	**612**	1 730.200	1 825.200	**672**	1 742.200	1 837.200	**732**	1 754.200	1 849.200
553	1 718.400	1 813.400	**613**	1 730.400	1 825.400	**673**	1 742.400	1 837.400	**733**	1 754.400	1 849.400
554	1 718.600	1 813.600	**614**	1 730.600	1 825.600	**674**	1 742.600	1 837.600	**734**	1 754.600	1 849.600
555	1 718.800	1 813.800	**615**	1 730.800	1 825.800	**675**	1 742.800	1 837.800	**735**	1 754.800	1 849.800
556	1 719.000	1 814.000	**616**	1 731.000	1 826.000	**676**	1 743.000	1 838.000	**736**	1 755.000	1 850.000
557	1 719.200	1 814.200	**617**	1 731.200	1 826.200	**677**	1 743.200	1 838.200	**737**	1 755.200	1 850.200
558	1 719.400	1 814.400	**618**	1 731.400	1 826.400	**678**	1 743.400	1 838.400	**738**	1 755.400	1 850.400
559	1 719.600	1 814.600	**619**	1 731.600	1 826.600	**679**	1 743.600	1 838.600	**739**	1 755.600	1 850.600
560	1 719.800	1 814.800	**620**	1 731.800	1 826.800	**680**	1 743.800	1 838.800	**740**	1 755.800	1 850.800
561	1 720.000	1 815.000	**621**	1 732.000	1 827.000	**681**	1 744.000	1 839.000	**741**	1 756.000	1 851.000
562	1 720.200	1 815.200	**622**	1 732.200	1 827.200	**682**	1 744.200	1 839.200	**742**	1 756.200	1 851.200
563	1 720.400	1 815.400	**623**	1 732.400	1 827.400	**683**	1 744.400	1 839.400	**743**	1 756.400	1 851.400
564	1 720.600	1 815.600	**624**	1 732.600	1 827.600	**684**	1 744.600	1 839.600	**744**	1 756.600	1 851.600
565	1 720.800	1 815.800	**625**	1 732.800	1 827.800	**685**	1 744.800	1 839.800	**745**	1 756.800	1 851.800
566	1 721.000	1 816.000	**626**	1 733.000	1 828.000	**686**	1 745.000	1 840.000	**746**	1 757.000	1 852.000
567	1 721.200	1 816.200	**627**	1 733.200	1 828.200	**687**	1 745.200	1 840.200	**747**	1 757.200	1 852.200
568	1 721.400	1 816.400	**628**	1 733.400	1 828.400	**688**	1 745.400	1 840.400	**748**	1 757.400	1 852.400
569	1 721.600	1 816.600	**629**	1 733.600	1 828.600	**689**	1 745.600	1 840.600	**749**	1 757.600	1 852.600

Table A.5

Channel	UL (MHz)	DL (MHz)	Channel	UL (MHz)	DL (MHz)	Channel	UL (MHz)	DL (MHz)
750	1 757.800	1 852.800	**810**	1 769.800	1 864.800	**870**	1 781.800	1 876.800
751	1 758.000	1 853.000	**811**	1 770.000	1 865.000	**871**	1 782.000	1 877.000
752	1 758.200	1 853.200	**812**	1 770.200	1 865.200	**872**	1 782.200	1 877.200
753	1 758.400	1 853.400	**813**	1 770.400	1 865.400	**873**	1 782.400	1 877.400
754	1 758.600	1 853.600	**814**	1 770.600	1 865.600	**874**	1 782.600	1 877.600
755	1 758.800	1 853.800	**815**	1 770.800	1 865.800	**875**	1 782.800	1 877.800
756	1 759.000	1 854.000	**816**	1 771.000	1 866.000	**876**	1 783.000	1 878.000
757	1 759.200	1 854.200	**817**	1 771.200	1 866.200	**877**	1 783.200	1 878.200
758	1 759.400	1 854.400	**818**	1 771.400	1 866.400	**878**	1 783.400	1 878.400
759	1 759.600	1 854.600	**819**	1 771.600	1 866.600	**879**	1 783.600	1 878.600
760	1 759.800	1 854.800	**820**	1 771.800	1 866.800	**880**	1 783.800	1 878.800
761	1 760.000	1 855.000	**821**	1 772.000	1 867.000	**881**	1 784.000	1 879.000
762	1 760.200	1 855.200	**822**	1 772.200	1 867.200	**882**	1 784.200	1 879.200
763	1 760.400	1 855.400	**823**	1 772.400	1 867.400	**883**	1 784.400	1 879.400
764	1 760.600	1 855.600	**824**	1 772.600	1 867.600	**884**	1 784.600	1 879.600
765	1 760.800	1 855.800	**825**	1 772.800	1 867.800	**885**	1 784.800	1 879.800
766	1 761.000	1 856.000	**826**	1 773.000	1 868.000			
767	1 761.200	1 856.200	**827**	1 773.200	1 868.200			
768	1 761.400	1 856.400	**828**	1 773.400	1 868.400			
769	1 761.600	1 856.600	**829**	1 773.600	1 868.600			
770	1 761.800	1 856.800	**830**	1 773.800	1 868.800			
771	1 762.000	1 857.000	**831**	1 774.000	1 869.000			
772	1 762.200	1 857.200	**832**	1 774.200	1 869.200			
773	1 762.400	1 857.400	**833**	1 774.400	1 869.400			
774	1 762.600	1 857.600	**834**	1 774.600	1 869.600			
775	1 762.800	1 857.800	**835**	1 774.800	1 869.800			
776	1 763.000	1 858.000	**836**	1 775.000	1 870.000			
777	1 763.200	1 858.200	**837**	1 775.200	1 870.200			
778	1 763.400	1 858.400	**838**	1 775.400	1 870.400			
779	1 763.600	1 858.600	**839**	1 775.600	1 870.600			
780	1 763.800	1 858.800	**840**	1 775.800	1 870.800			
781	1 764.000	1 859.000	**841**	1 776.000	1 871.000			
782	1 764.200	1 859.200	**842**	1 776.200	1 871.200			
783	1 764.400	1 859.400	**843**	1 776.400	1 871.400			
784	1 764.600	1 859.600	**844**	1 776.600	1 871.600			
785	1 764.800	1 859.800	**845**	1 776.800	1 871.800			
786	1 765.000	1 860.000	**846**	1 777.000	1 872.000			
787	1 765.200	1 860.200	**847**	1 777.200	1 872.200			
788	1 765.400	1 860.400	**848**	1 777.400	1 872.400			
789	1 765.600	1 860.600	**849**	1 777.600	1 872.600			
790	1 765.800	1 860.800	**850**	1 777.800	1 872.800			
791	1 766.000	1 861.000	**851**	1 778.000	1 873.000			
792	1 766.200	1 861.200	**852**	1 778.200	1 873.200			
793	1 766.400	1 861.400	**853**	1 778.400	1 873.400			
794	1 766.600	1 861.600	**854**	1 778.600	1 873.600			
795	1 766.800	1 861.800	**855**	1 778.800	1 873.800			
796	1 767.000	1 862.000	**856**	1 779.000	1 874.000			
797	1 767.200	1 862.200	**857**	1 779.200	1 874.200			
798	1 767.400	1 862.400	**858**	1 779.400	1 874.400			
799	1 767.600	1 862.600	**859**	1 779.600	1 874.600			
800	1 767.800	1 862.800	**860**	1 779.800	1 874.800			
801	1 768.000	1 863.000	**861**	1 780.000	1 875.000			
802	1 768.200	1 863.200	**862**	1 780.200	1 875.200			
803	1 768.400	1 863.400	**863**	1 780.400	1 875.400			
804	1 768.600	1 863.600	**864**	1 780.600	1 875.600			
805	1 768.800	1 863.800	**865**	1 780.800	1 875.800			
806	1 769.000	1 864.000	**866**	1 781.000	1 876.000			
807	1 769.200	1 864.200	**867**	1 781.200	1 876.200			
808	1 769.400	1 864.400	**868**	1 781.400	1 876.400			
809	1 769.600	1 864.600	**869**	1 781.600	1 876.600			

Index

Indoor Radio Planning: A Practical Guide for GSM, DCS, UMTS and HSPA Morten Tolstrup